职业教育双语教材

Bilingual Textbooks of Vocational Education

大型风力发电系统安装与调试

Installation and Commissioning of Large-scale Wind Power Generation System

王　欣　李良君　主编

李云梅　主审

Edited by Wang Xin Li Liangjun

Reviewed by Li Yunmei

·北京·

· Beijing ·

内容简介

本书基于大型风电机组常用机型，讲述风力发电的基础知识。在此基础上，依托大型风力发电系统实训平台，从安全操作规范、结构组成、特点和技术性能等方面对实训设备进行介绍，并安排风机对象结构的认知、叶片和变桨轴承的拆装、能源储存系统安装与调试、偏航功能的实现、变桨功能的实现、并网型逆变器工作原理实训、并网逆变器参数设置及电能质量分析、并网逆变控制系统安装与调试、发电系统整体运行共9个实训项目。在每个实训项目中先系统介绍相关理论知识，再进入实训环节，从而达到理实一体的效果。

本书是汉英双语职业教育的教材，可供相关专业院校师生使用。

图书在版编目（CIP）数据

大型风力发电系统安装与调试：汉、英/王欣，李良君主编．—北京：化学工业出版社，2023.7

ISBN 978-7-122-43280-3

Ⅰ．①大… Ⅱ．①王… ②李… Ⅲ．①风力发电系统-安装-汉、英②风力发电系统-调试方法-汉、英 Ⅳ．①TM614

中国国家版本馆CIP数据核字（2023）第065084号

责任编辑：韩庆利　　文字编辑：宋　旋　温潇潇
责任校对：李雨晴　　装帧设计：韩　飞

出版发行：化学工业出版社（北京市东城区青年湖南街13号　邮政编码100011）
印　　装：大厂聚鑫印刷有限责任公司
787mm×1092mm　1/16　印张16¼　字数420千字　2024年2月北京第1版第1次印刷

购书咨询：010-64518888　　售后服务：010-64518899
网　　址：http://www.cip.com.cn
凡购买本书，如有缺损质量问题，本社销售中心负责调换。

定　　价：58.00元

前言

为扩大与“一带一路”沿线国家的职业教育合作，贯彻落实天津市启动实施的将优秀职业教育成果输出国门与世界分享计划的要求，职业教育作为与制造业联系最紧密的一种教育形式，正发挥着举足轻重的作用。为了配合“鲁班工坊”的理论和实训教学，开展交流与合作，提高中国职业教育的国际影响力，创新职业院校国际合作模式，输出我国职业教育优秀资源，我们编写了《大型风力发电系统安装与调试》。

本教材采用项目导向、任务驱动的理念，构建风力发电技术基础知识与实际仿真系统实训相结合的课程内容。本教材共包含三部分：第一部分为风力发电技术，主要介绍风力发电的专业基础知识；第二部分为大型风力发电系统与结构，着重介绍大型风力发电系统实训平台的基本参数；第三部分为实训项目，是全书的重点内容。其中项目一重点掌握风力发电理论基础与风电机组基本结构；项目二通过利用常用工具对叶片和变桨轴承进行拆装，掌握变桨系统结构组成与变桨原理；项目三介绍了能源储存系统安装与调试；项目四和项目五重点介绍了偏航与变桨的工作过程和功能控制；项目六至项目八，掌握并网逆变器的工作原理、参数设置与并网逆变控制系统的整体安装与调试；项目九在掌握前面八个项目内容的基础上，实现发电系统的整体运行。

本教材由王欣、李良君主编，李云梅主审，张润华、袁进峰、马思宁、姚嵩、沈洁、李娜、崔立鹏参加了编写工作。王欣进行了框架设计，张润华负责整体统稿，李良君对整体框架和全部内容进行了审核把关。第一、第二部分由王欣、李良君编写；第三部分中，姚嵩编写项目一，李良君、袁进峰编写项目二，袁进峰、沈洁编写项目三，李良君编写项目四，沈洁、崔立鹏编写项目五，李娜编写项目六，马思宁编写项目七，张润华编写项目八，王欣编写项目九。

限于编者水平，书中定有不少疏漏之处，恳请读者批评指正。

编　者

目 录

第一部分　风力发电技术

一、认识风力发电

1. 风力发电的基本原理

风力发电是一个由风力发电机组（风机）将捕获到的风能转化为机械能，并通过主轴、齿轮箱等传动机构将机械能传递给发电机，再由发电机将机械能转换为电能的过程，如图1.1.1所示。发出来的电通过升压送到电网上，人们就可以利用了。

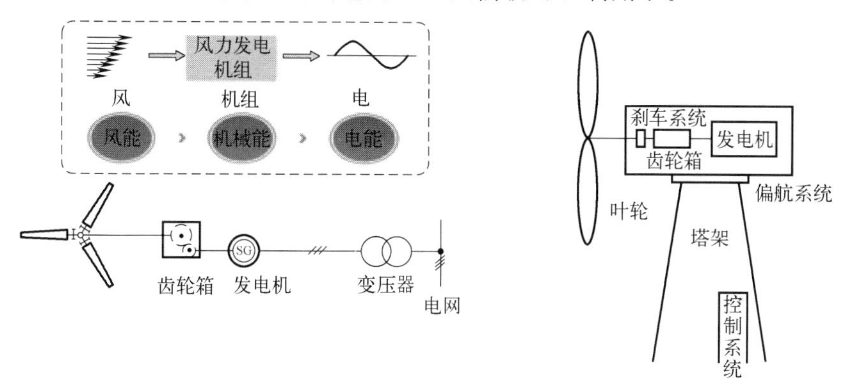

图 1.1.1　风力发电基本原理示意图

按照目前的风力发电技术，大约3m/s的风速，便可以开始发电。

由于MW级风力发电机组齿轮箱损坏率较高，因而采用直驱型风力发电机组（无齿轮箱）。

直驱型风力发电机组的风轮轴与发电机直接相连，采用多极低速同步发电机。而双馈型风力发电机组采用的是高速异步发电机。

2. 风力发电机组分类

按结构形式分类：水平轴、垂直轴；

按功率大小分类：微型（1kW以下）、小型（1～10kW）、中型（10～100kW）、大型（100kW以上）；

按塔架位置分类：上风式（迎风）、下风式（顺风）；

按叶片工作原理分类：升力型、阻力型；

按叶片数量分类：单叶片、双叶片、三叶片、多叶片；

按功率调节方式分类：定桨型（失速调节）、变桨型（变桨距调节）；

按风轮转速分类：定速型、变速型；

按传动机构分类：齿轮箱升速型、直驱型；

按发电机分类：异步型、同步型；

按并网方式分类：并网型、离网型、混合型。

大型风力发电机组常见类型：并网型水平轴上风式三叶片变速变桨距风力发电机组，其中双馈型即升速型（异步发电机）和直驱型（同步发电机）都比较多。

3. 风力发电模式

陆地风力发电，其主要技术方向是低风速发电，主要机型是 2～5MW 的大型风力发电机组。这种模式的关键是向电网输电。

海上风力发电，目前主要是在滩涂和近海海域布置大规模的风力发电场，一般安装 5MW 以上的大型风力发电机组。这种模式的主要制约因素是风力发电场的规划和建设成本。但是海上风力发电的优势是明显的，即不占用土地，海上风力资源较好。

4. 风力发电的发展历程和趋势

自 1887～1888 年冬，第一台自动运行且发电的风力机安装完成，至今，风力发电大致经历了试验研究、示范先行、商业开发、积累能量、竞争开发、规模领先、跨越发展的发展历程。目前，风力发电的发展趋势如下：

① 变桨距调节方式迅速取代失速调节方式；

② 变速运行方式迅速取代恒速运行方式；

③ 机组规模向大型化发展；

④ 直驱永磁、异步双馈两种形式共同发展，半直驱型机组迎面扑来；

⑤ 并网控制越来越友好、自动化；

⑥ 海上风力发电快速增长；

⑦ 全球风电行业市场高度集中，新兴市场未来发展迅速；

⑧ 风力发电成本越来越具备竞争优势。

风力发电技术进步很快，风力发电机组科技含量和可靠性高，虽然目前风力发电成本还比较高，但随着生产批量的增大和进一步的技术改进，成本将继续下降。

风力发电的突出优点是环境效益好，不排放任何有害气体和废弃物。陆地风力发电场虽然占了大片土地，但是风力发电机组基础使用的面积很小，不影响农田和牧场的正常生产。多风的地方往往是荒滩或者山地，建设风力发电场的同时也开发了旅游资源。

二、认知风力发电系统

1. 风力发电机组

(1) 双馈式机组

双馈异步风力发电机是一种绕线式感应发电机。双馈异步发电机的定子绕组直接与电网相连，转子绕组通过变频器与电网连接，转子绕组电源的频率、电压、幅值和相位按运行要求由变频器自动调节，机组可以在不同的转速下实现恒频发电，满足用电负载和并网的要求。由于采用了交流励磁，发电机和电力系统构成了“柔性连接”，即可以根据电网电压、电流和发电机的转速来调节励磁电流，精确地调节发电机输出电压，使其能满足要求。双馈式风力发电机组的叶轮通过多级齿轮增速箱驱动发电机，主要结构包括

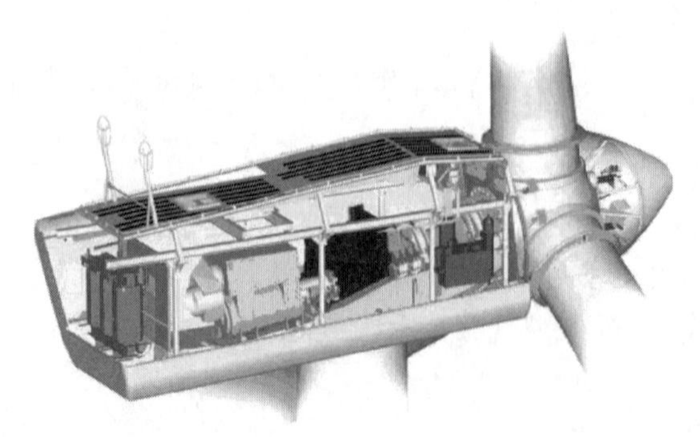

图 1.1.2　双馈式风力发电机组结构示意图

风轮、传动装置、发电机、变流器系统、控制系统、塔架等，如图 1.1.2 所示。

双馈式风力发电机组将齿轮箱传输到发电机转子轴的机械能转化为电能，通过发电机定子、转子传送给电网，如图 1.1.3 所示。发电机定子绕组直接和电网连接，转子绕组和频率、幅值、相位都可以按照要求进行调节的变流器相连。变流器控制发电机在亚同步和超同步转速下都保持发电状态。在超同步发电时，通过定、转子两个通道同时向电网馈送能量，这时变流器将直流侧能量馈送回电网。在亚同步发电时，通过定子向电网馈送能量、转子吸收能量产生制动力矩使发电机工作。变流系统能够双向馈电，故称双馈技术。

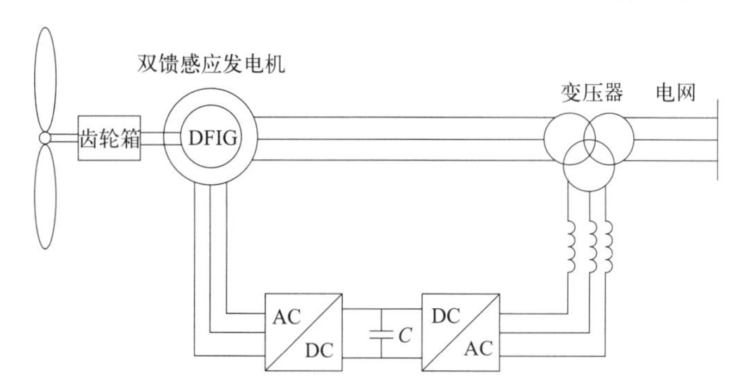

图 1.1.3　双馈变速恒频风力发电机组示意图

双馈式风力发电机组具有以下优点：

① 能控制无功功率，并通过独立控制转子励磁电流解耦有功功率和无功功率控制。

② 双馈感应发电机无需从电网励磁，而从转子电路中励磁。

③ 能产生无功功率，并可以通过网侧变流器传送给定子。

(2) 直驱式机组

直驱式风力发电机组由风力直接驱动发电机，亦称无齿轮风力发电机组。它采用多极电机与叶轮直接连接进行驱动的方式，免去齿轮箱这一传统部件，主要由风轮、传动装置、发电机、变流器、控制系统、塔架等组成，如图 1.1.4 所示。为了提高低速发电机效率，直驱式风力发电机组采用大幅度增加极对数的方式来提高风能利用率，采用全功率变流器实现风力发电机组的调速，如图 1.1.5 所示。

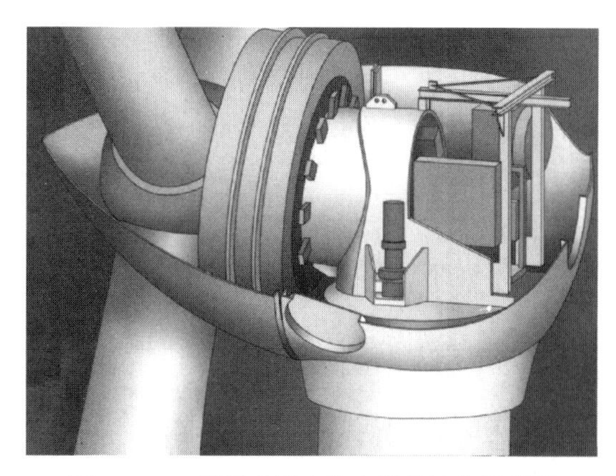

图 1.1.4　直驱式风力发电机组结构示意图

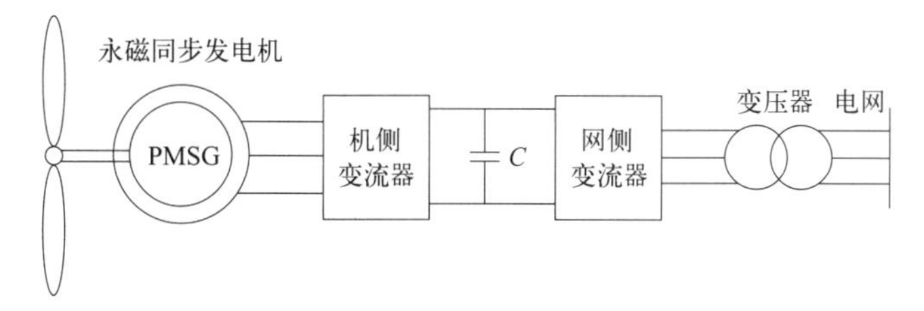

图 1.1.5　直驱变速恒频风力发电机组示意图

直驱式风力发电机组按照励磁方式可分为电励磁和永磁两种。

永磁同步发电机由于结构简单、无需励磁绕组、效率高的特点而在中小型风力发电机组中应用广泛，随着高性能永磁材料制造工艺的提高，大容量的风力发电机组也倾向于使用永磁同步发电机。永磁同步发电机通常用于变速恒频的风力发电机组中，发电机转子由风轮直接拖动，所以转速很低。由于去掉了增速齿轮箱，增加了机组的可靠性和寿命；利用许多高性能的永磁磁钢组成磁极，不像电励磁同步发电机那样需要结构复杂、体积庞大的励磁绕组，从而提高了气隙磁通密度和功率密度，在同功率等级下，减小了发电机体积。

直驱式风力发电机组有以下优点：

① 发电效率高。直驱式风力发电机组没有齿轮箱，减少了传动损耗，提高了发电效率，尤其是在低风速环境下，效果更加显著。

② 可靠性高。齿轮箱是风力发电机组运行出现故障频率较高的部件，直驱技术省去了齿轮箱及其附件，简化了传动结构，提高了机组的可靠性。同时，机组在低转速下运行，旋转部件较少，可靠性更高。

③ 运行及维护成本低。采用无齿轮直驱技术可减少风力发电机组零部件数量，避免齿轮箱油的定期更换，降低了运行维护成本。

④ 电网接入性能优异。直驱永磁风力发电机组的低电压穿越使得电网并网点电压跌落时，风力发电机组能够在一定电压跌落的范围内不间断并网运行，从而维持电网的稳定运行。

2. 风力发电机组基础

风力发电机组基础是支撑塔架与地面底部连接的平台。将风力发电机组支撑在 60～100m 甚至更高的高空，从而使其获得充足、稳定的风力来发电。风力发电机组基础的作用主要体现在：是风力发电机组的主要承载部件；用于安装、支撑风力发电机组；平衡风力发电机组在运行过程中所产生的各种载荷；保证机组安全、稳定地运行。

基础设计与基础所处的地质条件有着重要的关联，良好的地质条件可以为基础建造提供可靠的安全保证。从基础建设的要求可知，塔架基础的重要性及复杂性是不言而喻的。在复杂地质条件下如何确定安全合理的基础方案更是重中之重。风力发电机组基础包括陆上风电基础（图 1.1.6）和海上风电基础（图 1.1.7）。

图 1.1.6　陆上风电基础

图 1.1.7　海上风电基础

(1) 风力发电机组基础设计

① 基础的结构与类型。根据风力发电机组型号与容量自身特性，要求基础承载载荷也各不相同。根据风力发电机组的单机容量、轮毂高度和地基复杂程度，地基基础分为三个设计级别，见表 1.1.1。

机组地基基础设计应符合下列规定：所有机组地基基础，均应满足承载力、变形和稳定性的要求；1 级、2 级机组地基基础，均应进行地基基础变形计算；3 级机组地基基础，一般可不做变形验算，如有下列情况之一时，仍应做变形验算：其一，地基承载力特征值小于 130kPa 或压缩模量小于 8MPa；其二，软土等特殊性的岩土。

表 1.1.1　地基基础三个设计级别

设计级别	单机容量、轮毂高度和地基类型	设计级别	单机容量、轮毂高度和地基类型
1	单机容量大于 1.5MW 轮毂高度大于 80m 复杂地质条件或软土地基	3	单机容量小于 0.75MW 轮毂高度小于 60m 地质条件简单的岩土地基
2	介于 1 级、3 级之间的地基基础		

注：1. 地基基础设计级别按表中指标划分分属不同级别时，按最高级别确定；
2. 对 1 级地基基础，地基条件较好时，经论证基础设计级别可降低一级。

风力发电机组基础均为现浇钢筋混凝土独立基础，根据风电场场址工程地质条件和地基承载力以及基础载荷、尺寸大小不同，从结构的形式看，常用的可分为块状基础和框架式基础两种。块状基础，即实体重力式基础，应用广泛，对基础进行动力分析时，可以忽略基础的变形，并将基础作为刚性体来处理，而仅考虑地基的变形。按其结构剖面又可分为“凹”形和“凸”形两种：前者如图 1.1.8 所示，基础整个为方形实体钢筋混凝土，后者如图 1.1.9 所示；后者与前者相比，均属实体基础，区别在于扩展的底座盘上的回填土也成了基础重力的一部分，这样可节省材料，降低费用。

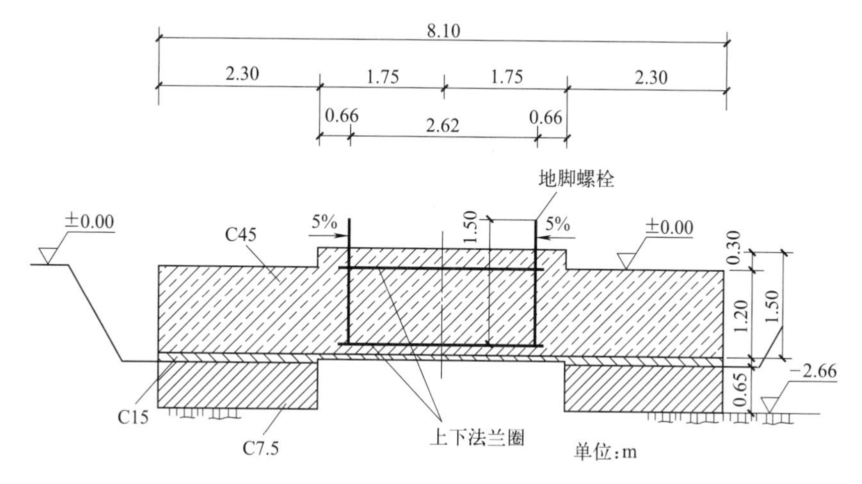

图 1.1.8　“凹”形基础结构

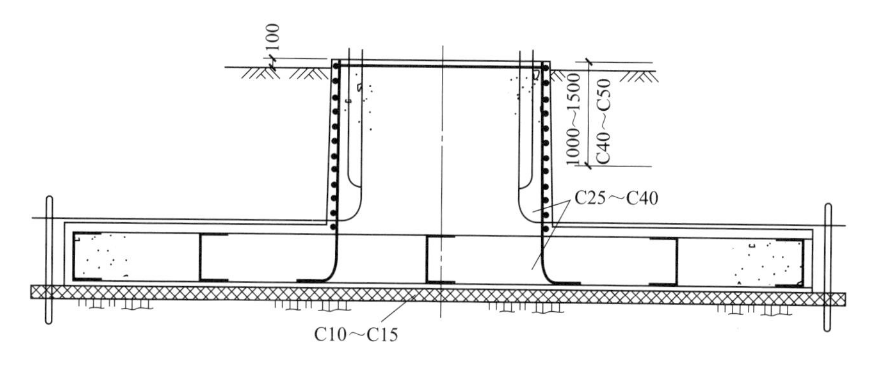

图 1.1.9　“凸”形基础结构

框架式基础实为桩基群与平面板梁的组合体，从单个桩基持力特性看，又分为摩擦桩基础和端承桩基础两种：桩上的载荷由桩侧摩擦力和桩端阻力共同承受的为摩擦桩基础；桩上载荷主要由桩端阻力承受的则为端承桩基础。

根据基础与塔架（机身）连接方式又可分为地脚螺栓式和法兰式两种类型基础。前者塔架用螺母与尼龙弹垫、平垫固定在地脚螺栓上，后者塔架法兰与基础段法兰用螺栓对接。地脚螺栓式又分为单排螺栓、双排螺栓、单排螺栓带上下法兰圈等。

② 风力发电机组基础设计有关注意事项。风力发电机组的基础用于安装、支撑风力发电机组，平衡风力发电机组在运行过程中所产生的各种载荷，以保证机组安全、稳定地运行。因此，在设计风力发电机组基础之前，必须对机组的安装现场进行工程地质勘察。充分了解、研究地基土层的成因、构造及其物理力学性质等，从而对现场的工程地质条件做出正确的评价，这是进行风力发电机组基础设计的先决条件。同时还必须注意到，由于风力发电机组的安装，将使地基中原有的应力状态发生变化，故还需应用力学的方法来研究载荷作用下地基土的变形和强度问题，以使地基基础的设计满足以下两个基本条件：

a. 要求作用于地基上的载荷不超过地基容许的承载能力，以保证地基在防止整体破坏方面有足够的安全储备。

b. 控制基础的沉降，使其不超过地基容许的变形值，以保证风力发电机组不因地基的变形而损坏或影响机组的正常运行。因此，风力发电机组基础设计的前期准备工作是保证机组正常运行必不可少的重要环节。

③ 风力发电机组对基础的要求及基础的受力状况。当风力发电机组运行时，机组除承受自身的重量 Q 外，还要承受由风轮产生的正压力 P、风载荷 q 以及机组调向时所产生的扭矩 M_n 等载荷的作用。这些载荷主要是靠基础予以平衡，以确保机组安全、稳定运行。

图 1.1.10 显示了上述载荷在基础上的作用状况，图中 Q 和 G 分别为机组及基础的自重。倾覆力矩 M 是由机组自重的偏心、风轮产生的正压力 P 以及风载荷 q 等因素所引起的合力矩。M_n 为机组调向时所产生的扭矩。剪力 F 则由风轮产生的正压力 P 以及风载荷 g 所引起。

一般情况下，由于剪力 F 及风力发电机组在调向过程中所产生的扭矩 M_n 都不是很大，且与其他载荷相比要小得多，所以在考虑到不影响计算效果的同时，又能满足工程要求的前提下，编者认为，在实际计算中，此两项可以略去不计。因此在对风力发电机组基础的设计中，风力发电机组对基础所产生的载荷主要应考虑机组自重 Q 与倾覆力矩 M 两项。经上述简化后，风力发电机组基础的力学模型如图 1.1.11 所示。

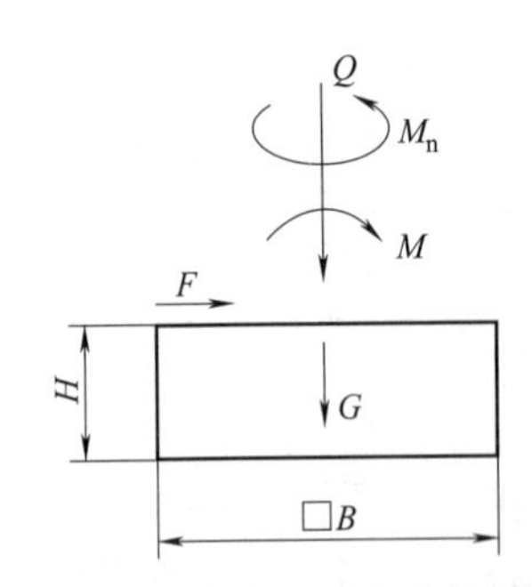

图 1.1.10　载荷在基础上的作用状态

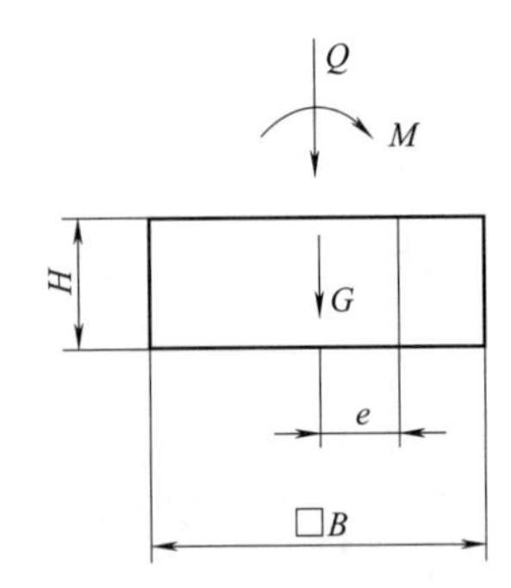

图 1.1.11　风力发电机组基础的力学模型

(2) 海上风力发电机组的基础

根据地理位置及地质条件的不同，海上风力发电机组基础不同的设计模式类型与场址条件密切相关，其成本占投资的20%～30%。海上风力发电机组基础设计模式主要有以下几种类型：单桩基础、重力基础、三脚架基础、导管架基础、漂浮式基础。

① 单桩基础。单桩基础（图1.1.12）结构最简单，所以应用较为广泛。它由焊接钢管组成，桩的直径一般为3～5m，壁厚约为桩直径的1%，安装时需要打入海床10～20m以下深度进行固定。所以单桩基础适用于较浅且海床较为坚硬的水域，但是，如果海床有岩石则不适用。

单桩基础具有的优点是制造简单，不需要做任何海床准备；不足之处则是受海底地质条件和水深的约束较大，施工安装费用较高，需要做好防冲刷防护。

② 重力基础。重力基础（图1.1.13）是利用基础的重力使整个系统固定，不适合流沙型的海底情况。基础的重力可以通过向基础内部填充钢筋、沙子、水泥和岩石等来获得，重力基础一般为钢筋混凝土结构。

重力基础所具有的优点：结构简单，造价低；稳定性和可靠性高。缺点：需要预先进行海床准备，体积和重量都比较大，安装不方便，适用水深范围太小。

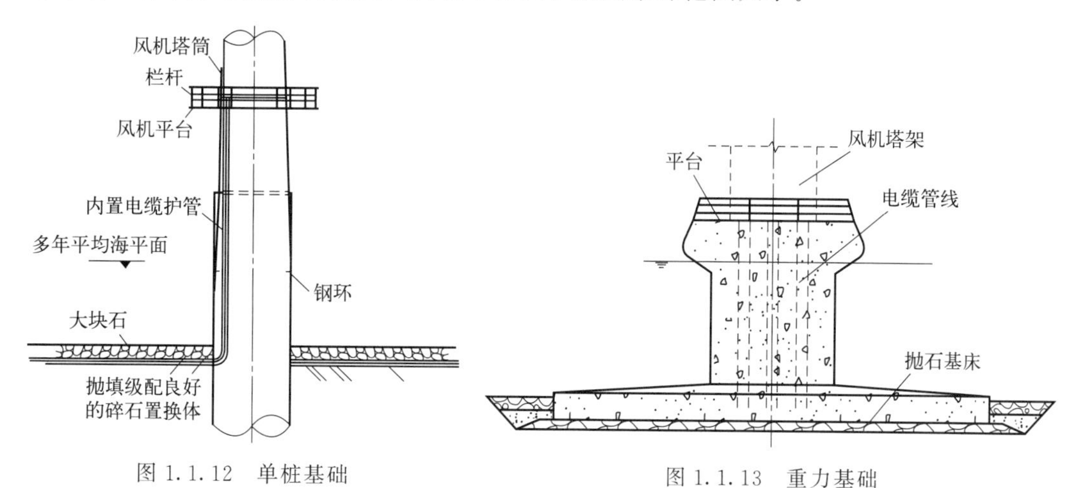

图1.1.12　单桩基础

图1.1.13　重力基础

③ 三脚架基础。三脚架基础（多桩基础）采用标准的三腿支撑结构，其结构由中心柱、三根插入海床一定深度的圆柱钢管和斜撑结构构成，此结构能够很好地抵抗波浪、水流力等，如图1.1.14所示。中心柱提供风机塔架的基本支撑，增强了周围结构的刚度和强度。

三脚架基础的优点：制造简单，不需要做任何海床准备，可用于深海域，不需要冲刷防护。缺点：受地质条件约束较大，不适于浅海域，建造与安装成本较高。

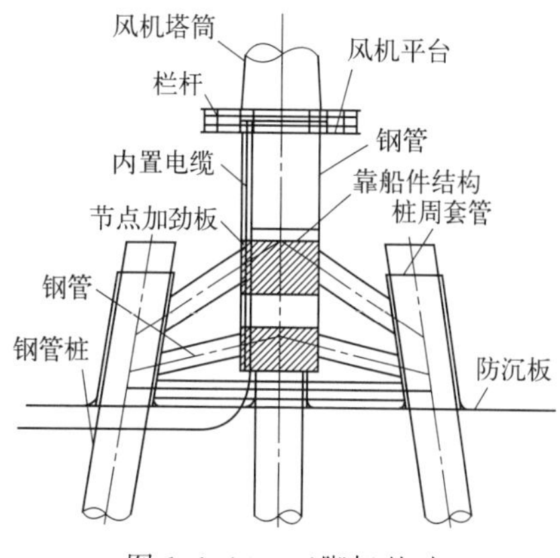

图1.1.14　三脚架基础

④ 导管架基础。导管架基础（图1.1.15）从外形看像一个锥台形空间框架，其适用的水深范围比较大。它具有的优点：建造和施工方便，受到波浪和水流的作用载荷比较小，对地质条件要求不高。缺点：造价随着水深的增加增长很快。

⑤ 漂浮式基础。漂浮式基础（图1.1.16）必须有浮力支撑风力发电机组的重量，并且在可接受的限度内能够抑制倾斜、摇晃和法向移动。在一定的水深区域，漂浮式基础可能是

最好的选择。

优点：安装与维护成本低，在其寿命终止时拆除费用也低；对水深不敏感，安装深度可达 50m 以上；波浪荷载较小。缺点：稳定性差，平台与锚固系统的设计有一定难度。

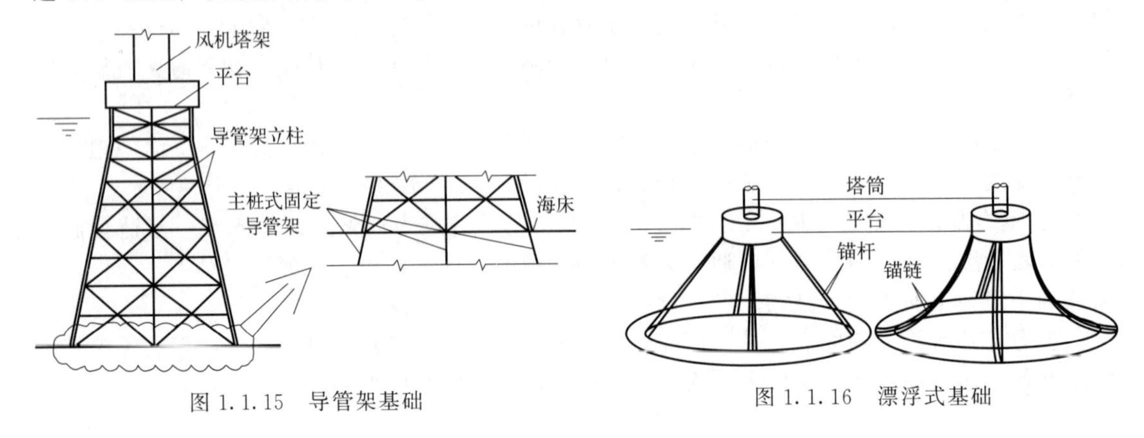

图 1.1.15　导管架基础　　　　图 1.1.16　漂浮式基础

3. 风电场集电系统

风电场电力汇集系统，亦称风电场集电系统，是汇集风力发电机组电能并输送到风电场升压变电站或用电负荷的电力连接系统。可根据电压等级分类，如 10kV 集电系统、35kV 集电系统等。

目前，风力发电技术的成熟度、规模化开发应用情况和市场发展的趋势都是相当卓越的。2020 年，新建风电容量达 97GW，2022 年之后陆上风电建设进入平台期，大部分增长来自海上风电项目。2021 年，新建海上风电容量首次突破 10GW 大关，并将在 2030 年达到 30GW。

(1) 陆上风电场集电系统

陆上风电场一般处于环境恶劣的地区，风力发电机组的年均负载率较低，因此对风电集电系统要求耐候性强、可靠性高和空载损耗低。

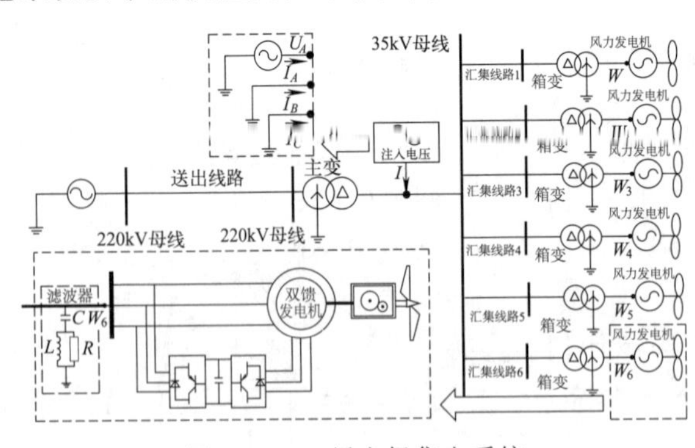

图 1.1.17　风电场集电系统

① 电气一次系统。根据用电的层级分为一次系统和二次系统，其中一次系统包括风力机、发电机和变频器、机组升压变压器（发电机输出 690V，升压至 10kV 或 35kV，再接入风电场升压变电站），其风电场集电系统如图 1.1.17 所示。

风电场集电系统的主接线原则为：

a. 可靠性。任一断路器检修，不影响其所在回路供电；断路器或母线故障、母线检修时，尽量减少停运回路数和时间，并保证对一级负荷及全部或大部分二级负荷的供电；尽量减小发电厂、变电所全部停电的可能性。

b. 灵活性。调度时，灵活切入和切出机组、变压器和线路，灵活调配电源和负载；检修时，方便停运断路器、母线及其继电保护装置，不致影响电力系统的运行和供电；扩建时，从初期接线容易过渡到最终接线。在不影响连续供电和停电最短的条件下，投入新的机组或变压器、线路而互不干扰，且对系统的改建工作量最小。

c. 经济性。要求投资省，占地面积小，电能损失少。

在风电场集电系统中的电气一次系统设计过程中，应重点围绕主接线设计、初选风电场的主接线、机组分组与连接、集电线路方案、风电场自用、主要电力设备等，按选定的接入电力系统方式和电气主接线方案等方向进行设计。

② 电气二次系统。风电场的电气二次系统包含继电器、接触器、控制开关、自动开关、接线端子、成套保护和测控装置。二次系统是风电场集电系统不可缺少的重要组成部分，它对一次设备的工作进行监测、控制、调节、保护，以及为运维人员提供运行工况或生产指挥信号所需的低压电气设备，使一次系统能安全经济地运行。在风电场保护装置中，我国普遍采用的成套保护装置和测控装置如图1.1.18所示。

图 1.1.18　成套保护装置和测控装置

目前，由于大规模集中式开发模式以及风电出力的随机性，风电场内部的无功调度及电压控制一直是相关领域研究的热点。当风电场以分散接入的方式加入电网时，对其系统的整体电压稳定性影响很小，当风电场以大规模集中接入的方式加入电网时，由于电压等级高且传输距离远，使其对区域电网系统运行影响较大。为降低风电并网对区域电网的负面作用，风电并网导则规定各风电场需具备并网点的电压及功率因数调节控制能力，用于在风况出现波动时保持并网点各参数稳定。

(2) 海上风电场集电系统

海上风电作为拥有巨大潜力的大规模可再生能源，在世界各国节能减排的要求以及各国政府有力的财政政策支持下，已经成为未来风能利用的必然趋势。从世界各国海上风电的发展与规划来看，海上风电发展表现出风电场容量逐渐增加与离岸距离不断扩大等特点。自1991年丹麦建成全球首个海上风电场开始，海上风电场的规模呈现逐年递增的状态。Horns Rev作为全球首个大型海上风电场，装机容量为160MW，离岸距离为15km。2021年，中国全年海上风电新增装机1690万千瓦，是此前累计建成总规模的1.8倍。截止到2021年底，中国累计装机规模达到2639万千瓦，跃居世界第一。

大规模远距离海上风电场可能意味着更多数量的风力发电机组和更长距离的电能传输要求。众所周知，海上环境恶劣，电气设备需要专门的防护措施，价格也远远高于陆上。海上条件特殊，施工需要借助专门的工具与设备，因此，建设与运行维护成本也大大高于陆上。为了实现海上风电场经济可靠地并网运行，就需要对海上风电场的电气系统提出一些特殊要求。

为了收集散布在风电场各处的风力发电机组发出的电能，海上风电场电气系统通过海底电缆以一定的形式将机组连接起来，并将电能输送至电网。从结构上看，通常可以划分为集电系统、海上升压平台与输电系统三个部分。集电系统通过中压海底电缆将各风力发电机组相互连接，并接入相应的升压站。海上升压平台将各“串”风机以一定的主接线形式连接，并根据需要将电压等级升高。输电系统通过高压海底电缆将风电场接入系统并网点。图1.1.19为一个典型大规模海上风电场电气系统结构图。目前常用的结构形式主要有三种：交流系统、交直流混合系统以及直流系统。

① 交流系统。由于目前海上风力发电机组大都采用690V的机端电压，为了减少风电场

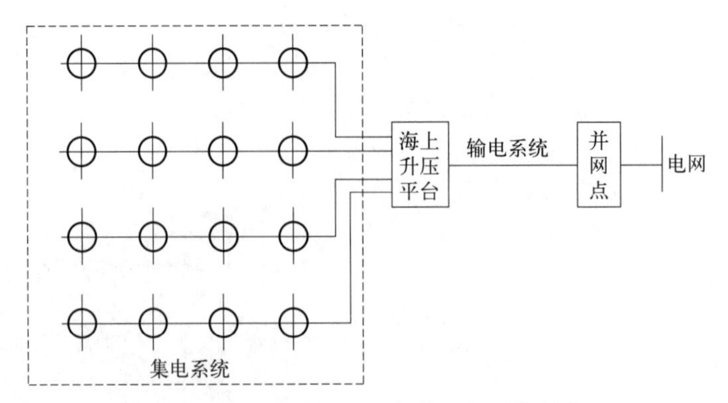

图 1.1.19　大规模海上风电场电气系统结构图

内部电能传输的损耗，常见的做法是在风力发电机组出口装设箱式变压器将电压等级升高。综合设备成本与传输损耗的因素，普遍认为 30～36kV 是交流电气系统中风机之间连接的最佳电压等级。当海上风电场容量小于 100MW、离岸距离小于 15km 时，通常无需装设海上变电站，直接通过中压线路连接至陆上变电站后，接入电网。当风电场规模较大、离岸距离较远时，可以采用海上变电站将电压等级升高，经高压输电线路连接至并网点。

② 交直流混合系统。大规模的海上风电场并网不仅需要考虑数十公里的海底电缆电能输送的经济性问题，还需要充分考虑其对电网运行的稳定性影响。采用高压直流输电联网是一种既能满足风电场并网导则，又具有较高经济性的风电场并网方式。交直流混合电气系统，即采用交流集电系统将海上风力发电机组连接成串，接入海上换流站，然后采用高压直流输电的方式，将风电场接入电网。其接线方式具体如图 1.1.20 所示。基于当前直流输电技术的发展情况，当海上风电场容量大于 100MW，离岸距离超过 90km 时，采用基于电压源换流的柔性直流输电方式（VSC-HVDC）更经济。当风电场容量大于 350MW、离岸距离超过 100km 时，则可以考虑采用传统的 HVDC 输电方式。

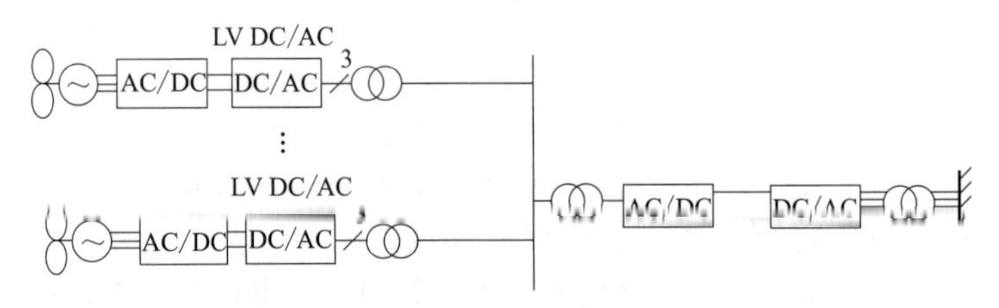

图 1.1.20　交直流混合系统

③ 直流系统。随着柔性直流输电技术的快速发展，直流输电在较低电压等级和较短输电距离时也具备了一定的竞争力。尤其是直流输电方式在海上风电场实现工程化以后，直流方式在集电系统中也获得了越来越多的关注。在直流集电系统中，风力发电机组是通过一组 AC/DC/DC 变换器将电压升高至中压水平。为了与海上风电场高压直流输电线路相互连接，直流集电系统目前主要有两种设计思路：并联连接与串联连接。并联连接采用 DC/DC 换流站将中压直流升高至高压水平，如 150kV，然后通过直流输电线路，经陆上 DC/AC 换流站接入电网。而串联连接则采用海底电缆将风力发电机组相互串联，以获得 N 倍的直流电压，达到升压的目的。然后同样经过高压直流输电线路和陆上 DC/AC 换流站接入电网。具体接线如图 1.1.21 和图 1.1.22 所示。直流系统目前还处于设想与研究阶段。

并联直流集电系统与交流集电系统，主要在两方面有区别：

a. 线路回路数。海上风电场集电系统采用电缆线路，即交流集电系统通常采用三芯海

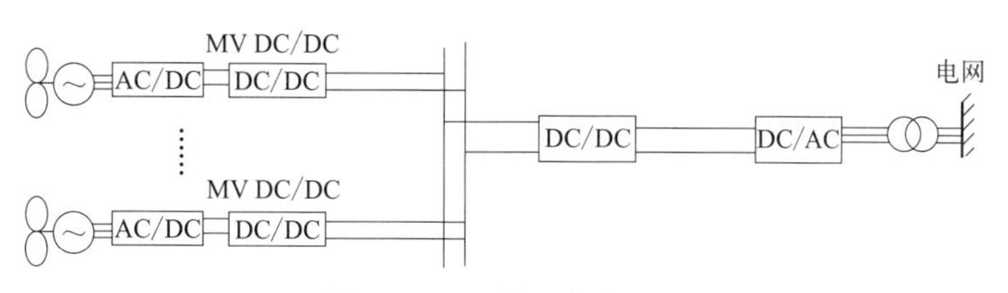

图 1.1.21 并联直流系统

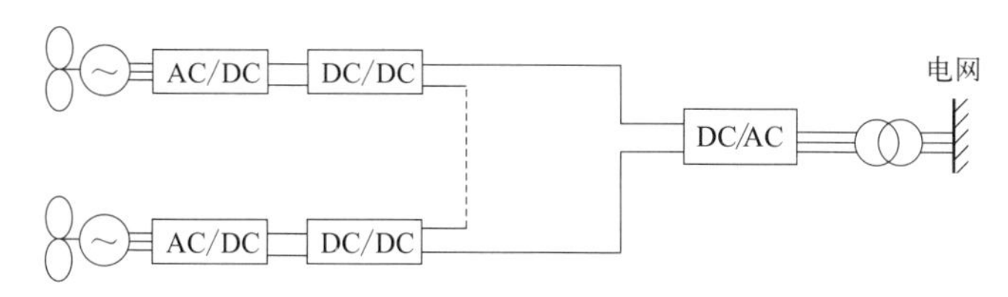

图 1.1.22 串联直流系统

缆，而直流集电系统则采用单芯海缆。因此，交流集电系统只需要 1 根电缆，而并联直流集电系统则需要 2 根电缆。这不仅影响海底电缆成本，也将对海缆的敷设费用造成影响。

在电缆成本方面，与直流电缆相比，交流电缆通常为三相四线制或三相五线制，绝缘安全要求更高，结构较复杂，因此，成本也高出许多。但是，对于相同截面的交直流电缆来说，30kV 的直流海底电缆与 35kV 的交流海底电缆能够承载的负荷相差不大。也就是说，对于相同的风力发电机组来说，采用 30kV 直流系统时输出的电流几乎是 35kV 交流系统输出电流的 2 倍，即相同数量的风力发电机组构成的串型结构所需要的直流电缆的截面可能远远大于所需要的交流电缆。

在安装费用方面，海缆的敷设费用高昂。而相同导体截面的交流海底电缆，其外径远远大于相同导体截面的直流电缆，使得交流电缆在船只上的盘绕与搬运都更加困难与复杂。同时相同长度的交流海底电缆的重量也远远大于直流电缆，因此，交流海底电缆的敷设费用一般要高于直流海底电缆。但是，考虑到海底电缆敷设需要 25～50m 的间距，与交流集电系统相比，直流集电系统的海底电缆敷设工作量可能较大，所需要的风电场海域租赁费用也相对较高。

b. 中压变流器。并联 DC 集电系统中需采用 DC/DC 变压器将风机出口较低的电压等级升高至中压水平。虽然 DC/DC 变流器有许多设计方式，但是，在目前的电力电子技术条件下，当直流升压变比大于 10 时，则需要采用具有电隔离的 DC/DC 变流器结构，类似 DC/AC-变压器-AC/DC 的形式。

4. 风电场并网控制

(1) 并网型风力发电系统工作原理

风力发电有两种不同的类型：独立运行的离网型和接入电力系统运行的并网型。离网型的风力发电规模较小，通过蓄电池等储能装置或者与其他能源发电技术相结合可以解决偏远地区的供电问题。规模较大的风力发电场由几十台甚至成百上千台风力发电机组构成，通常都会接入电网。

并网型风力发电系统（图 1.1.23）通常单机容量大，且由多台风力发电机组组建成风电场，集中向电网输送电能。并网型风力发电的频率应恒等于电网频率。按其发电机运行方式可分为恒速恒频风力发电系统和变速恒频风力发电系统两大类。

① 恒速恒频风力发电系统。恒速恒频风力发电系统中主要采用三相同步发电机、笼型

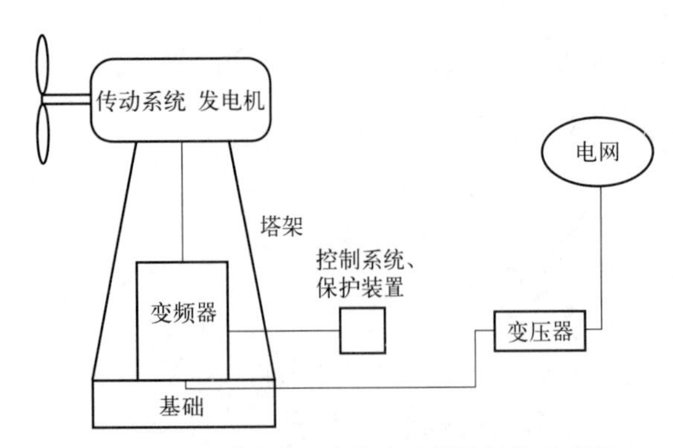

图 1.1.23 并网型风力发电系统的结构示意图

异步发电机（SCIG）。在定桨距并网型风力发电机组中，一般采用 SCIG，通过定桨距失速控制的风轮使其以略高于同步转速 n 的转速［一般为（1～1.05）n］稳定发电运行。图 1.1.24 所示为采用 SCIG 的恒速恒频风力发电系统结构示意图，由于 SCIG 在向电网输出有功功率的同时，需从电网吸收滞后的无功功率，以建立转速为 v_1 的旋转磁场，这加重了电网无功功率的负担，导致电网功率因数下降。为此，在 SCIG 机组与电网之间设置合适容量的并联电容器组以补偿无功功率。在整个运行风速范围内（3～25m/s），气流的速度是不断变化的，为了提高中低风速运行时的效率，定桨距风力发电机组普遍采用三相（笼型）异步双速发电机。

恒速恒频风力发电系统具有电机结构简单、成本低、可靠性高等优点。其主要缺点为：运行范围窄；不能充分利用风能（其风能利用系数不可能保持在最大值）；风速跃升时会导致主轴、齿轮箱和发电机等部件承受很大的机械应力。

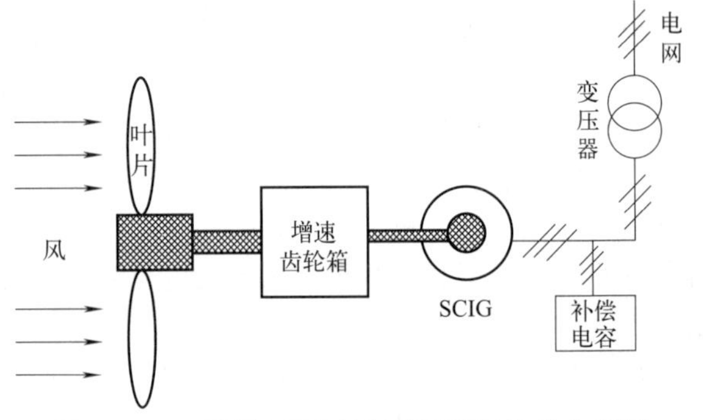

图 1.1.24 采用 SCIG 的恒速恒频风力发电系统

② 变速恒频风力发电系统。为了克服恒速恒频风力发电系统的缺点，20 世纪 90 年代中期，基于变桨距技术的各种变速恒频风力发电系统开始进入市场，其主要特点为：低于额定风速时，调节发电机转矩使转速跟随风速变化，使风轮的叶尖速比保持在最佳值，维持风力发电机组在最大风能利用率下运行；高于额定风速时，调节桨距角以限制风力机吸收的功率不超过最大值；恒频电能的获得是通过发电机与电力电子变换装置相结合实现的。目前，变速恒频风力发电机组主要采用绕线转子双馈异步发电机，低速同步发电机直驱型风力发电系统亦受到广泛重视。

a. 基于绕线转子双馈异步发电机的变速恒频风力发电系统。绕线转子双馈异步发电机（DFIG）的转子侧通过集电环和电刷加入交流励磁，既可输入电能也可输出电能。图 1.1.25 为基于绕线转子双馈异步发电机的变速恒频风力发电系统结构示意图，其中，DFIG 的转子绕组通过可逆变换器与电网相连，通过控制转子励磁电流的频率实现宽范围变速恒频发电运行。其工作原理为：

假定定子磁场转速是 n_0，即同步转速是 n_0，转子通入三相低频励磁电流形成低速旋转磁场，发电机转子机械转速 n_1 与该磁场的旋转速度 n_2 相叠加，等于定子同步转速 n_0，即

$$n_1 \pm n_2 = n_0$$

从而在 DFIG 定子绕组中感应出相应于同步转速 n_0 的工频电压。当发电机转速 n_1 随风速变化而变化时，调节转子励磁电流的频率，改变 n_2，即可调节以补偿 n_1 的变化（一般的变化范围为 n_0 的 30%，可双向调节），从而保持输出电能频率恒定。

图 1.1.25 所示变速恒频方案，由于是在转子电路中实现的，而流过转子电路的功率是由 DFIG 转速运行范围所决定的，转差功率一般只为额定功率的 1/4～1/3，故显著降低了

变换器的容量和成本。此外，调节转子励磁电流的有功、无功分量，可独立调节发电机的有功、无功功率，以调节电网的功率因数、补偿电网的无功需求。事实上，由于 DFIG 转子采用了可调节频率、幅值、相位的交流励磁，发电机和电力系统构成了“柔性连接”。

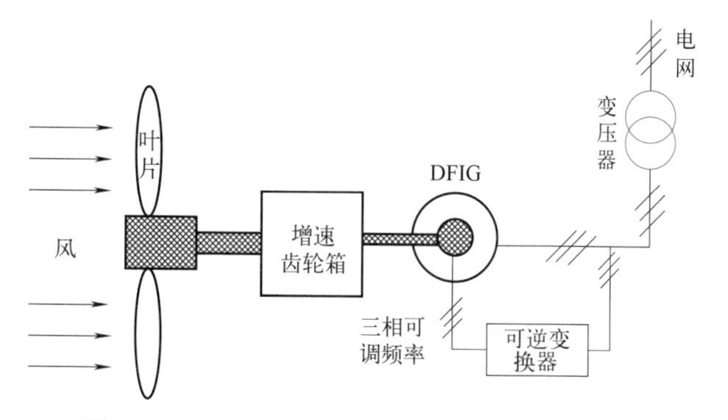

图 1.1.25　基于 DFIG 的变速恒频风力发电系统

b. 基于低速同步发电机的直驱型风力发电系统。直驱型风力发电系统中，风轮与永磁式（或电励磁式）同步发电机直接连接，省去了常用的升速齿轮箱。图 1.1.26 为永磁直驱型变速恒频风力发电系统结构示意图，风能通过风力机和永磁同步发电机（PMSG）转换为 PMSG 定子绕组中频率、幅值变化的交流电，输入到全功率变换器中（其通常采用可控 PWM 整流或不可控整流后接 DC/AC 变换），先经整流为直流，然后经三相逆变器变换为三相工频交流电输出。该系统通过定子侧的全功率变换器对系统的有功、无功功率进行控制，并控制发电机的电磁转矩以调节风轮转速，实现最大功率跟踪。与基于 DFIG 的风力发电系统相比，该系统可在较宽的转速范围内并网，但其全功率变换器的容量要求较大。与带齿轮箱的风力发电系统相比，该系统提高了效率与可靠性，降低了运行噪声，但发电机转速低，为获得一定的功率，发电机应具备较大的电磁转矩，故其体积大、成本高。

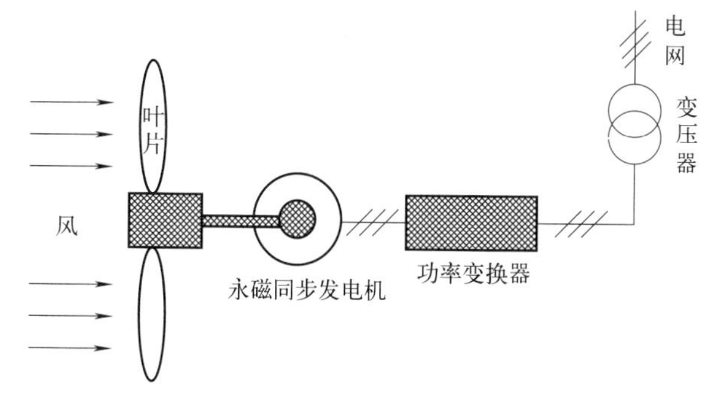

图 1.1.26　永磁直驱型变速恒频风力发电系统

（2）风电并网方式概述

目前风力发电的并网方式大致可以分为异步发电机组并网、同步发电机组并网和双馈发电机组并网三种方式。

① 异步发电机组并网。因为风力机为低速运转的动力机械，在风力机与异步发电机转子之间经增速齿轮传动来提高转速，以达到适合异步发电机运转的转速。一般与电网并联运行的异步发电机多选用 4 极或 6 极发电机，因此异步发电机转速必须超过 1500r/min 或 1000r/min，才能运行在发电状态向电网送电。根据发电机理论，异步发电机并入电网运行时，是靠滑差率来调整负荷的，其输出的功率与转速近乎成线性关系。因此对机组的调速要求，不像同步发电机那么严格精确，不需要同步设备和整步操作，只要转速接近同步转速时就可并网。但异步发电机在并网瞬间会出现较大的冲击电流（约为异步发电机额定电流的 4～7 倍），并使电网电压瞬时下降。随着风力发电机组单机容量的不断增大，这种冲击电流对发电机自身部件的安全及对电网的影响也愈加严重。过大的冲击电流，有可能使发电机与电网连接的主回路中的自动开关断开；而电网电压的较大幅度下降，则可能会使低压保护动作，从而导致异步发电机根本不能并网。

当前在风力发电系统中采用的异步发电机并网方法有以下几种：

a. 直接并网。这种并网方法要求在并网时发电机的相序与电网的相序相同，当风力驱动的异步发电机转速接近同步转速时即可自动并入电网。但如上所述，直接并网时会出现较大的冲击电流及电网电压的下降，因此这种并网方法只适合用于异步发电机容量在百千瓦级以下而电网容量较大的情况下。

b. 降压并网。降压并网方法是在异步发电机与电网之间串接电阻或电抗器，或者接入自耦变压器，以达到降低并网合闸瞬间冲击电流幅值和电网电压下降的幅度。因为电阻、电抗器等元件要消耗功率，在发电机并入电网以后，进入稳定运行状态时，必须将其迅速切除。这种并网方法适用于百千瓦级以上、容量较大的机组。

c. 通过晶闸管软并网。晶闸管软并网方法是在异步发电机定子与电网之间，通过每相串入一只双向晶闸管连接起来，三相均由晶闸管控制，双向晶闸管的两端与并网自动开关的动合触头并联。接入双向晶闸管的目的是将发电机并网瞬间冲击电流控制在允许的限度内。其并网的过程如下：当风力发电机组接收到由控制系统内微处理器发出的启动命令后，先检查发电机的相序与电网的相序是否一致，若相序正确，则发出松闸命令，风力发电机组开始启动。当发电机转速接近同步转速时（99%～100%同步转速），双向晶闸管的控制脚同时由180°到0°逐渐同步打开；与此同时，双向晶闸管的导通角则同时由0°到180°逐渐增大，此时并网自动开关未动作，动合触头未闭合，异步发电机即通过晶闸管平稳地并入电网；随着发电机转速继续升高，发电机的滑差率渐趋于零。当滑差率为零时，并网自动开关动作，动合触头闭合，双向晶闸管被短接，异步发电机的输出电流将不再经双向晶闸管，而是通过已闭合的自动开关触头流入电网。在发电机并网后，应立即在发电机端并入补偿电容，将发电机的功率因数提高到0.95以上。

② 同步发电机组并网。同步发电机在运行时，由于它既能输出有功功率，又能提供无功功率，周波稳定，电能质量高，已被电力系统广泛应用。

直驱交流永磁同步发电机组的并网。由风力机直接驱动低速交流发电机，通过工作速度快、驱动功率小、导通压降低的IGBT逆变器并网。这种系统并网运行的特点如下：

a. 由于不采用齿轮箱，机组水平轴向的长度大幅减小，电能生产的机械传动路径缩短，避免了因齿轮箱旋转而产生的损耗、噪声等。

b. 由于发电机具有大的表面，散热条件更有利，使发电机运行时的温升降低，减小发电机温升的起伏。

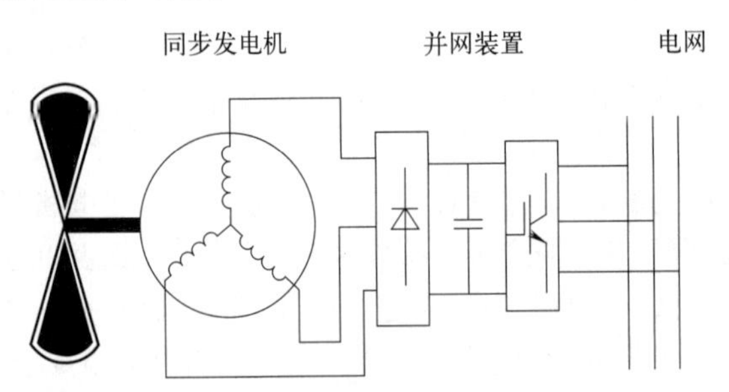

图 1.1.27 交流同步发电机并网电路

图1.1.27为采用交流同步发电机的典型电能装置转换电路。整个并网发电系统主要由同步发电机、并网装置等组成。

三相同步发电机输出的交流电流采用不可控整流器整流为直流以后，经过直流滤波环节，送入到DC/AC逆变器的输入端，逆变为电压、频率、相角、功率因数和谐波都符合电网要求的电能，再经过交流滤波环节后并入电网。

③ 双馈发电机组并网。图1.1.28为交流双馈发电机的典型电能转换电路。整个并网发电系统主要由双馈发电机、双脉冲整流器等组成。

这种并网方案的特点是在发电机侧和电网侧分别加入脉冲整流器，在低风速的情况下，发电机输出的交流电压经过发电机侧脉冲整流器升压后，可以满足电网侧脉冲整流器的正常工作。

(3) 风电场变电站

风电场变电站是把一些设备组装起来，用以切断或接通、改变或者调整电压的场所。风电场变电站主要装设有变压器、配电装置、无功补偿设备和控制设备等，用以升高风电场集电线路汇集的电能电压、控制风力发电机组发出电力的汇集和输送，如图1.1.29所示。

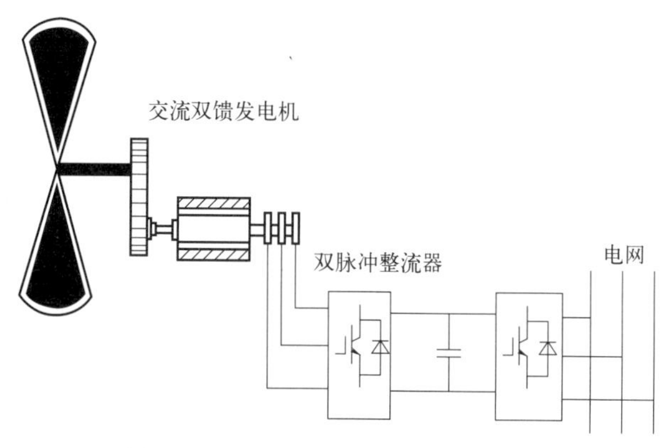

图 1.2.28　交流双馈发电机并网电路

图 1.1.29　风电场升压变电站

变电站主要结构包括：主控室、户内配电装置、户外配电装置、无功补偿、防雷接地系统、主变压器、风电场变电站继电保护、风电场变电站综合自动化系统等，如图 1.1.30 所示。

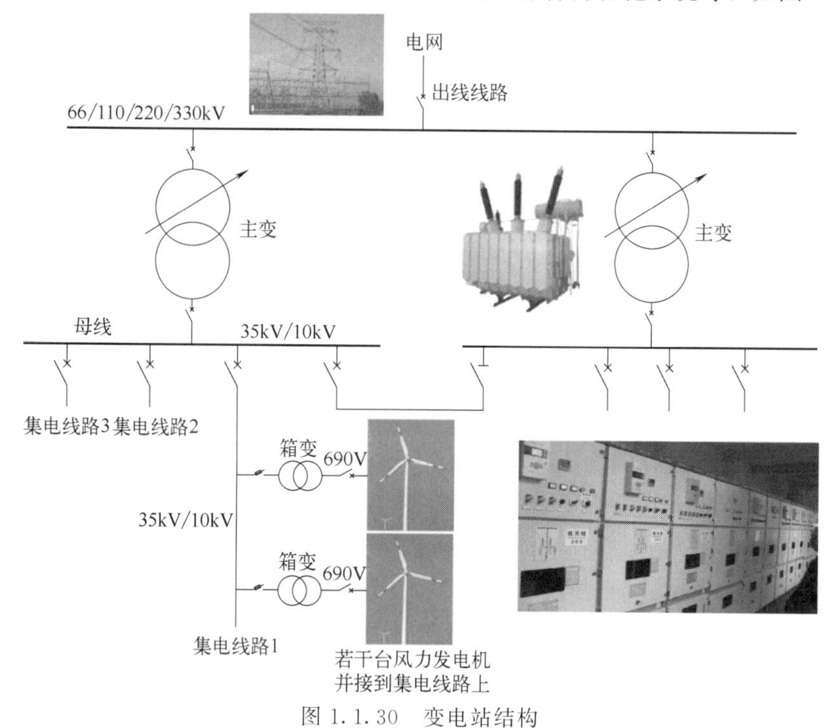

图 1.1.30　变电站结构

第二部分 大型风力发电系统实训平台

一、产品概述

THWPWG-3B 型 大型风力发电系统实训平台如图 2.1.1 所示。该实训平台参照 MW 级风机控制系统设计，具有形象的物理仿真对象，创造了良好的实训环境，包括一套风力发电机组模拟对象和四台控制柜，分别是：风机对象模型（MW 级风机结构，可实现主动偏航和独立变桨功能）、能源控制・监控管理・气象站控制柜、偏航变桨控制系统控制柜、能源转换储存控制系统控制柜、并网逆变控制系统控制柜。可以实现风机变桨系统、电气系统、偏航系统的控制过程实训和风机的整机运行演示。可用于职业院校学生的风力发电电动变桨偏航装置电气控制操作及电气故障排除、风电并网控制技术培训，也可用于风电检修工技能鉴定培训。

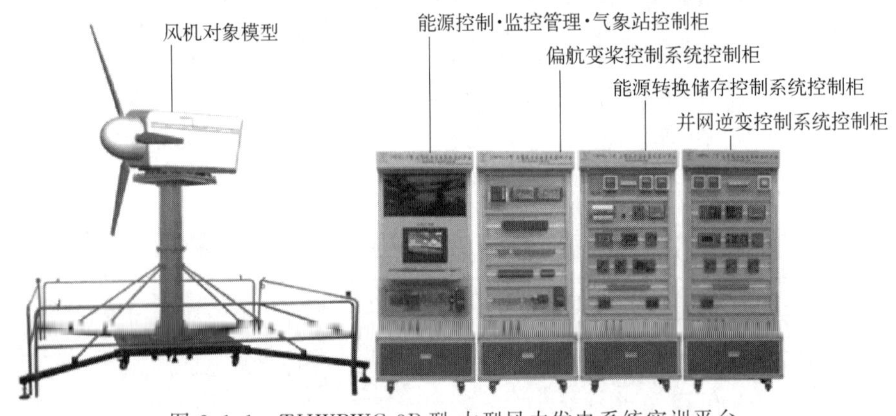

图 2.1.1 THWPWG-3B 型 大型风力发电系统实训平台

二、产品特点

1. 形象生动、方便灵活

风力发电机组的变桨、偏航、刹车装置机械结构采用半实物模型，形象生动，电气控制部分元器件采用网孔板安装结构模式，安装方便灵活。

2. 温度监测

对象系统安装有环境模拟系统，可模拟－40～＋80℃环境温度。发电机和变频器及增速齿轮箱均装有多个温度传感器用于监测温度。

3. 模拟风速风向

有模拟风速风向功能，根据风速大小及风向变化反映变桨、偏航机构的工作特点。

4. 并网逆变

并网逆变控制系统将直流24V逆变成交流36V、50Hz（安全电压），通过升压变压器升至交流220V、50Hz，与单相市电相连实现并网发电功能。主控制器采用TI公司32位定点TMS320F2812芯片，逆变电源的输出功率因数接近于1，输出电流为正弦波。控制策略采用双闭环控制结构，内环为并网电流环，外环为直流电压环。并网同步采用数字锁相技术，锁相精度高，易于实现，同时实现了输出端的功率因数校正控制。

5. 电量测量、最大功率跟踪、电池管理

能源转换储存控制系统完成电量测量、最大功率跟踪、储能和蓄电池管理等功能。最大功率跟踪微处理器采用51系列单片机，具有通用性和在线下载功能，方便用户编程调试，硬件完全开放，用户可以编写不同的MPPT算法实现最大功率跟踪，并将调节参数发送给PWM驱动模块进行调节。PWM驱动CPU采用PIC系列单片机，接收调节参数并输出不同占空比的PWM信号，控制主电路，实现功率调节。智能充放电控制器可以根据蓄电池电压高低，调节充电状态和电流的大小，保护蓄电池过充或过放，延长蓄电池使用寿命。

6. 安全防护

完善的安全装置，外围安装有可拆卸式安全栏杆，可保证使用者不进入危险区域，同时安装有红外反射警示开关，当使用者意外进入危险区域时，系统主动断电刹车。

三、技术性能

1. 额定工作电压

三相四线AC 380V±10%，50Hz。

2. 工作环境

温度－10～＋40℃，相对湿度＜85%（25℃），海拔＜4000m。

3. 装置容量

小于2.5kV·A。

4. 外形尺寸

3400mm×3400mm×3800mm（风机对象模型）；
880mm×600mm×2100mm（能源控制·监控管理·气象站控制柜）；
880mm×600mm×2100mm（偏航变桨控制系统控制柜）；
880mm×600mm×2100mm（能源转换储存控制系统控制柜）；
880mm×600mm×2100mm（并网逆变控制系统控制柜）。

四、产品结构和组成

1. 风机对象模型

主要由变桨电机、偏航电机、绝对值编码器、风力发电机、原动机、齿轮箱、叶片、风轮轴、机舱、塔架、轮毂等组成。

2. 能源控制·监控管理·气象站控制柜

主要由LED显示屏、工控机、PLC、低压电器、开关等器件组成。LED屏可显示风速、风向、偏航角度、变桨角度等信息。工业平板电脑上安装有风力发电系统监控软件，可实现风力发电系统的监控与管理、风力发电机组模拟和对风力发电系统的控制。

3. 偏航变桨控制系统控制柜

主要由从 PLC、变频器、绝对值编码器、交流减速电机、控制按钮等器件组成。可完成变桨偏航控制系统的安装、手动与自动变桨偏航 PLC 控制程序的编写与调试，以及从（slave）PLC 与监控管理主（master）PLC 的通信。

4. 能源转换储存控制系统控制柜

主要由直流电压电流采样模块、温度告警模块、PWM 驱动模块、CPU 核心模块、人机交互模块、通信模块、防雷器、智能型充放电控制器、蓄电池组、开关电源、直流电压表、直流电流表等组成。

5. 并网逆变控制系统控制柜

主要由核心模块、接口模块、液晶显示模块、键盘接口模块、驱动模块、直流电压升压模块、直流电压采样模块、交流电压采样模块、交流电流采样模块、温度告警模块、通信模块、开关电源、直流电机、方形指示灯、直流电压表、直流电流表、多功能数显表、变压器等组成。

五、安全操作规范

为了顺利完成实训项目，确保实训时设备的安全、可靠及长期的运行，实训人员要严格遵守如下安全规程。

1. 实训前的准备

① 实训前仔细阅读使用说明书，熟悉系统的相关部分。

② 实训前仔细阅读系统操作说明及实训的注意事项。

③ 实训前仔细阅读变频器的用户手册，了解变频器的用法。

④ 实训前确保各系统控制柜电源处于断开状态。

⑤ 实训前根据实训指导书中相关内容熟悉此次实训的操作步骤。

2. 实训中的注意事项

① 使用前先检查各电源是否正常。

② 接线前务必熟悉装置的各单元模块的功能及接线位置。

③ 实训接线前必须先断开总电源，严禁带电接线。

④ 接线完毕，检查无误后方可通电。

⑤ 熟练掌握偏航变桨电机控制原理及程序控制方法。

⑥ 控制柜中存在 AC 220V 的接入点，实训时要注意安全。

⑦ 实训始终，实训台要保持整洁，不可随意放置杂物，以免发生短路等故障。

⑧ 实训完毕，应及时关闭电源开关，并及时清理实训台。

⑨ 严格按照正确的操作步骤给系统上电和断电，以免误操作给系统带来损坏。

⑩ 在操作系统的过程中，能源转换储存控制系统蓄电池开关打开之后，有一个等待智能充放电控制器自检初始化的过程，必须等到智能充放电控制器的“红灯”灭掉后才能进行下一步操作。

⑪ 在实训过程中，设备安装时注意防止高处跌落、挤压受伤。

⑫ 在实训过程中，有“危险”标志的地方为强电注意安全。

3. 实训的步骤

实训时要做到以下几点：

(1) 预习报告详细完整，熟悉设备

实训开始前，指导老师要对学生的预习报告做检查，要求学生了解本次实训的目的、内容和安全实训操作步骤，只有满足此要求后，方能允许开始实训。

指导老师要对实训装置做详细介绍，学生必须熟悉该次实训所用的各种设备，明确这些设备的功能与使用方法。

(2) 建立小组，合理分工

每次实训都以小组为单位进行，每组由 2～3 人组成。

(3) 试运行

在正式实训开始之前，先熟悉装置的操作，然后按一定安全操作规范接通电源，观察设备是否正常。如果设备出现异常，应立即切断电源，并排除故障；如果一切正常，即可正式开始实训。

(4) 认真负责，实训有始有终

实训完毕后，应请指导老师检查实训资料。经指导老师认可后，按照安全操作步骤关闭所有电源，并把实训中所用的物品整理好，放回原位。

4. 实训总结

这是实训的最后、最重要阶段，应分析实训现象并撰写实训报告。每位实训参与者要独立完成一份实训报告，实训报告的编写应持严肃认真、实事求是的态度。

实训报告是根据实训中观察发现的问题，经过自己分析研究或组员之间分析讨论后写出的实训总结和心得体会，应简明扼要、字迹清楚、结论明确。

实训报告应包括以下内容：

① 实训名称、专业、班级、学号、姓名、同组者姓名等。

② 实训目的、实训内容、实训步骤。

③ 实训设备的型号、规格。

④ 实训资料的整理。

⑤ 用理论知识对实训结果进行分析总结，得出正确的结论。

⑥ 对实训中出现的现象、遇到的问题进行分析讨论，写出心得体会，并提出自己的建议和改进措施。

⑦ 实训报告应写在一定规格的报告纸上，保持整洁。

⑧ 每次实训每人独立完成一份报告，按时送交指导老师批阅。

第三部分　实训项目

项目一　风机对象结构的认知

项目描述

查阅 THWPWG-3B 型 大型风力发电系统实训平台的说明书，了解风机对象模型结构组成，运用所学的知识和相关安装手册，完成风机对象模型的认知，理解风力发电基本原理，并掌握各种风电设备在运行与控制中的作用。

能力目标：

① 了解风力发电机组结构。

② 认识常见风电设备并记录设备型号及参数。

③ 理解风力发电基本原理，并掌握各种风电设备在运行与控制中的作用。

项目环境

THWPWG-3B 型 大型风力发电系统实训平台，如图 2.1.1 所示。

① 风力发电机组的变桨、偏航、刹车装置机械结构采用半实物模型，形象生动，电气控制部分元器件采用网孔板安装结构模式，安装方便灵活。

② 对象系统安装有环境模拟系统，可模拟外部环境温度。发电机和动力电机及增速齿轮箱均装有多个温度传感器用于监测温度。

③ 装置具有模拟风速风向功能，根据风速大小及风向变化反映变桨、偏航机构工作特点。

项目原理及基础知识

风力发电是利用风能来发电，风力发电机组是将风能转化为电能的机械，风轮是风力发电机组的最主要部件，由叶片和轮毂组成。叶片具有良好的空气动力外形，在气流作用下能产生空气动力使风轮旋转，将风能转换成机械能，再通过齿轮箱增速驱动发电机，将机械能转换成电能。在理论上，最好的风轮只能将约 60％的风能转换成机械能。现代风力发电机组风轮的效率可达到 50％以上，在机组输出达到额定功率之前，其功率与风速的立方成正比，即风速增加 1 倍，输出功率增加 7 倍，可见风力发电的效率与当地的风速关系极大。

风力发电机组根据应用场合的不同又分为并网型和离网型。离网型风力发电机组亦称独立运行风力发电机组，是应用在无电网地区的风力发电机组，一般功率较小。独立运行风力发电机组一般需要与蓄电池和其他控制装置共同组成独立运行风力发电系统。这种独立运行系统可以是几千瓦乃至上几十千瓦以解决一个村落的供电系统，也可以是几十到几百瓦的小型风力发电机组以解决一家一户的供电。并网型风力发电机组主要由两大部分组成：风力机部分——将风能转换为机械能；发电机部分——将机械能转换为电能。

一、风力发电机组的基本结构

风力发电机组是由风轮、变桨系统、传动系统、偏航系统、液压系统、制动系统、发电机、控制与安全系统、机舱、塔架等组成。

各主要组成部分功能简述如下：

① 风轮：风轮由叶片和轮毂构成。叶片是吸收风能的单元，用于将空气的动能转换为叶轮转动的机械能。轮毂的作用是将叶片固定在一起，并且承受叶片上传递的各种载荷，然后传递到发电机转动轴上。

② 变桨系统：变桨系统通过改变叶片的桨距角，使叶片在不同风速时处于最佳的吸收风能的状态，当风速超过切出风速时，使叶片顺桨刹车。

③ 传动系统：传动系统将机械能传递给发电机。一般包括低速轴、高速轴、齿轮箱、联轴器和制动器等。但不是每一种风力发电机组都必须具备所有这些环节。有些机组的轮毂直接连接到齿轮箱上，不需要低速传动轴。也有一些机组设计成无齿轮箱的，叶轮直接连接到发电机。

④ 偏航系统：偏航系统采用主动对风齿轮驱动形式，与控制系统相配合，使叶轮始终处于迎风状态，充分利用风能，提高发电效率。同时提供必要的锁紧力矩，以保障机组安全运行。

⑤ 液压系统：风力发电机组的液压系统的主要功能是刹车（高、低速轴，偏航刹车）。液压系统一般由电动机、油泵、油箱、过滤器、管路及各种液压阀等组成。

⑥ 制动系统：目前获得广泛应用的水平轴风力发电机刹车系统一般由空气制动系统和机械制动系统两部分组成。空气制动系统主要分为定桨距风力发电机的叶尖扰流器和变桨距风力发电机的变桨距控制两类。空气制动系统能够使风轮速度降下来，但却不能使风轮完全停止转动，机械制动系统起着使风机停机的作用。

⑦ 发电机：发电机是将叶轮转动的机械动能转换为电能的部件。转子与变频器连接，可向转子回路提供可调频率的电压，输出转速可以在同步转速±30%范围内调节。

⑧ 控制与安全系统：控制与安全系统包含一台不断监控风电机状态的计算机，并控制偏航、变桨装置。为防止任何故障（即齿轮箱或发电机的过热），该控制器可以自动停止风电机的转动，并发出警报。

⑨ 机舱：机舱包容着风电机的关键设备，包括齿轮箱、发电机。维护人员可以通过风电机塔进入机舱。机舱左端是风电机转子，即转子叶片及轴。

⑩ 塔架：风电机塔载有机舱及转子。通常高的塔具有优势，因为离地面越高，风速越大。

二、风力发电机组主要参数

① 风轮直径：通常风力发电机组的功率越大，风轮直径越大；

② 叶片数目：高速发电机组为2～4片，低速机组大于4片；

③ 叶片材料：现代常采用高强度低密度的复合材料；
④ 风能利用系数：一般为 0.15～0.5；
⑤ 启动风速：一般为 3～5m/s；
⑥ 停机风速：通常为 15～35m/s；
⑦ 输出功率：现代风力发电机组一般为几百千瓦到几兆瓦；
⑧ 发电机：分为直流发电机和交流发电机；
⑨ 其他：塔架高度，等等。

项目实施

一、仪表、设备、工具清单

① THWPWG-3B 型 大型风力发电系统实训平台；
② 梯子；
③ 手机；
④ Word 文档。

二、安全操作规范

① 认识风力发电机组设备时，需利用梯子进行高空作业，要求梯子必须平稳安放在平整的地面上；

② 上下梯子和站在梯子上作业时，要踩实踩稳，专心致志，不得嬉戏打闹；

③ 作业时 4 人一组，2 人分别站在 2 个梯子上协作拆装，另 2 人在地面负责安全保护和传递工具、零部件；

④ 当心坠落、踩空，当心工具或零部件掉落。

三、实训步骤

① 列出你所知道的风力发电机组基本结构，并叙述各部分器件的主要功能。

② 风机对象模型主要由变桨电机、偏航电机、绝对值编码器、风力发电机、原动机、齿轮箱、叶片、风轮轴、机舱、塔筒、轮毂等组成。

以 THWPWG-3B 型 大型风力发电系统实训平台为例，根据设备技术参数及其在风力发电机组运行与控制中的作用，找到列表中各种设备，并用手机拍摄图片，将图片插入到表 3.1.1 设备位置。

表 3.1.1 风机对象模型主要设备技术参数

序号	名称	规格	数量	设备位置及照片	备注
1	风机对象模型	3400mm×3400mm×3800mm	1 台		
2	交流减速机	变桨电机 80YB25GV22-GM10-GK75RC 输出功率：单相 25W 额定电压：220V 额定电流：0.25A 额定转速：1300r/min 中间减速箱减速比：1∶10 直角中控减速箱减速比：1∶75	3 台		

续表

序号	名称	规格	数量	设备位置及照片	备注
3	交流减速机	偏航电机 90YB90GY38-90GM10-90GF60HE 输出功率：三相 90W 额定电压：380V 额定电流：0.41A 额定转速：1300r/min 减速比：1∶600	3 台		
4	卧式减速机	GH28-0.75-70S/380 输出功率：三相 750W 额定电压：380V 额定电流：1.97A 额定转数：1400r/min 减速箱减速比：1∶70	1 台		
5	回转支撑	011.20.280	1 个		
6	行星齿轮增速箱	NB300L2-20-PC-V01A 增速比：19.75∶1 输出类型：V01A 连续输入转矩：850N·m 最大输入转矩：1200N·m	1 台		
7	导电滑环	SRH50120-6P/36S 380V AC/24V DC 额定电流 6 环 10A/36 环 5A 绝缘电阻≥500MΩ/500V DC 工作速度：0～500r/min 工作温度：－20～＋80℃ 工作湿度：60％RH	1 个		
8	三相永磁式发电机	24V-300W-400r 额定输出电压：24V 额定功率：300W 额定转速：400r/min	1 台		
9	绝对值编码器	桨叶角度绝对值编码器 TRD-MA-512N 输出信号：格雷码(最大 9bit，本设备采用 8 位 256 线) 最高相应频率：30kHz 容许最高转速：3000r/min 电源电压：DC 10.8～26.4V 输出形式：NPN 或 PNP 集电极开路	3 个		
10	增量式编码器	J38S-6G-600BZ-C5-24	1 个		

续表

序号	名称	规格	数量	设备位置及照片	备注
11	霍尔传感器	SM12-31010NA 开关类别:电感式(S) 外形:圆柱形(M) 工作电压:24V 检测距离:10mm 输出方式:三线直流 NPN 负逻辑常开输出 输出状态:常开(A) 标准检测物:永磁体 工作环境温度:－25～＋70℃	5个		
12	U形光电开关	测模拟风速 SU003-3K 开关类别:电感式(S) 工作电压:24V 输出方式:三线直流 NPN 负逻辑常开输出 输出状态:常开(A) 工作环境温度:－25～＋70℃	1个		
13	信号灯	工作电压 DC/24V(绿色)	36个		
14	反射传感器	ZR-350N 开关类别:电感式(S) 工作电压:24V 检测距离:1200mm 输出方式:三线直流 NPN 负逻辑常开输出 输出状态:常开(A) 工作环境温度:－25～＋70℃	4个		
15	贴片式 PT100	模拟量输出 测量温度:－50～＋420℃	6个		
16	抗震压力表	YN 100ZQ/10MPa 测量压力范围 0～10MPa	1只		
17	二位三通电磁阀	3WE6E61B/CG24N9Z6L 24V DC 二位三通阀	1只		
18	直动式溢流阀	DBDH6P10B/100 7个压力级可调溢流阀	1只		

续表

序号	名称	规格	数量	设备位置及照片	备注
19	二位四通电磁阀	24V DC 二位四通阀	1只		
20	微型液压动力单元	一体化液压站电机 两个回油口、一个进油口 额定电压：380V 50Hz 额定功率：750W 压力：10MPa 容量：6L	1套		整体使用，确保密封性完好
21	压力传感器	TP212-10M423 模拟量输出：0～5V 测量压力：10MPa	1只		
22	储液罐	330 121 407 储藏冷却液 2L	1只		
23	水泵	电源：单相 220V 额定功率：95W 额定电流：0.46A 额定流量：16L/min，进口水压 1m 以上	1台		
24	电磁阀	24V DC 水流截止阀	1只		

项目作业

根据 PLC/单片机对设备的使用和控制情况，尝试将部分器件分成两大组：输入设备和输出设备，并填写完成表 3.1.2 和表 3.1.3。

表 3.1.2　输入设备名称及作用

输入设备名称	作用	输入设备名称	作用

表 3.1.3　输出设备名称及作用

输出设备名称	作用	输出设备名称	作用

续表

输出设备名称	作用	输出设备名称	作用

项目二　叶片和变桨轴承的拆装

项目描述

查阅风力发电机组叶轮、变桨系统的安装手册，了解叶片与变桨轴承的结构组成，以及相关零部件的装配工艺，运用所学的知识和相关安装手册，完成1.5MW风力发电机组模拟机叶片和变桨轴承的拆装。

能力目标：

① 掌握风力发电机组叶片与变桨轴承的装配工具及检测工具的应用；

② 掌握工艺拆、装的对角拧紧法；

③ 具备工艺要求阅读能力和部件装配能力。

项目环境

本实训任务主要涉及此实训平台的风机对象模型的叶轮部分，完成该任务需要参考THWPWG-3B型大型风力发电系统实训平台设备说明手册，认识并了解叶片和变桨系统各部件及结构组成，了解相关工具的使用规范和注意事项。

一、风轮结构

风轮又称叶轮，主要由叶片、轮毂和变桨系统组成，有三个叶片。叶片又称桨叶，限位开关即限位传感器。风轮的主要部件和结构如图3.2.1～图3.2.3所示。

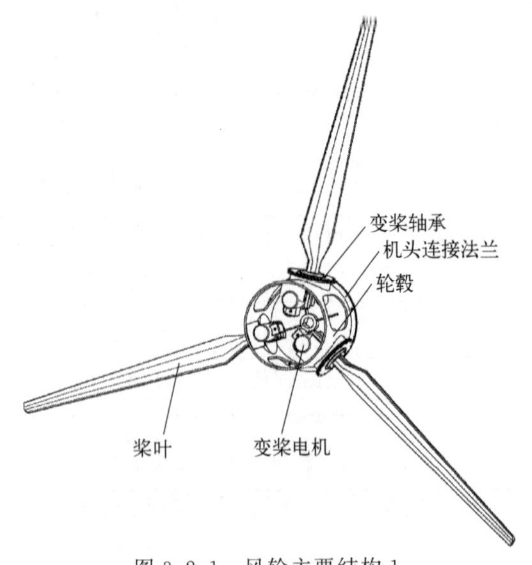

图3.2.1　风轮主要结构1

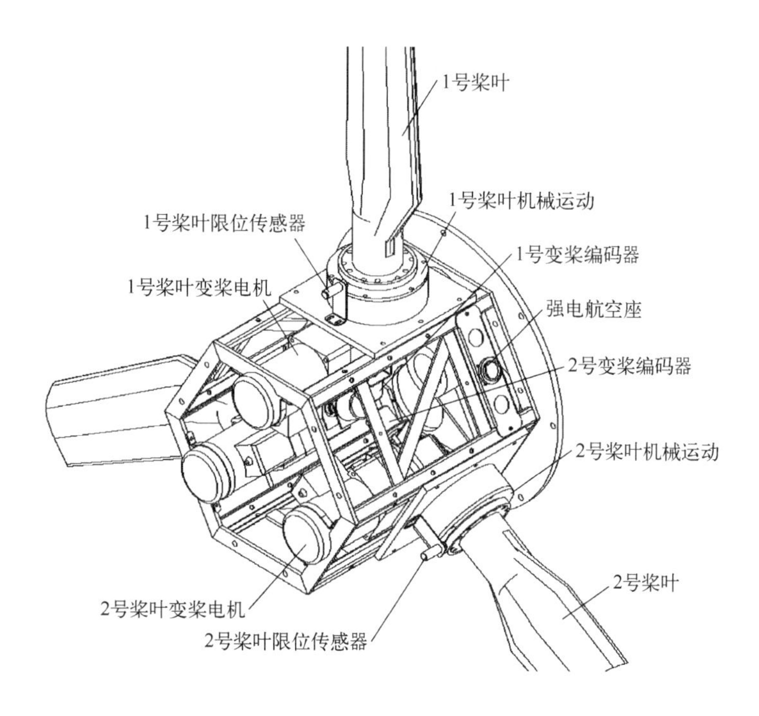

图 3.2.2　风轮主要结构 2

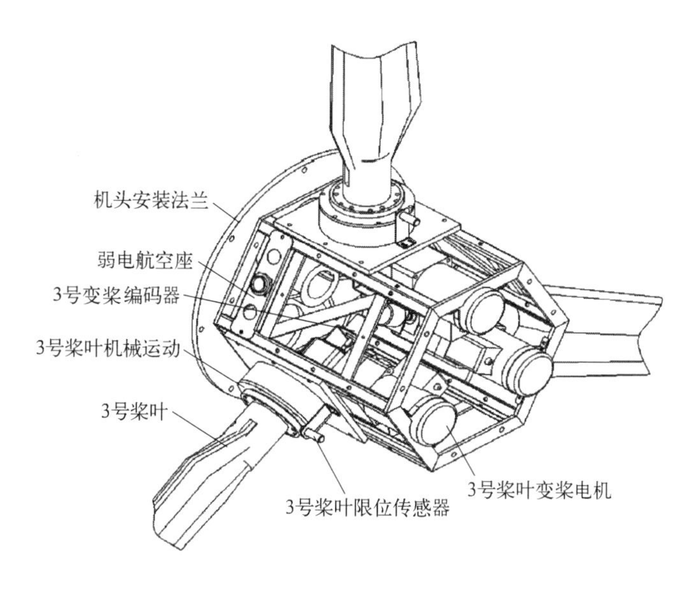

图 3.2.3　风轮主要结构 3

二、变桨系统总装图

变桨系统总装图如图 3.2.4 所示。

三、控制原理图

变桨系统的控制原理如图 3.2.5 所示，每个桨叶电机 U_2、Z_2 之间接运行电容。

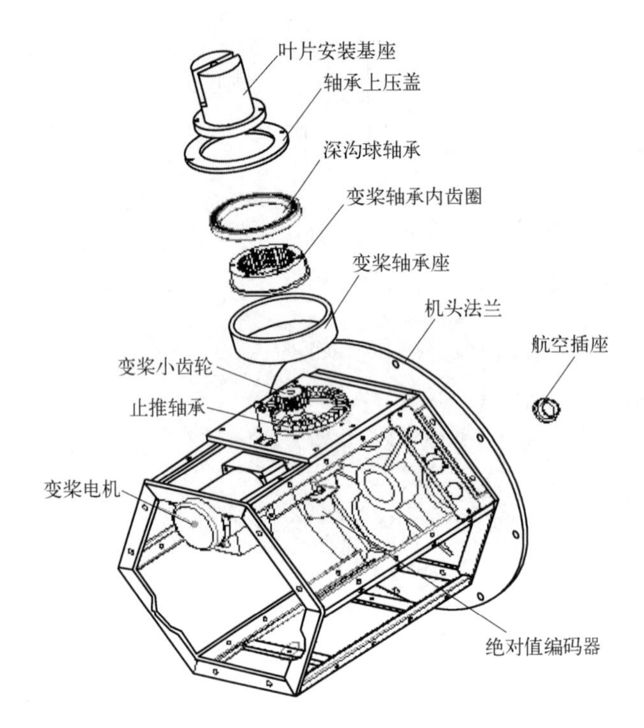

图 3.2.4　变桨系统总装图

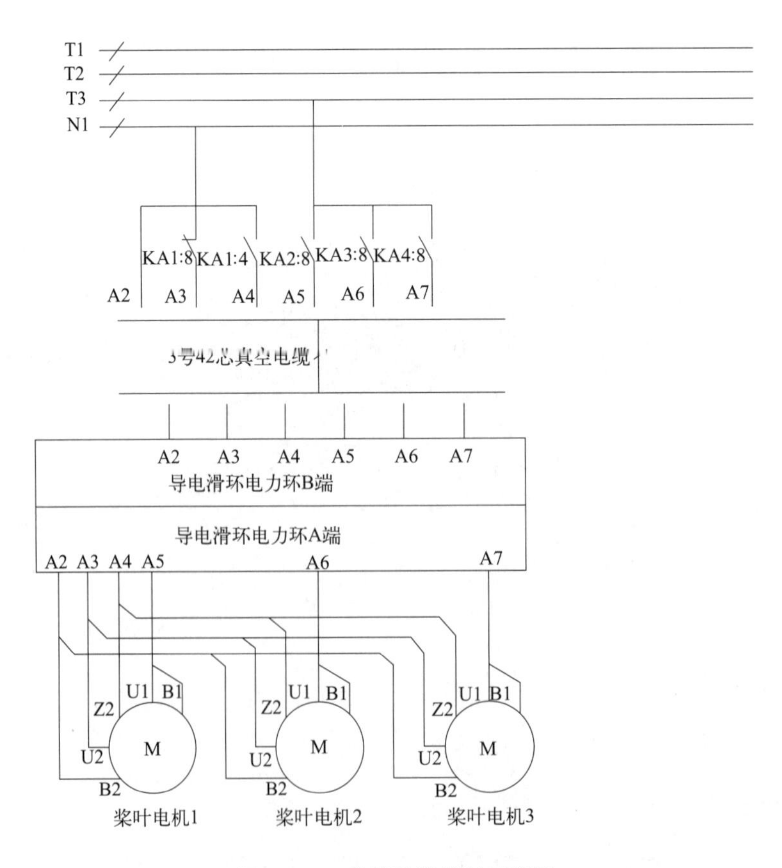

图 3.2.5　变桨系统控制原理图

项目原理及基础知识

一、叶片

1. 叶片的作用

风力发电机组是一种将风能转化为机械能，再由机械能转化为电能的机组和系统，前一种转化是由风轮实现的，后一种转化是由发电机实现的。风轮主要由两部分组成：叶片（一般为3片）和轮毂（图3.2.6），叶片是将风能转化为机械能的唯一关键部件。叶片有一个空气动力外形，它的外形决定了整个机组的空气动力性能，一个具有良好空气动力外形的叶片，可以使机组的能量转换效率更高，获得更多的风能。叶片在转换能量的同时，又承受着很大的载荷（风力和质量力）。在自然界中风况是复杂多变的，因此叶片上承载的载荷也是复杂的，整个风力发电机组主要载荷的来源是叶片，所以叶片必须有足够的强度和刚度。

由此可见，叶片的材料、结构和工艺是非常关键的。材料和结构保证叶片的强度和刚度，并且重量要轻，还要有合适的工艺和方法，保证能够做出带有复杂的外形、符合空气动力学原理的外形的大尺寸构件。

图3.2.6　叶片和轮毂

2. 叶片的性能指标

叶片的主要经济技术指标包括：叶片应用范围（风力发电机组类型、适用风场级）、风轮直径、叶片长度、最大转速、叶片重量、风轮最大风能利用系数、安全风速（极端风速）、设计寿命等。

例：某1.5MW、40.25m风力发电机组叶片主要技术参数见表3.2.1。

表3.2.1　某1.5MW、40.25m风力发电机组叶片技术参数

环境条件		额定转速	17.3r/min
运行温度	−30～+50℃	切入风速	3m/s
环境温度	−40～+50℃	切出风速	25m/s
防雷要求	Ⅰ级(IEC 61400-24)	额定风速	10～11m/s
系统参数		安全风速	52.5m/s
设计等级	IEC ClassⅢ	叶片技术参数	
叶片数目	3	叶片长度	40.25m
工作寿命	20年	额定功率	1500kW
风轮直径	82～83m	叶片质量	6400kg
旋转方向	顺时针(从上风向看)	最大风能利用系数 C_p	0.4935(D=82.5m)
功率控制	变速变桨距	一阶固有频率(挥舞方向)	0.7～0.9Hz
风轮运行转速范围	≤20r/min	一阶固有频率(摆振方向)	1.4～1.65Hz

续表

叶片技术参数		连接方式	
表面防护	叶片外表面为喷漆	螺栓孔位置公差	ϕ0.5mm
叶片重心位置	～12.0m(距叶根)	螺栓孔分布圆直径	ϕ1800mm
叶片材质	GFRP(环氧)	螺栓孔数量	54/64
最大弦长	3.24m	螺纹规格	M30
最大扭角	12.8°	叶片圆柱段外径	ϕ1895mm

3. 叶片的生产工艺

决定叶片质量的主要是叶片的空气动力性能，即最大风能利用系数 $C_{p_{max}}$、推力系数 C_T，再就是叶片的结构强度和重量。这些关键性能除了与气动外形设计及性能和载荷计算、材料选择、结构设计与强度和刚度计算、模具设计与制造等密切相关外，还取决于生产过程中的工艺控制。叶片结构形式如图 3.2.7 所示。

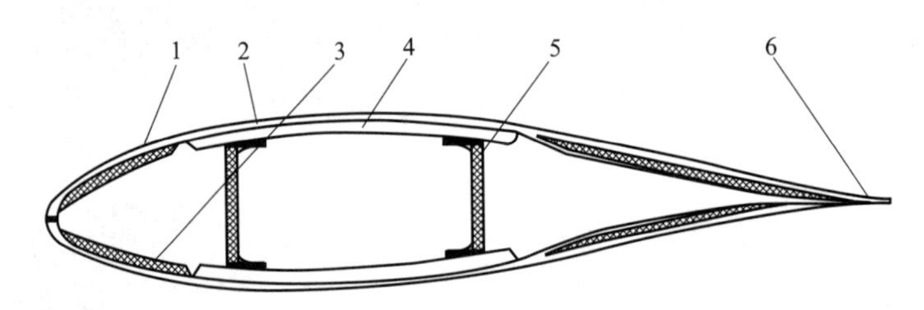

图 3.2.7　叶片剖面结构图

1—胶衣层；2—蒙皮层；3—泡沫层；4—单向纤维层；5—大梁；6—结构胶层

目前复合材料叶片的生产工艺，一般是在各专用模具和工具上分别成型叶片蒙皮、主梁、腹板及其他部件，然后在主模具上把这些部件胶接组装在一起，合模加压固化后制成整体叶片。总体的生产工艺路线如图 3.2.8 所示。

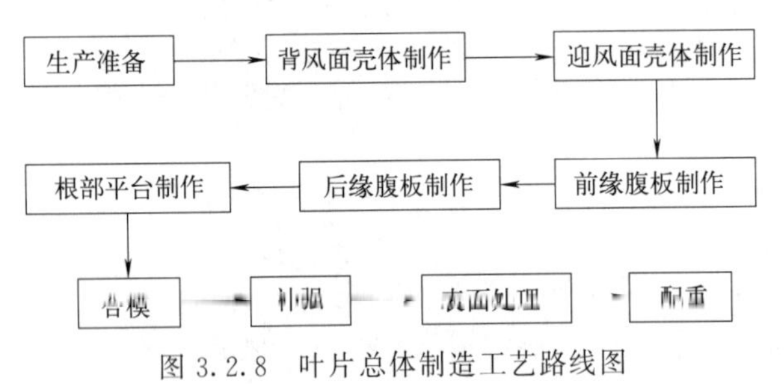

图 3.2.8　叶片总体制造工艺路线图

叶片生产中的关键工艺是成型工艺，大致可分为手糊成型、模压成型、拉挤成型、纤维缠绕树脂传递模塑、预浸料成型以及真空吸注成型，模压成型、树脂传递模塑及真空吸注成型为闭模模塑工艺。目前比较可靠的典型成型工艺有两种：一种是预浸料（图 3.2.9），一种是真空吸注（也称真空导注）。预浸料成型工艺（图 3.2.10）的可控性更好，质量稳定性也更好，但在技术人员和工人培训相对成熟的情况下，真空吸注工艺（图 3.2.11）对质量的稳定性控制也能达到要求，而且真空吸注工艺的经济性要比预浸料工艺好很多。以真空吸注生产工艺为例，叶片的生产过程大致有下料、大梁和翼梁制作、层铺、真空吸注成型、合模和起模、型修、检验配平出厂等工序。

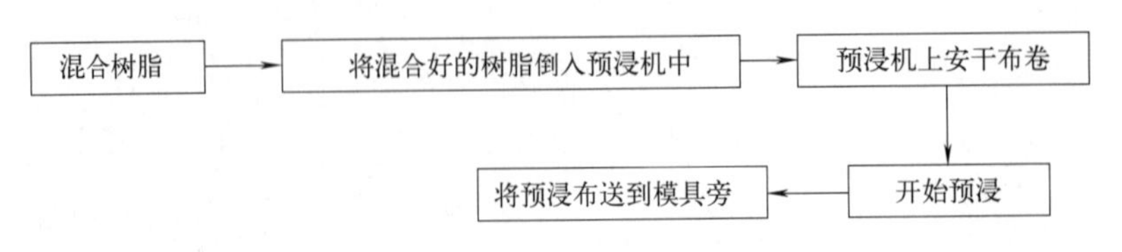

图 3.2.9　预浸料的制作

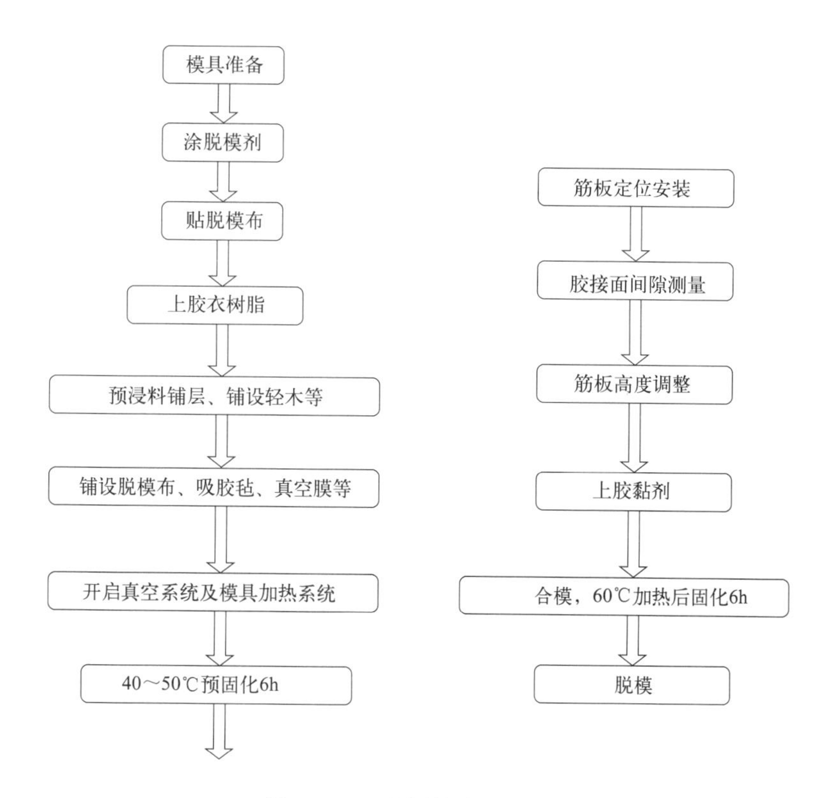

图 3.2.10　预浸料成型工艺图

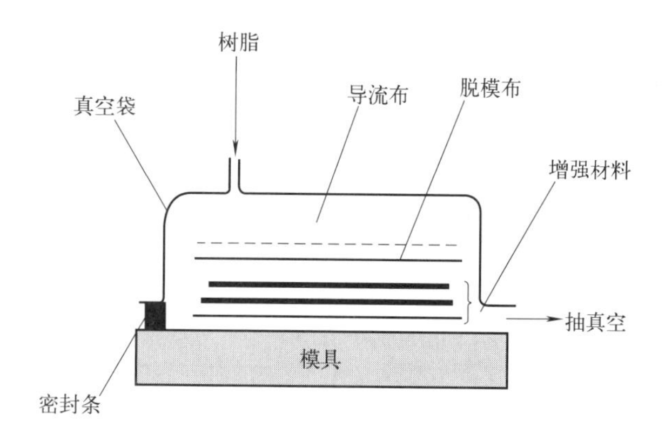

图 3.2.11　真空吸注成型工艺图

二、变桨系统

1. 概述

变桨系统的所有部件都安装在轮毂上。风力发电机组正常运行时，所有部件都随轮毂以一定的速度旋转。

变桨系统通过控制叶片的角度来控制风轮的转速，进而控制风力发电机组的输出功率，并能够通过空气动力制动的方式使风力发电机组安全停机。

风力发电机组的叶片（根部）通过变桨轴承与轮毂相连，每个叶片都要有自己的相对独立的电控同步的变桨驱动系统。变桨驱动系统通过一个小齿轮与变桨轴承内齿啮合联动。

风力发电机组正常运行期间，当风速超过机组额定风速时（风速在12m/s到25m/s之间时），为了控制功率输出，变桨角度限定在0°到30°之间（变桨角度根据风速的变化进行自动调整），通过控制叶片的角度使风轮的转速保持恒定。任何情况引起的停机都会使叶片顺桨到90°位置（执行紧急顺桨命令时叶片会顺桨到90°限位位置）。

变桨系统有时需要由备用电池供电进行变桨操作（比如变桨系统的主电源供电失效后），因此变桨系统必须配备备用电池以确保机组发生严重故障或重大事故的情况下可以安全停机（叶片顺桨到90°限位位置）。此外，还需要一个冗余限位开关（用于95°限位），在主限位开关（用于90°限位）失效时确保变桨电机的安全制动。

由机组故障或其他原因而导致备用电源长期没有使用时，风力发电机组主控就需要检查备用电池的状态和备用电池供电变桨操作功能的正常性。

每个变桨驱动系统都配有一个绝对值编码器安装在电机的非驱动端（电机尾部），还配有一个冗余的绝对值编码器安装在叶片根部变桨轴承内齿旁，它通过一个小齿轮与变桨轴承内齿啮合联动记录变桨角度。

风力发电机组主控接收所有编码器的信号，而变桨系统只应用电机尾部编码器的信号，只有当电机尾部编码器失效时风力发电机组主控才会控制变桨系统应用冗余编码器的信号。

2. 变桨系统的作用

根据风速的大小自动调整叶片与风向之间的夹角，实现风轮最大效率利用风能（额定风速以下），或有一个恒定转速（额定风速以上）；利用空气动力学原理可以使桨叶顺桨90°与风向平行，使风力发电机组停机。

3. 主要部件组成

变桨系统的主要部件组成见表3.2.2。

表3.2.2 变桨系统的主要部件组成

部件名称	数量
电控箱(中控箱、轴控箱)	1套(4个)
变桨电机(配有变桨系统主编码器:A编码器)	3套
备用电池箱	3套
机械式限位开关	3套(6个)
限位开关支架相关连接件	3套
冗余编码器:B编码器	3套
冗余编码器支架、测量小齿轮及相关连接件	3套
各部件间的连接电缆及电缆连接器	1套

注：随着风力发电技术和电力电子技术的发展，目前一些机组的变桨控制已将备用电池箱改用超级电容，并将其放置在轴控箱中，因此，轮毂内只有四个控制箱。还有一些厂家不仅用超级电容代替备用电池箱，也不再使用中控箱，滑环线缆直接与三个轴控箱分别连接。

4. 变桨系统各部件的连接

变桨系统各部件连接框图如图3.2.12所示，变桨机构机械连接如图3.2.13所示。

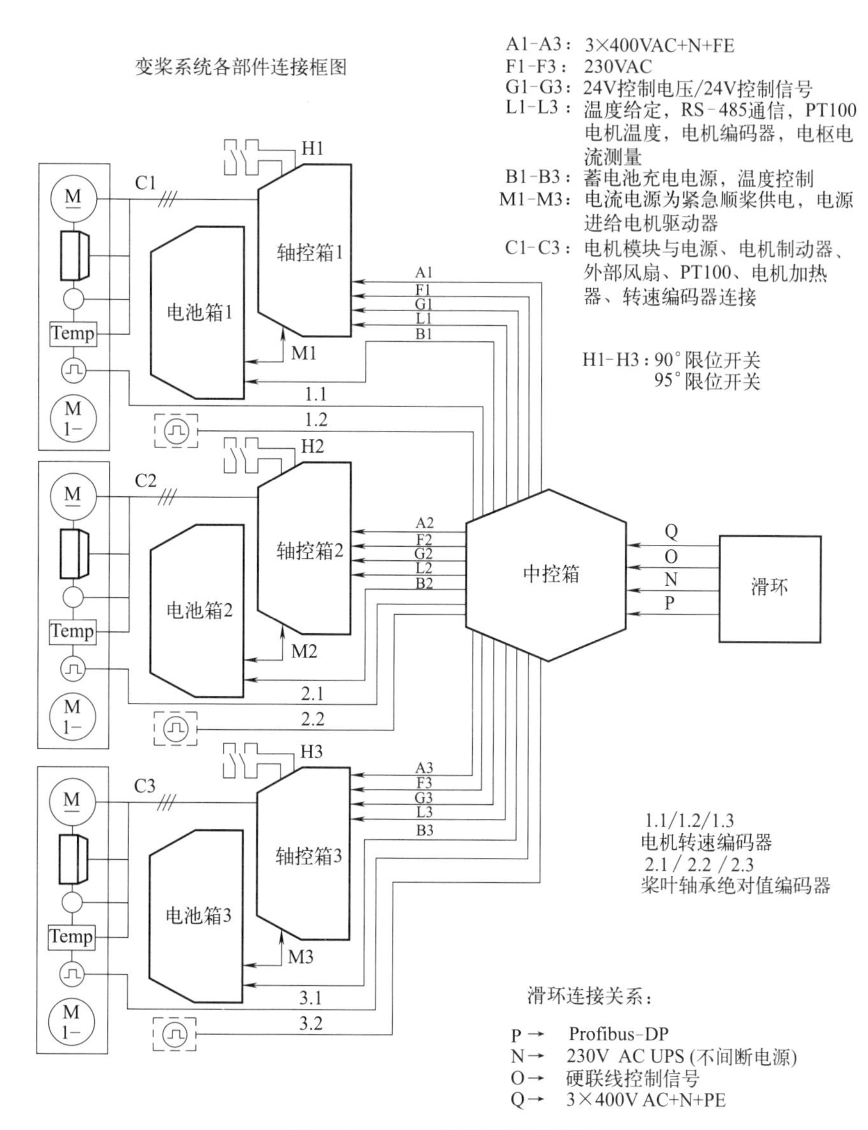

图 3.2.12　各部件间连接框图

(1) 中控箱

变桨中央控制箱执行轮毂内的轴控箱和位于机舱内的机舱控制柜之间的连接工作，如图 3.2.14 所示。

机舱控制柜通过滑环向变桨中央控制柜提供电能和控制信号。另外，风力发电机组控制系统和变桨控制器之间用于数据交换的 Profibus-DP 的连接也通过这个滑环实现。

变桨控制器位于变桨中央控制箱内，用于控制叶片的位置。另外，三个电池箱内的电池组的充电过程由安装在变桨中央控制箱内的中央充电单元控制。

在变桨系统内有三个轴控箱，每个叶片分配一个轴控箱，如图 3.2.15 所示。箱内的变流器控制变桨电机速度和方向。

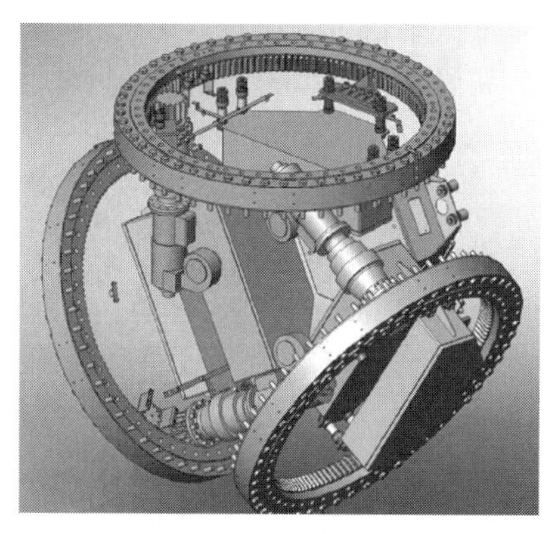

图 3.2.13　变桨机构机械连接

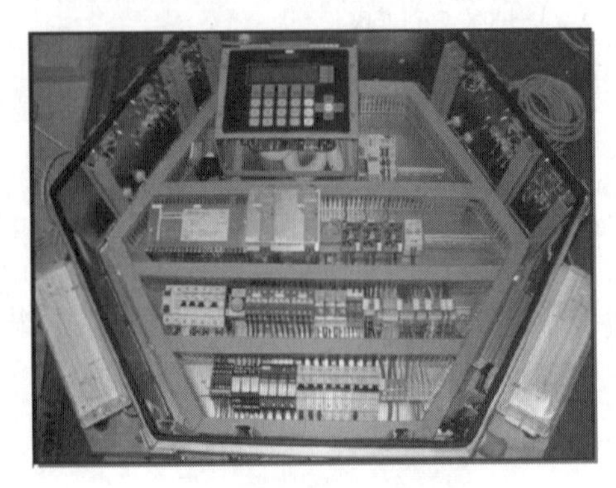

图 3.2.14　中控箱

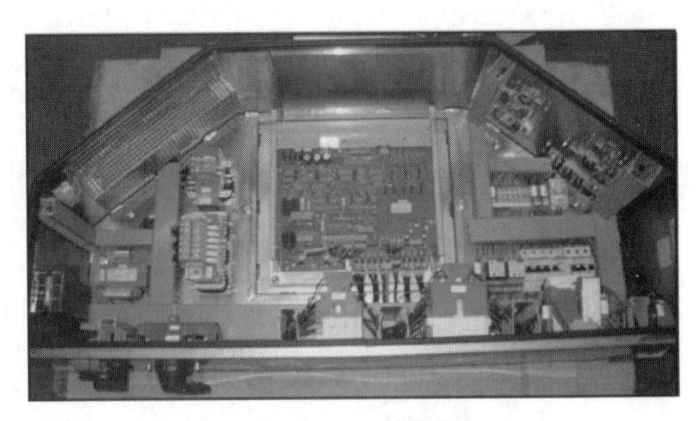

图 3.2.15　轴控箱

(2) 电池箱

和轴控箱一样，每个叶片分配一个电池箱，如图 3.2.16 所示。在供电故障或 EFC 信号（紧急顺桨控制信号）复位的情况下，电池供电控制每个叶片转动到顺桨位置。

(3) 变桨电机

变桨电机（图 3.2.17）是直流电机，正常情况下电机受轴控箱变流器控制转动，紧急顺桨时电池供电电机动作。

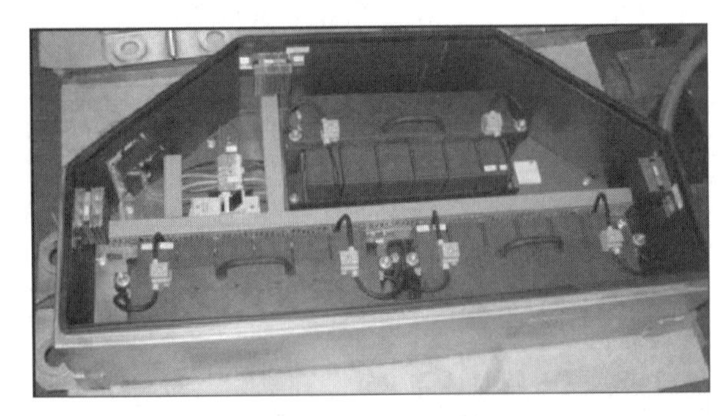

图 3.2.16　电池箱

图 3.2.17　变桨电机

(4) 绝对值编码器和冗余编码器

MW 级以上大型风力发电机组大多采用电动变桨系统，在每个变桨电机后部安装一个绝对值编码器（A 编码器），连接到变桨控制箱作为叶片位置反馈信号。在叶根处安装一个冗余编码器（B 编码器），冗余编码器连接到变桨控制箱作为叶片位置的冗余反馈信号，如图 3.2.18 所示。

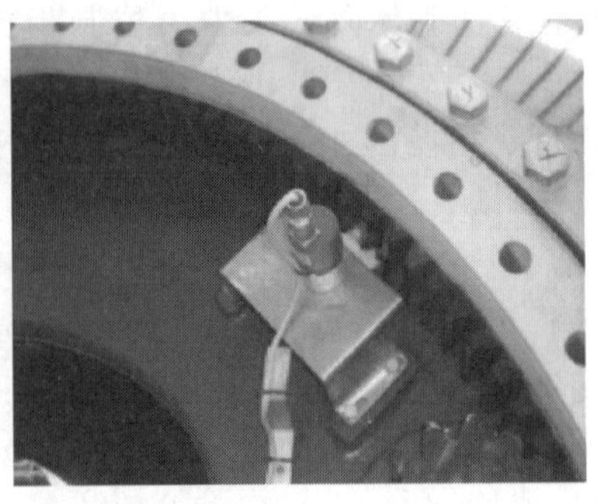

图 3.2.18　绝对值编码器和冗余编码器

(5) 限位开关

每个叶片对应两个限位开关：90°限位开关和95°限位开关。95°限位开关作为冗余开关使用，如图3.2.19所示。

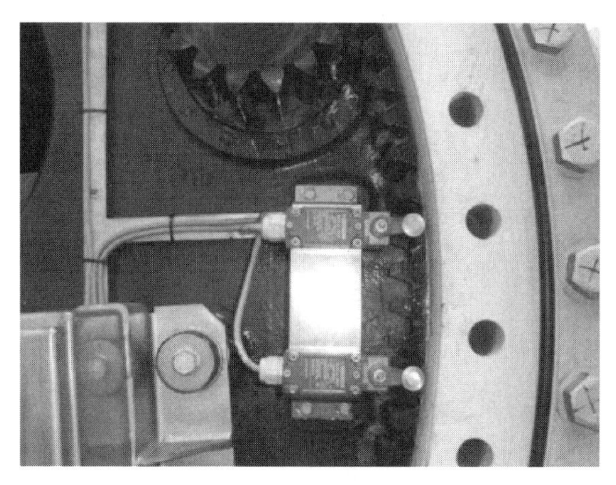

图3.2.19　限位开关

(6) 各部件间连接电缆

变桨中央控制箱、轴控箱、电池箱、变桨电机、冗余编码器和限位开关之间通过电缆进行连接。为了防止连接电缆时产生混乱，电缆有各自的编号。

项目实施

一、仪表、设备、工具清单

① THWPWG-3B型大型风力发电系统实训平台风机对象模型；

② 内六方扳手1套；

③ 尖嘴钳1把；

④ 十字螺丝刀、一字螺丝刀各1把。

二、安全操作规范

① 确保所有电源处于关闭状态；

② 拆装叶片与变桨轴承时，需利用梯子进行高空作业，要求梯子必须平稳安放在平整的地面上；

③ 上下梯子和站在梯子上作业时，要踩实踩稳，专心致志，不得嬉戏打闹；

④ 作业时4人一组，2人分别站在2个梯子上协作拆装，另2人在地面负责安全保护和传递工具、零部件；

⑤ 当心坠落、踩空，当心工具或零部件掉落，工具和零部件不得放在设备上面；

⑥ 负责保护的人员，禁止站在待拆除部件下面；

⑦ 拆装过程中注意零部件和工具的放置，要求分门别类、整齐、不妨碍操作，要特别注意连接螺栓、螺母、垫片等小件的放置，避免丢失和损坏。

三、实训步骤

① 负责拆装的2人，正确使用工具，遵照对角拧松法，依次将叶片、叶片安装座、变

桨限位开关、轴承上压盖、深沟球轴承、变桨轴承内齿圈、变桨轴承座、变桨小齿轮等零部件拆下，拆叶片时注意叶片的角度位置；

② 遵照对角拧紧法，将拆下的零部件组装还原，注意叶片的角度位置也要还原；

③ 在上述 2 人拆装的过程中，另外 2 人负责扶好梯子，随时注意在梯子上作业的人员和自身的安全，同时为负责拆装的人员递送工具盒零部件，如图 3.2.20 所示；

图 3.2.20　协同作业

④ 两组交换任务，再拆装一遍；

⑤ 撰写实训报告。

项目作业

（1）根据实训操作过程中的切身体会，描述什么是对角拧紧法。

（2）超级电容与备用电池箱（蓄电池）相比，有什么优势？

项目三　能源储存系统安装与调试

项目描述

能源转换储存控制系统主要由直流电压电流采样模块、温度告警模块、PWM 驱动模块、CPU 核心模块、人机交互模块、通信模块、防雷器、智能型充放电控制器、蓄电池组、开关电源、直流电压表、直流电流表等组成。

能力目标：

① 具备掌握各个功能模块原理的能力；

② 具备读懂电路原理图与接线图的能力；

③ 具备在完成系统安装后能够实现调试的能力。

项目环境

能源储存控制单元主要由直流电压表、电流表、断路器、开关电源、MPPT 控制器、蓄电池组、智能充放电控制器组成。

一、各模块功能说明

① 直流电压表、电流表：监测电量参数，如风机输出电压、风机输出电流等。

② 断路器：主要完成各支路的断开与接入。

③ 连接模块：在控制对象与控制器之间起连接作用，方便插拔连接线。

④ 开关电源：为电路和设备提供稳定的直流电源。

⑤ MPPT 控制器：主要完成最大功率跟踪（MPPT）算法。

⑥ 蓄电池组：无风时可以供给电力，也可以在发电电力突变时起缓冲器作用、储存电能、电力调峰等扩大系统的使用范围，提高并网系统的附加经济值。

⑦ 智能充放电控制器：根据蓄电池电压高低，调节充电状态和电流的大小，防止蓄电池过充或过放，延长蓄电池使用寿命。

其外形如图 3.3.1 所示。

能源转换储存控制系统主要设备技术参数见表 3.3.1。

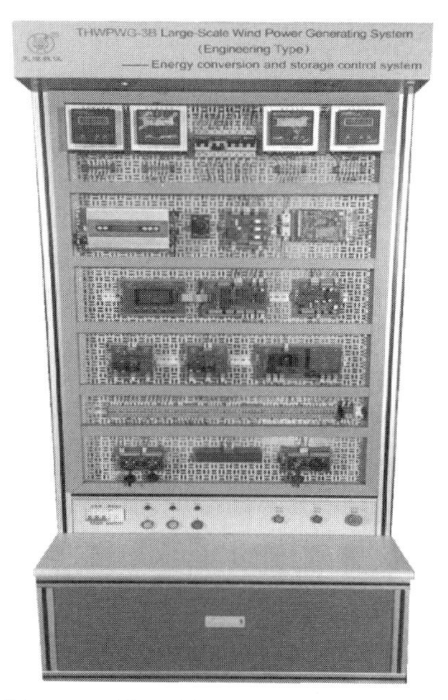

图 3.3.1　能源转换储存控制系统控制柜

表 3.3.1　能源转换储存控制系统主要设备技术参数

序号	名称	规格	数量
1	能源转换储存控制系统控制柜	880mm×620mm×2120mm	1 台
2	直流电压电流采样模块	输入：0～60V，0～5A　输出：0～5V	1 个
3	PWM 驱动模块	PWM 波频率：19.2kHz 占空比调节范围：0～90%	1 个
4	人机交互模块	分辨率：128×64	1 个
5	智能型充放电控制器	功率：600W 蓄电池额定电压：12/24V 自动切换	1 台
6	蓄电池组	电压：12V　容量：24A・h	4 块
7	开关电源	额定输入电压：AC 220V 额定输出电压：DC 24V 额定功率：35W	1 个
8	直流电压表	输入电压范围：0～500V 精度：0.5%±5 个字 通信：RS485 通信接口	2 个
9	直流电流表	输入电流范围：0～5A 精度：0.5%±5 个字 通信：RS485 通信接口	2 个

二、电路框图

能源储存控制单元电路框图如图 3.3.2 所示。

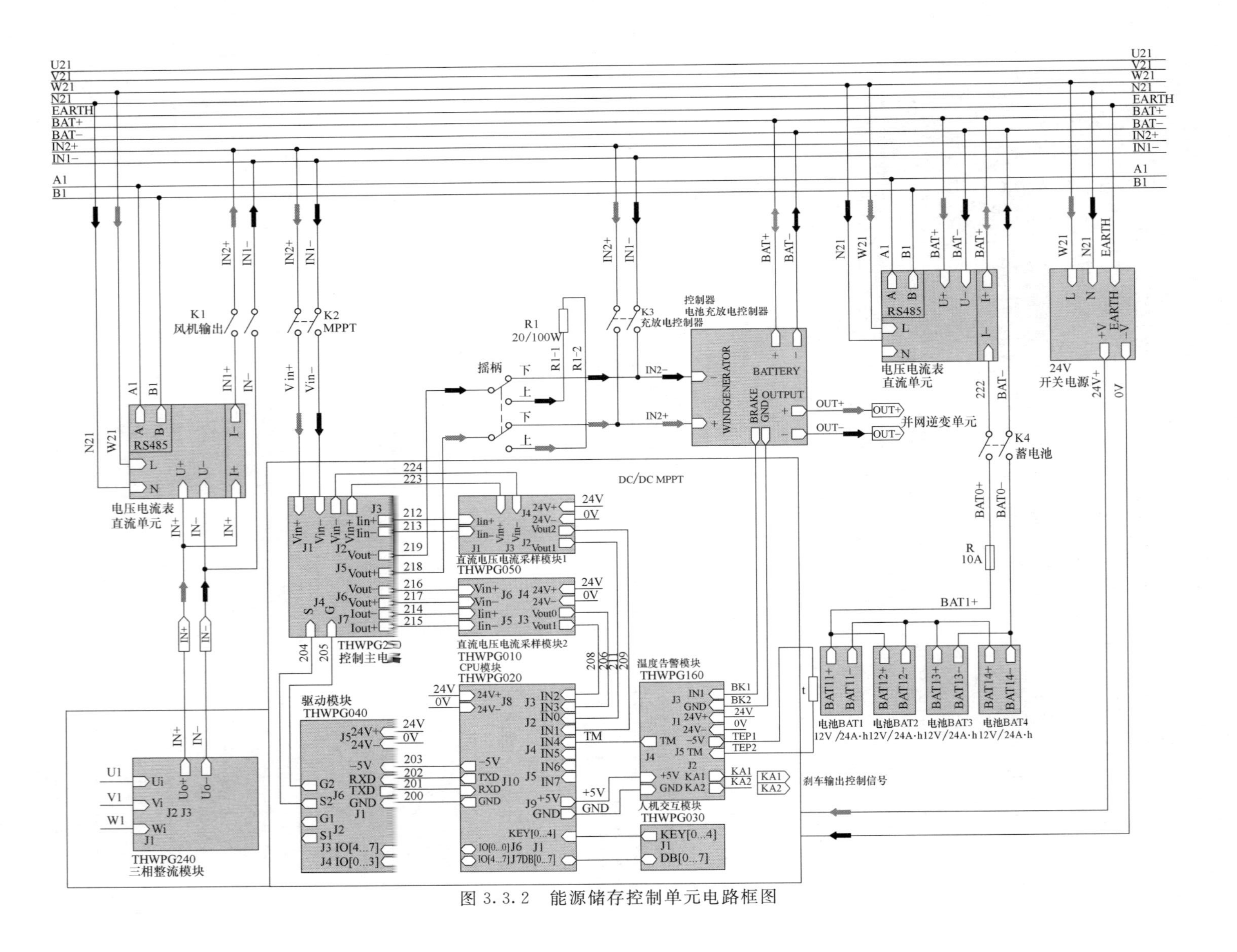

图 3.3.2　能源储存控制单元电路框图

项目原理及基础知识

一、直流单元模块

直流电压、电流表：监测电量参数，如风机输出电压、风机输出电流等。

(1) 主要功能

① 直流电压（电流）测量、显示。

② RS485、串行输出、模拟量输出和上下限告警、控制（接点输出）。

③ 可手动或上位机设定地址、波特率、显示数值、上下限报警阈值和回差。

(2) 技术指标

① 特点。

模块化设计，可根据用户需要选择单显示、显示加其他功能。

采用最新 PIC 芯片，抗干扰能力强。

② 常规说明。

外壳材料：阻燃塑料。

隔离：电源/输入/输出相互隔离。

③ 性能指标。

最大显示：±19999。

显示分辨率：0.001。

最大输入范围：电压：DC 0～600V　电流：DC 0～5A。

输入方式：单端输入。

精度等级：0.5 级。

吸收功率：＜1.2V·A。

测量速度：约 5 次/s。

输出负载能力：≤300Ω。

电源：AC 220V±20%，50/60Hz。

工频耐压：电源/输入/输出间：AC 2kV/min·1mA。

脉冲串干扰（EFT）：2kV 5kHz。

浪涌冲击电压：2kV 1.2/50μs。

绝缘电阻：≥100MΩ。

(3) 按键操作说明

按键操作说明见表 3.3.2。

表 3.3.2　按键操作说明表

显示字符	对应定义	数据范围
PASS	精度调节	0～9999
Baud	波特率	1200、2400、4800、9600、19200
dISP	显示数值	0～19999
ADDr	本机地址	1～32
rHNu	继电器输出上限阈值	上限阈值＞下限阈值时，区间外告警 上限阈值＜下限阈值时，区间内告警 上限阈值＝下限阈值时，关闭告警功能
rLHu	继电器输出下限阈值	
SAVE	保存退出	
E	不保存直接退出	
r_rE	回差	
DECi	小数点位置	

选定的功能不同，菜单也会有相应的增减。可操作按键有 set、←和→键。按“set”键进入调试程序主菜单，LED 显示 PASS，按“←”键显示 dISP、DECi、SAVE、E，按“→”键反向循环显示，按“←”或“→”键到需要设定项，按“set”键进入修改定项菜单。

当以下每项调整到所需数值后，可按“set”键退回到主菜单。

① dISP：LED 显示 00000（若已调整小数点显示位则相应位置显示小数点）。同时修改位闪动，按“→”修改位加 1，按“←”键修改位向左移动一位，如此循环，调至所需数值（显示数值对应满输入时显示数值，例：输入 DC 100V，可设置显示为 100.00）。

② DECi：LED 显示当前小数点位置，按“→”键小数点右移一位，移至最后一位单位指示翻转（若无单位显示则忽略），按“←”键。小数点左移一位，移至最高位单位指示翻转（若无单位则忽略）。

③ 本机地址（ADDr）：LED 闪动显示本机地址，按动“←”键数值加 1，按动“→”键数值减 1，调至相应数值（1～32）。

④ 波特率（Baud）：LED 闪动显示当前设置波特率（1200～19200），按动“←”键或“→”键更改数值。

⑤ rHNu/rLNu：LED 显示当前动作阈值，同时修改位闪动，按动“→”键修改位加一，按动“←”键修改位向左移动一位，如此循环，调至实际动作阈值。

⑥ 继电器动作回差设置（r_rE）：LED 显示当前动作回差（回差值为实际测量的最后两位数值），同时修改位闪动，按动“→”键修改位加 1，按动“←”键修改位向左移动一位，如此循环，调至实际动作回差。

⑦ 精度调节（PASS）：显示精度调整，修改精度时需相应的密码，无需用户调整。

显示 S 时按“set”保存修改并退出。显示 E 时按“set”忽略修改直接退出。说明：在任何一级菜单下，按键后，无操作时间大于 60s 后系统将自动退出设置菜单。

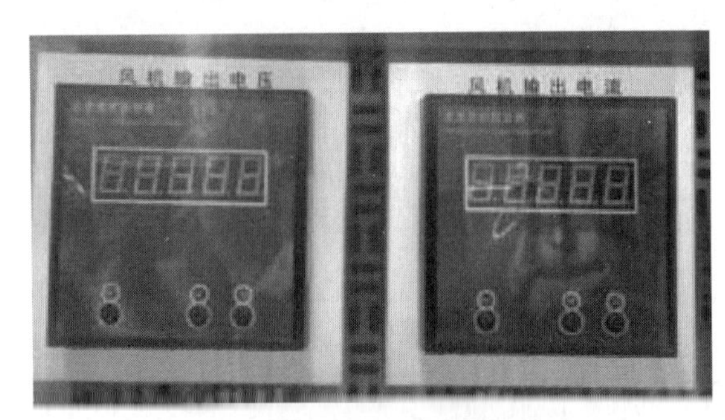

图 3.3.3　直流单元模块

(4) 实物图

直流单元模块如图 3.3.3 所示。

(5) 端口定义

直流电压表和直流电流表端口定义分别见表 3.3.3 和表 3.3.4。

表 3.3.3　直流电压表

序号	定义	说明
1	U+	被测电压输入
2	U−	
3	A	RS485
4	B	
5	L	交流 220V 电源输入
6	N	

表 3.3.4　直流电流表

序号	定义	说明
1	I+	被测电流输入
2	I−	
3	A	RS485
4	B	
5	L	交流 220V 电源输入
6	N	

二、开关电源

利用现代电力技术控制开关晶体管开通和关断时间比例，维持稳定输出电压，一般由脉冲宽度调制（PWM）控制 IC 和 MOSFET 构成。该系统中使用的开关电源（图 3.3.4）为

MPPT控制器功能模块提供+24V直流电源。

图3.3.4　开关电源

(1) 特点

开关：电力电子器件工作在开关状态。

高频：电力电子器件工作在高频。

直流：开关电源输出的是直流电。

(2) 结构组成

大部分常用的开关电源是为电子设备提供直流电源，电子设备所需要的直流电压一般都在几伏到十几伏，而交流市电电源供给的电压为220V（110V），频率为50Hz（60Hz）。开关电源的作用就是把一个高电压等级的工频交流电变换成一个低电压等级的直流电。工频交流电进入开关电源后被直接整流，因此省去了体积大、重量大的工频整流变压器。

整流器输出为电压很高的直流电，整流后的电压经电容滤波，电压的平均值为300～310V。高电压等级的直流电送往逆变器的输入端，经逆变器变换，变为高电压、高频交流电，目前开关电源逆变器的变换工作频率在几十千赫兹到几百千赫兹。逆变器输出的交流电能接高频降压变压器的原边，由于经逆变器产生的高频交流电的频率比工频高得多，所以高频变压器的体积要比同容量的工频变压器小得多，从根本上减小了整个电源的体积和重量。

逆变器产生的高频交流电经高频变压器降压后，再经过整流、稳压等环节，变换出符合负载要求的低压直流电能，供给负载。从开关电源的结构中可以看出，电路中没有调整管，不会消耗额外的能量，所有电子器件都工作在开关状态，如果忽略开关器件的导通压降和电路的杂散电阻，电路的效率应为1。工频整流电路一般为不可控整流电路，根据电源容量的大小，可以是单相整流，一般选用单相桥式结构，大容量的开关电源可用三相交流电源。

(3) 分类

在开关电源技术领域是边开发相关电力电子器件，边开发开关变频技术，两者相互促进推动着开关电源每年以超过两位数字的增长率向着轻、小、薄、低噪声、高可靠、抗干扰的方向发展。

开关电源可分为AC/DC和DC/DC两大类，DC/DC变换器现已实现模块化，且设计技术及生产工艺在国内外均已成熟和标准化，并已得到用户的认可，但AC/DC的模块化，因其自身的特性使得在模块化的进程中，遇到较为复杂的技术和工艺制造问题。

(4) 选用

开关电源在输入抗干扰性能上，由于多级串联，一般的输入干扰如浪涌电压很难通过。在输出电压稳定度这一技术指标上与线性电源相比有较大优势，其输出电压稳定度可达(0.5～1)%。开关电源作为一种电力电子集成器件，在选用中应注意：

① 输出电流的选择。因开关电源工作效率高，一般可达到80%以上，故在其输出电流的选择上，应准确测量或计算用电设备的最大吸收电流，以使被选用的开关电源具有高的性能价格比，通常输出计算公式为：$I_s=KI_f$。式中，I_s为开关电源的额定输出电流；I_f为用电设备的最大吸收电流；K为裕量系数，一般取1.5～1.8。

② 接地。开关电源比线性电源会产生更多的干扰，对共模干扰敏感的用电设备，应采取接地和屏蔽措施，按ICE1000、EN61000、FCC等EMC限制，开关电源均采取EMC电磁兼容措施，因此开关电源一般应带有EMC电磁兼容滤波器。如利德华福的HA系列开关

电源，将其 FG 端子接大地或接用户机壳，方能满足上述电磁兼容的要求。

③ 保护电路。开关电源在设计中必须具有过流、过热、短路等保护功能，故在设计时应首选保护功能齐备的开关电源，并且其保护电路的技术参数应与用电设备的工作特性相匹配，以避免损坏用电设备或开关电源。

(5) 技术指标

输入：115V AC 0.8A；230V AC 0.45A；50/60Hz。

输出：+24V 1.5A

尺寸：130mm×98mm×36mm。

(6) 端口定义

开关电源端口定义见表 3.3.5。

表 3.3.5 端口定义表

序号	定义	说明	序号	定义	说明
1	L	交流电源输入	4	V−	0V
2	N		5	V+	+24V 输出
3	PE	地线			

三、密封铅酸蓄电池

1. 结构特点

这种电池虽然也是铅酸蓄电池，但与原来的铅酸蓄电池相比具有很多优点，特别是受那些需要将电池配套设备安装在一起的用户青睐。

这是因为 VRLA 电池是全密封的，不会漏酸，而且在充放电时不会像老式铅酸蓄电池那样会有酸雾放出来而腐蚀设备，污染环境，所以从结构特性上人们把 VRLA 电池又叫作密闭（封）铅酸蓄电池。为了区分，把老式铅酸蓄电池叫作开口铅酸蓄电池。由于 VRLA 电池从结构上来看，它不但是全密封的，而且还有一个可以控制电池内部气体压力的阀，所以 VRLA 铅酸蓄电池的全称便成了“阀控式密闭铅酸蓄电池”。

2. 技术参数

铅酸蓄电池的电性能用下列参数量度：电池电动势、开路电压、终止电压、工作电压、放电电流、容量、内阻、使用寿命（浮充寿命、充放电循环寿命）、能量、储存性能等。

(1) 电池电动势、开路电压、工作电压

当蓄电池用导体在外部接通时，正极和负极的电化学反应自发地进行，倘若电池中电能与化学能转换达到平衡时，正极的平衡电极电势与负极平衡电极电势的差值，便是电池电动势，它在数值上等于达到稳定值时的开路电压。

电池电动势与单位电量的乘积，表示单位电量所能做的最大电功。但电池电动势与开路电压意义不同。电池电动势可依据电池中的反应利用热力学计算或通过测量计算，有明确的物理意义；电池在开路状态下的端电压称为开路电压。电池的开路电压等于电池正极电极电势与负极电极电势之差。电池工作电压是指电池有电流通过（闭路）的端电压。电池放电初始的工作电压称为初始电压。电池在接通负载后，由于欧姆电阻和极化过电位的存在，电池的工作电压低于开路电压。

(2) 容量

电池容量是指电池储存电量的数量，以符号 C 表示。常用的单位为安时（A·h）或毫安时（mA·h）。电池容量可以分为额定容量（标称容量）、实际容量。

放电率：放电率针对蓄电池放电电流大小，分为时间率和电流率。放电时间率指在一定放电条件下，放电至终了电压时间长短。依据 IEC 标准，放电时间率有 20、10、5、3、1、0.5 小时率等，分别表示为：20Hr、10Hr、5Hr、3Hr、2Hr、1Hr、0.5Hr 等。

放电终止电压：铅蓄电池以一定的放电率在 25℃ 环境温度下放电至能再反复充电使用的最低电压称为放电终止电压。大多数固定型电池规定以 10Hr 放电时（25℃）终止电压为 1.8V/只。终止电压值视放电速率和需要而定。通常，为使电池安全运行，以小于 10Hr 的小电流放电时，终止电压取值稍高，以大于 10Hr 的大电流放电时，终止电压取值稍低。在通信电源系统中，蓄电池放电的终止电压，由通信设备对基础电压要求而定。

放电电流率：是为了比较标称容量不同的蓄电池放电电流大小而设的，通常以 10Hr 电流为标准，用 I10 表示，3Hr 及 1Hr 放电电流则分别以 I3、I1 表示。

额定容量：固定铅酸蓄电池规定在 25℃ 环境下，以 10Hr 电流放电至终了电压所能达到的额定容量。10Hr 额定容量用 C10 表示。10Hr 的电流值为 C10/10。其他小时率下容量表示方法为：3Hr 容量（A·h）用 C3 表示，在 25℃ 环境温度下实测容量（A·h）是放电电流与放电时间（h）的乘积，阀控铅酸固定型电池 C3 和 I3 值应该为 C3＝0.75×C10（A·h），I3＝2.5×I10（A），1 小时定容量（A·h）用 C1 表示，实测 C1 和 I1 值应为 C1＝0.55×C10（A·h），I1＝5.5×I10（A）。

实际容量：电池在一定条件下输出电量。等于放电电流与时间乘积，单位为 A·h。

(3) 内阻

电池内阻包括欧姆内阻和极化内阻，极化内阻又包括电化学极化与浓差极化。内阻的存在，使电池放电时的端电压低于电池电动势和开路电压，充电时端电压高于电池电动势和开路电压。电池的内阻不是常数，在充放电过程中随时间不断变化，因为活性物质的组成、电解液浓度和温度都在不断地改变。欧姆电阻遵守欧姆定律；极化电阻随电流密度增加而增大，但不是线性关系，常随电流密度的对数增大而线性增大。

(4) 循环寿命

蓄电池经历一次充电和放电，称为一次循环。在一定放电条件下，电池工作至某一容量规定值前，电池所能承受的循环次数，称为循环寿命。各种蓄电池使用循环次数有差异，传统固定型铅酸电池为 500～600 次，起动型铅酸电池为 300～500 次。阀控式密封铅酸电池循环寿命为 1000～1200 次。影响循环寿命的因素：一是厂家产品的性能，二是维护工作质量。固定型铅电池使用寿命，还可以用浮充寿命（年）来衡量。对于起动型铅酸蓄电池，按我国颁布的国家标准，采用过充电耐久能力及循环耐久能力单元数来表示寿命，而不采用循环次数表示寿命。

(5) 能量

电池的能量是指在一定放电制度下，蓄电池所能给出的电能，通常用瓦时（W·h）表示。电池的能量分为理论能量和实际能量。理论能量 W 可用理论容量和电池电动势（E）的乘积表示，即 $W_{理}=C_{理}E$，电池的实际能量为一定放电条件下的实际容量“$C_{实}$”与平均工作电压“$U_{平}$”的乘积，即 $W_{实}=C_{实}U_{平}$。

常用比能量来比较不同的电池系统。比能量是指电池单位质量或单位体积所能输出的电能，单位分别是 W·h/kg 或 W·h/L。比能量有理论比能量和实际比能量之分。前者指 1kg 电池反应物质完全放电时理论上所能输出的能量。实际比能量为 1kg 电池反应物质所能输出的实际能量。由于各种因素的影响，电池的实际比能量远小于理论比能量。

实际比能量和理论比能量的关系可表示如下：

$$W_{实}=W_{理}\times K_V\times K_R\times K_m$$

式中，K_V 为电压效率；K_R 为反应效率；K_m 为质量效率。电压效率是指电池的工作电压与电池电动势的比值。电池放电时，由于电化学极化、浓差极化和欧姆压降，工作电压小于电池电动势。反应效率表示活性物质的利用率。电池的比能量是综合性指标，它反映了电池的质量水平，也表明生产厂家的技术和管理水平。

(6) 储存性能

蓄电池在储存期间，由于电池内存在杂质，这些杂质可与负极活性物质组成微电池，发生负极金属溶解和氢气的析出现象。又如溶液中及从正极板栅溶解的杂质，若其标准电极电位介于正极和负极标准电极电位之间，则会被正极氧化，又会被负极还原。所以有害杂质的存在，使正极和负极活性物质逐渐被消耗，而造成电池容量丧失，这种现象称为自放电。电池自放电率用单位时间内容量降低的百分数表示，即用电池储存前（C10′）（C10″）容量差值和储存时间 T（天、月）的容量百分数表示。

3. 技术指标

单节蓄电池容量：12V 24A·h。单节蓄电池尺寸：165mm×125mm×175mm。

4. 实物图

蓄电池组如图 3.3.5 所示。

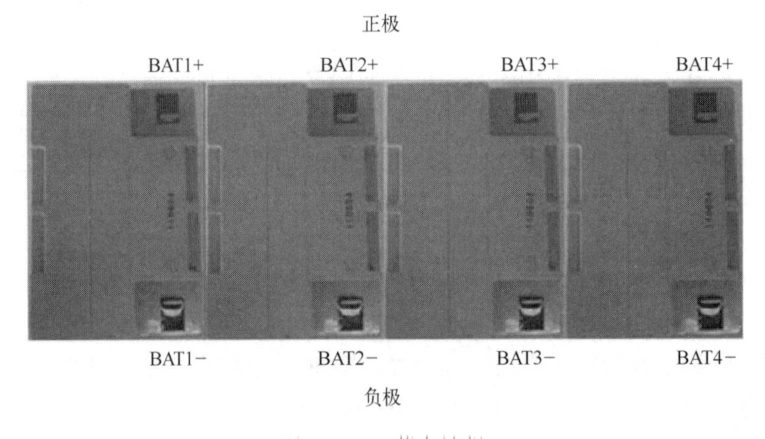

图 3.3.5 蓄电池组

5. 端口定义

蓄电池组端口定义见表 3.3.6。

表 3.3.6 端口定义表

序号	定义	说明
1	BATX+	蓄电池正极输出
2	BATX−	蓄电池负极输出

四、MPPT 控制器

MPPT 控制器完成对风机的最大功率跟踪，有效地提高了风机的工作效率，同时也改善了系统的工作性能，如图 3.3.6 所示。

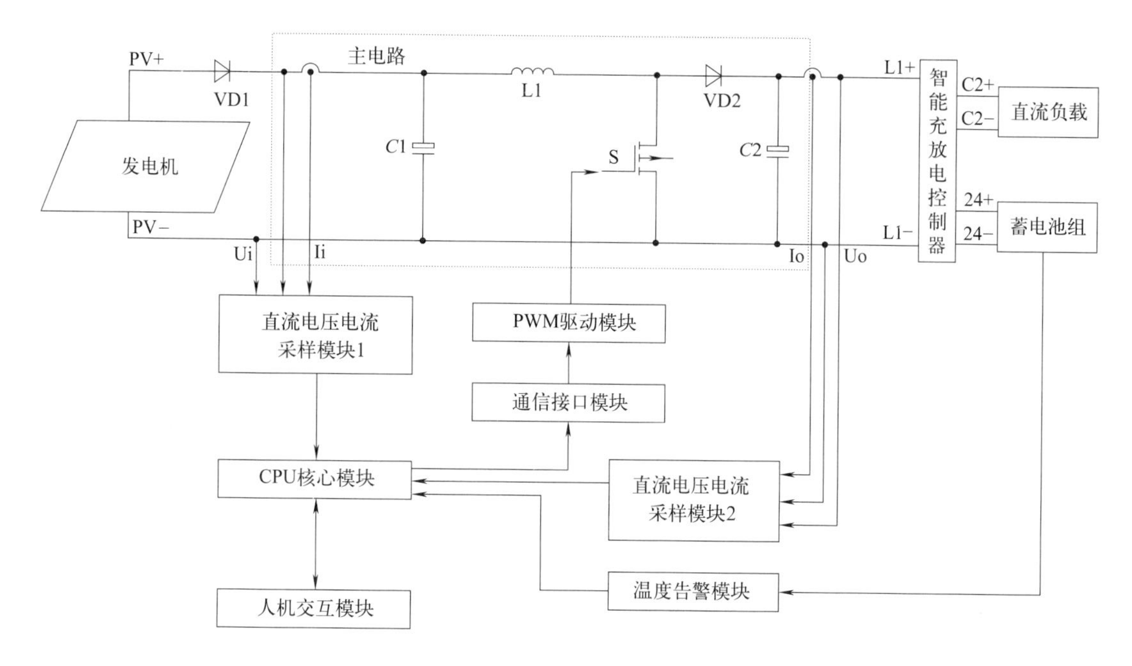

图 3.3.6　MPPT 控制器功能框图

1. MPPT 控制器各功能模块说明

(1) 直流电压电流采样模块 1

通过电压霍尔传感器和电流传感器将风机阵列电池输出的电压和电流转换成满足单片机输入端要求的电压信号。

(2) 人机交互模块

CPU 核心模块的输入、输出终端。

(3) 温度告警模块

检测蓄电池温度，超过参考温度时，液晶显示电池温度过高。

(4) PWM 驱动模块

接收最大功率跟踪微处理器输出的功率调节参数（占空比参数）并输出不同占空比的 PWM 信号，将 PWM 微处理输出的 PWM 信号与 DC/DC 电路隔离，并把 PWM 信号转换成满足开关管需求的驱动信号，提高驱动能力。

(5) 直流电压电流采样模块 2

通过电压霍尔传感器和电流传感器将输出的电压和电流转换成满足单片机输入端要求的电压信号。

(6) 主电路

通过实时采集风机阵列电池的功率来调整主电路的占空比，等效调节了负载的阻抗，使负载取用的功率得以改变，可始终跟随风机阵列电池输出的最大功率点，等效为一个阻抗变换器。

(7) 智能充放电控制器

根据蓄电池电压高低，调节充电状态和电流的大小，防止蓄电池过充或过放，延长蓄电池使用寿命。

2. CPU 核心模块

根据风机阵列电压、电流采样信号进行最大功率跟踪的程序设计和调试。这部分硬件对用户完全开放，用户可以编写不同的 MPPT 算法实现最大功率跟踪，并将调节参数（占空比参数）通过串口发送给 PWM 驱动模块进行调节。微处理器采用 51 系列单片机，具有在线下载功能，方便用户编程调试，实现 MPPT 控制算法。

（1）原理图

CPU 核心模块原理如图 3.3.7 所示。

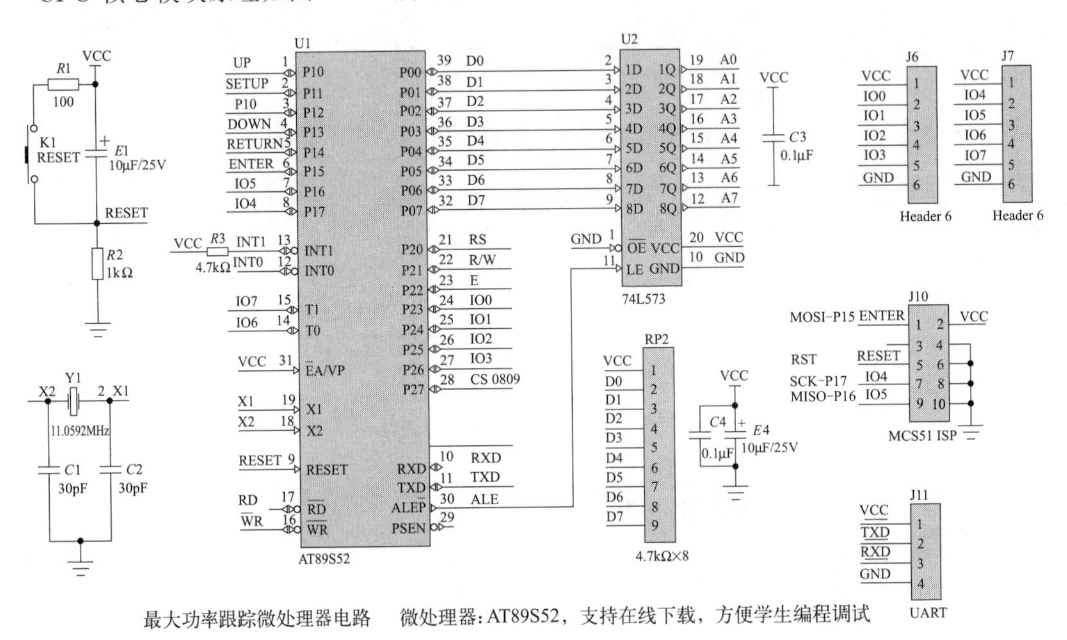

图 3.3.7　CPU 核心模块原理图

（2）实物照片

CPU 核心模块如图 3.3.8 所示。

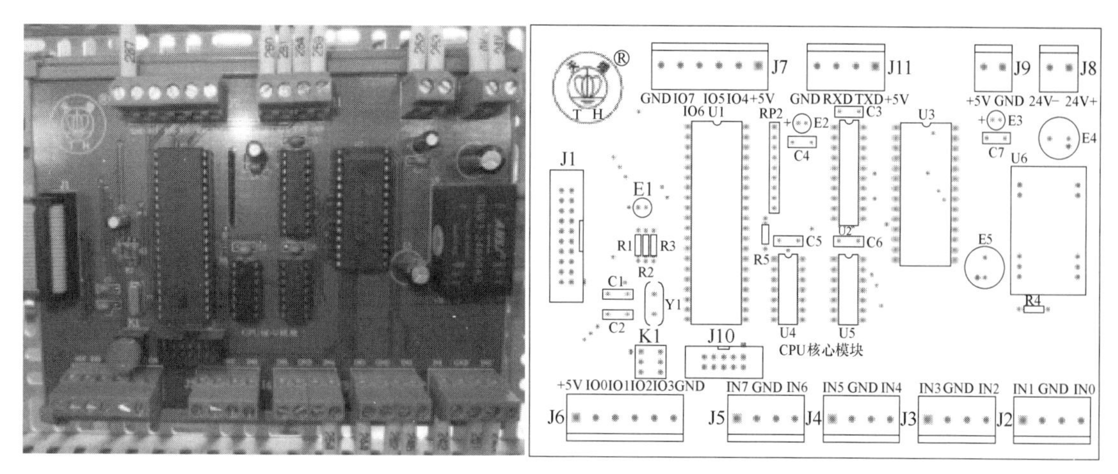

图 3.3.8 CPU 核心模块实物图

(3) 端口定义

CPU 核心模块端口定义如表 3.3.7 所示。

表 3.3.7 端口定义表

序号	名称	说明	扩展接口	备注
1	J2:IN0	AD 采样输入 0,5V 以内电压信号		
2	J2:GND			
3	J2:GND	AD 采样输入 1,5V 以内电压信号		
4	J2:IN1			
5	J3:IN2	AD 采样输入 2,5V 以内电压信号		
6	J3:GND			
7	J3:GND	AD 采样输入 3,5V 以内电压信号		
8	J3:IN3			
9	J4:IN4	AD 采样输入 4,5V 以内电压信号		
10	J4:GND			
11	J4:GND	AD 采样输入 5,5V 以内电压信号		
12	J4:IN5			
13	J5:IN6	AD 采样输入 6,5V 以内电压信号	√	
14	J5:GND		√	
15	J5:GND	AD 采样输入 7,5V 以内电压信号	√	
16	J5:IN7		√	
17	J6:+5V	+5V 电源输出	√	
18	J6:IO0	开关量输入输出接口 0,5V 以内电压信号	√	
19	J6:IO1	开关量输入输出接口 1,5V 以内电压信号	√	
20	J6:IO2	开关量输入输出接口 2,5V 以内电压信号	√	
21	J6:IO3	开关量输入输出接口 3,5V 以内电压信号	√	
22	J6:GND	地	√	
23	J7:+5V	+5V 电源输出	√	
24	J7:IO4	开关量输入输出接口 4,5V 以内电压信号	√	
25	J7:IO5	开关量输入输出接口 5,5V 以内电压信号	√	
26	J7:IO6	开关量输入输出接口 6,5V 以内电压信号	√	
27	J7:IO7	开关量输入输出接口 7,5V 以内电压信号	√	
28	J7:GND	地	√	
29	J11:+5V	+5V 电源输出		
30	J11:TXD	串行口发送端		

续表

序号	名称	说明	扩展接口	备注
31	J11:RXD	串行口接收端		
32	J11:GND	地		
33	J9:+5V	隔离电源 DC/DC 输出 5V 电源正极	√	
34	J9:GND	隔离电源 DC/DC 输出 5V 电源负极	√	
35	J8:24V+	隔离电源 DC/DC 输入 24V 正极		
36	J8:24V-	隔离电源 DC/DC 输入 24V 负极		
37	J1	至人机交互模块		20 排线

(4) 编程注意事项

MPPT 控制器内含有智能充放电控制器，由于智能充放电控制器在上电之后需要至少 1min 才能完成启动工作，因此在设计最大功率跟踪程序时需要延时 1min，以确保智能充放电控制器正常工作。

CPU 核心模块微处理器与 PWM 驱动模块微处理器之间的通信采用串口通信，波特率：9600bit/s；8 位数据位；1 位停止位；无校验位；1 个 8Bit 表示占空比，数据范围：0x00～0xFF（0x00：占空比为 0%；0xFF：占空比为 99%）。

四路待转换的模拟信号分别接在 ADC0809 的 IN0、IN1、IN2、IN3 输入通道上：汇编采用累加器 A 与外部数据存储器传送指令选择通道和读取 AD 值：MOVX @DPTR，A；MOVX A，@DPTR。各通道的访问地址如下所示：

① 风机输出电压（IN0）地址：0x7FF8；风机输出电池（IN1）地址：0x7FF9；

② Boost 变换器输出电流（IN2）地址：0x7FFA；Boost 变换器输出电压（IN3）地址：0x7FFB；

③ 选择通道用汇编指令：MOVX @DPTR，A，其中 DPTR 为地址，A 可以为任何值【0x00～0xFF】；

④ 读取 AD 值用汇编指令：MOVX A，@DPTR，其中 DPTR 为地址，A 为 AD 转换值。

四路待转换的模拟信号分别接在 ADC0809 的 IN0、IN1、IN2、IN3 输入通道上：采用 C 语言需要预定义各通道为外部数据区，定义如下：

① #define　II_CURRENT　XBYTE[0x7FF8]　/* 风机输出电流 */

② #define　UI_VOL　XBYTE[0x7FF9]/* 风机输出电压 */

③ #define　IO_CURRENT　XBYTE[0x7FFA]　/* Boost 变换器输出电流 */

④ #define　UO_VOL　XBYTE[0x7FFB]/* Boost 变换器输出电压 */ 选择通道用 C 语句：XXXX=X，其中 XXXX 为预定义名，X 可以为任意值【0x00～0xFF】。

⑤ 读取 AD 值用 C 语句：变量=XXXX，其中 XXXX 为预定义名，变量的内容为 AD 转换值。

五、人机交互模块

人机交互模块是 CPU 核心模块的输入、输出终端。

1. 原理图

人机交互模块原理如图 3.3.9 所示。

2. 12864 液晶屏

液晶是一种高分子材料，因为其特殊的物理、化学、光学特性，20 世纪中叶被广泛应用在轻薄型显示器上。

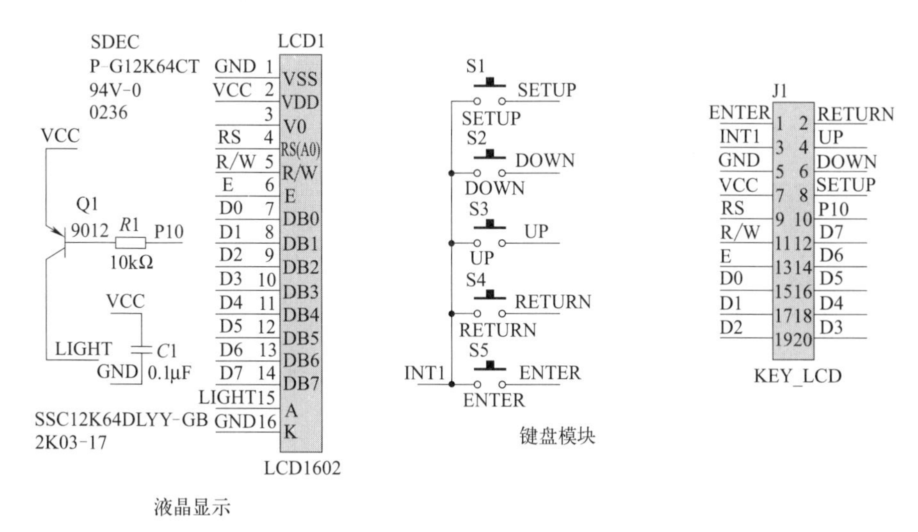

图 3.3.9　人机交互模块原理图

液晶显示器的主要原理是以电流刺激液晶分子产生点、线、面并配合背部灯管构成画面。为叙述简便，通常把各种液晶显示器都直接叫作液晶。

各种型号的液晶通常是按照显示字符的行数或液晶点阵的行、列数来命名的。比如：1602 的意思是每行显示 16 个字符，一共可以显示两行；类似的命名还有 0801、0802、1601 等，这类液晶通常都是字符型液晶，即只能显示 ASCII 码字符，如数字、大小写字母、各种符号等。12232 液晶属于图形型液晶，它的意思是液晶由 122 列、32 行组成，即共有 122×32 个点来显示各种图形，可以通过程序控制这 122×32 个点中的任一个点显示或不显示。类似的命名还有 12864、19264、192128、320240 等，根据客户需求，厂家可以设计出任意数组合的点阵液晶。

液晶体积小、功耗低、显示操作简单，但是它有一个致命的弱点，其使用的温度范围很窄，通用型液晶正常工作温度范围为 0～+55℃，存储温度范围为−20～+60℃，即使是宽温度级液晶，其正常工作温度范围也仅为−20～+70℃，存储温度范围为−30～+80℃，因此在设计相应产品时，务必要考虑周全，选取合适的液晶。

3. 实物照片

人机交互模块如图 3.3.10 所示。

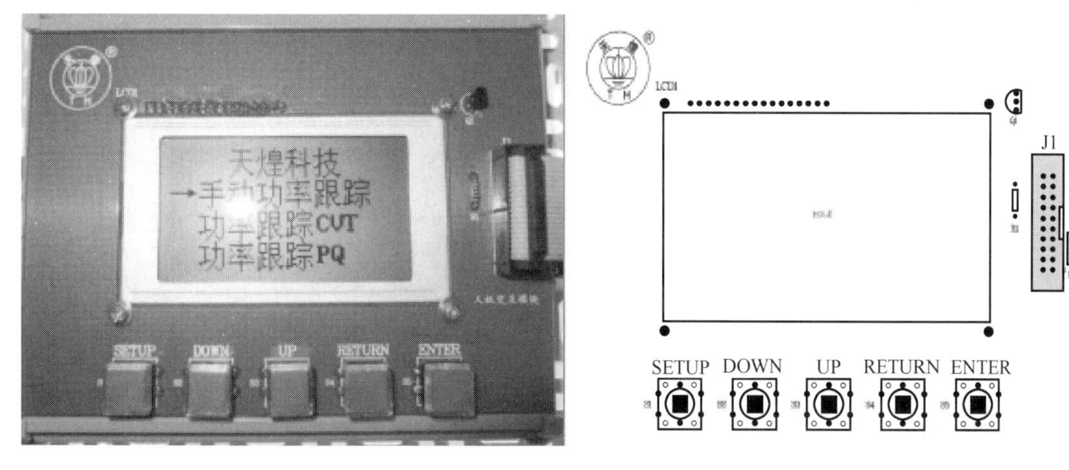

图 3.3.10　人机交互模块

4. 端口定义

人机交互模块端口定义表如表 3.3.8 所示。

表 3.3.8 端口定义表

序号	名称	说明	序号	名称	说明
1	ENTER	确定按键	2	RETURN	返回按键
3	INT1	按键公共端	4	UP	向上按键
5	GND	地	6	DOWN	向下按键
7	VCC	电源	8	SETUP	设置按键
9	RS	寄存器选择	10	P10	背光控制
11	R/W	读/写信号	12	D7	数据位 7
13	E	使能信号	14	D6	数据位 6
15	D0	数据位 0	16	D5	数据位 5
17	D1	数据位 1	18	D4	数据位 4
19	D2	数据位 2	20	D3	数据位 3

六、PWM 驱动模块

接收 CPU 模块输出的功率调节参数（占空比参数），经 CPU 处理并输出 PWM 信号，再经过光耦隔离给主电路模块。

1. 原理图

PWM 驱动模块原理如图 3.3.11 所示。

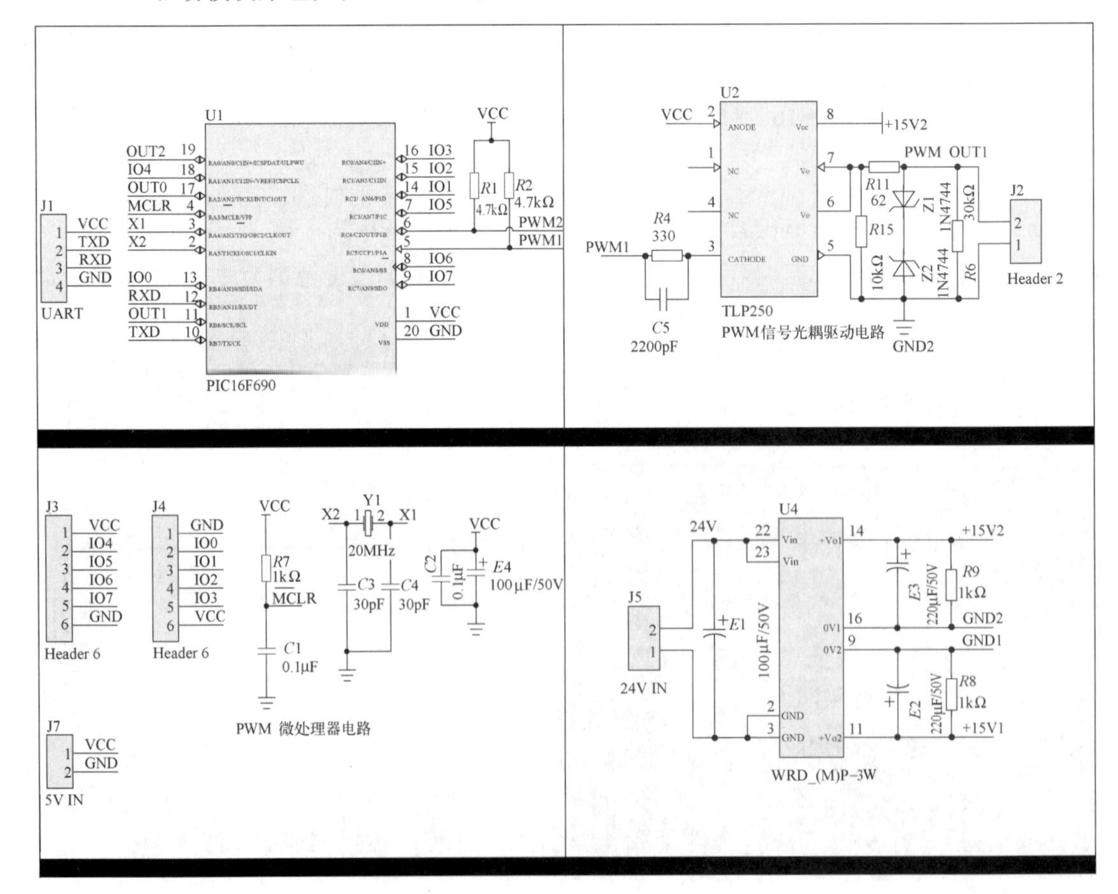

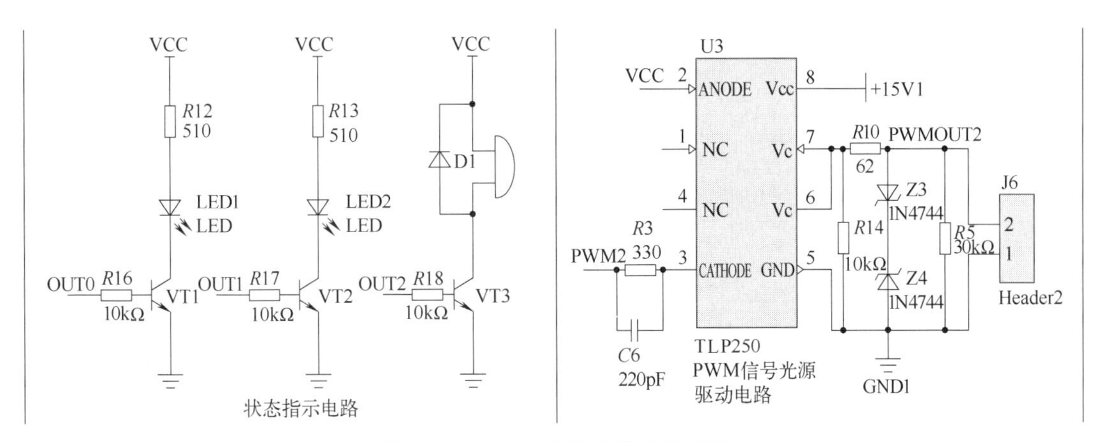

图 3.3.11 PWM 驱动模块原理图

2. 实物照片

PWM 驱动模块如图 3.3.12 所示。

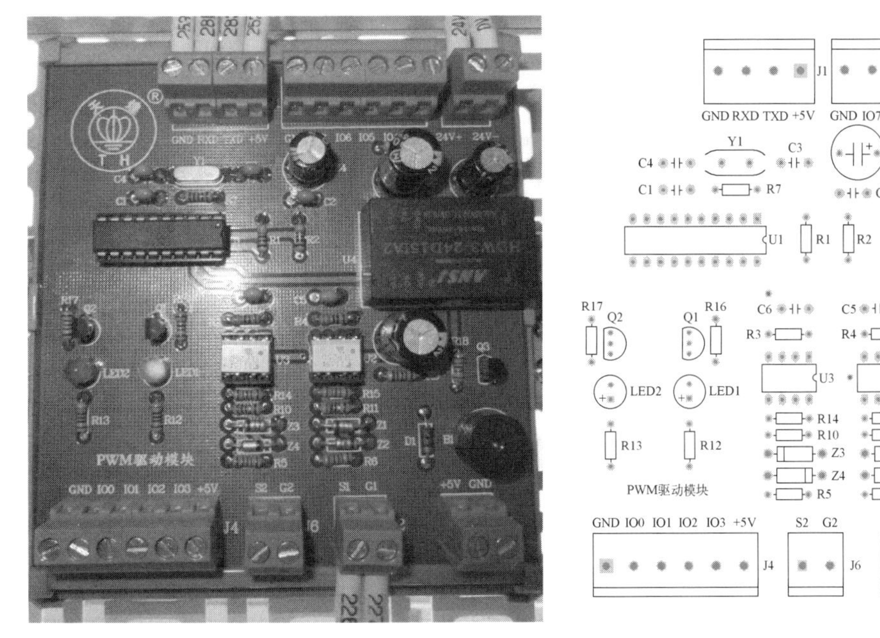

图 3.3.12 PWM 驱动模块

3. 端口定义

PWM 驱动模块端口定义如表 3.3.9 所示。

表 3.3.9 端口定义表

序号	名称	说明	扩展接口	备注
1	J4:GND	地	√	
2	J4:IO0	开关量输入输出接口 0，5V 以内电压信号	√	
3	J4:IO1	开关量输入输出接口 1，5V 以内电压信号	√	
4	J4:IO2	开关量输入输出接口 2，5V 以内电压信号	√	
5	J4:IO3	开关量输入输出接口 3，5V 以内电压信号	√	
6	J4:+5V	+5V 电源输出	√	
7	J3:+5V	+5V 电源输出	√	
8	J3:IO4	开关量输入输出接口 4，5V 以内电压信号	√	

续表

序号	名称	说明	扩展接口	备注
9	J3:IO5	开关量输入输出接口 5，5V 以内电压信号	√	
10	J3:IO6	开关量输入输出接口 6，5V 以内电压信号	√	
11	J3:IO7	开关量输入输出接口 7，5V 以内电压信号	√	
12	J3:GND	地	√	
13	J1:+5V	+5V 电源输入		
14	J1:TXD	串行口发送端		
15	J1:RXD	串行口接收端		
16	J1:GND	地		
17	J7:+5V	+5V 电源输出	√	
18	J7:GND	地	√	
19	J5:24V+	隔离电源 DC/DC 输入 24V 正极		
20	J5:24V−	隔离电源 DC/DC 输入 24V 负极		
21	J6:S1	隔离 PWM 驱动信号 1 负端		
22	J6:G1	隔离 PWM 驱动信号 1 正端		
23	J2:S1	隔离 PWM 驱动信号 2 负端	√	
24	J2:G1	隔离 PWM 驱动信号 2 正端	√	

七、温度告警模块

1. 原理图

温度告警模块原理如图 3.3.13 所示。

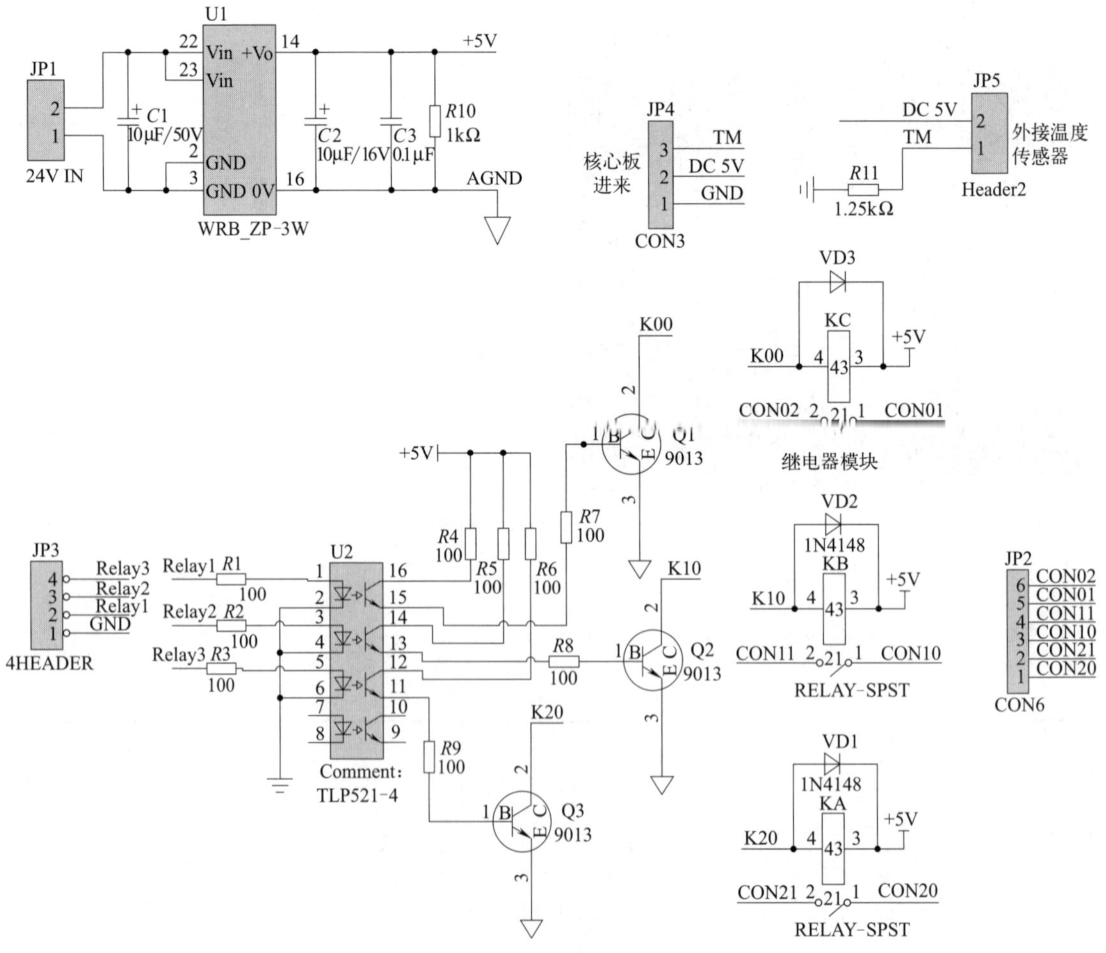

图 3.3.13　温度告警模块原理图

2. 实物照片

温度告警模块如图 3.3.14 所示。

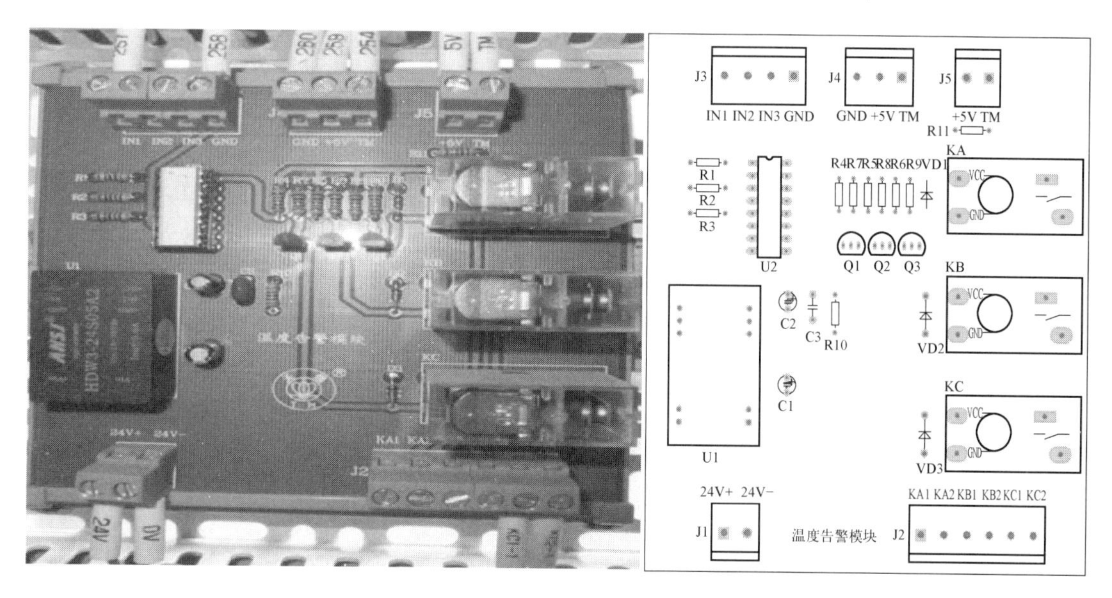

图 3.3.14 温度告警模块

3. 端口定义

温度告警模块端口定义如表 3.3.10 所示。

表 3.3.10 端口定义表

序号	名称	说明	扩展接口	备注
1	J1:24V+	隔离电源 DC/DC 24V 输入		
2	J1:24V−			
3	J2:KA1	KA 继电器	√	
4	J2:KA2		√	
5	J2:KB1	KB 继电器	√	
6	J2:KB2		√	
7	J2:KC1	KC 继电器		
8	J2:KC2			
9	J3:IN1	KA 继电器控制信号输入	√	
10	J3:IN2	KB 继电器控制信号输入	√	
11	J3:IN3	KC 继电器控制信号输入		
12	J3:GND	地		
13	J4:GND	+5V 电源输入		
14	J4:+5V			
15	J4:TM	温度模拟量输出		
16	J5:+5V	接温度传感器		
17	J5:TM			

八、直流电压电流采样模块 1

通过电压霍尔传感器和电流传感器将风机阵列输出的电压、电流转换成满足单片机输入端要求的电压信号。

1. 原理图

直流电压电流采样模块 1 原理如图 3.3.15 所示。

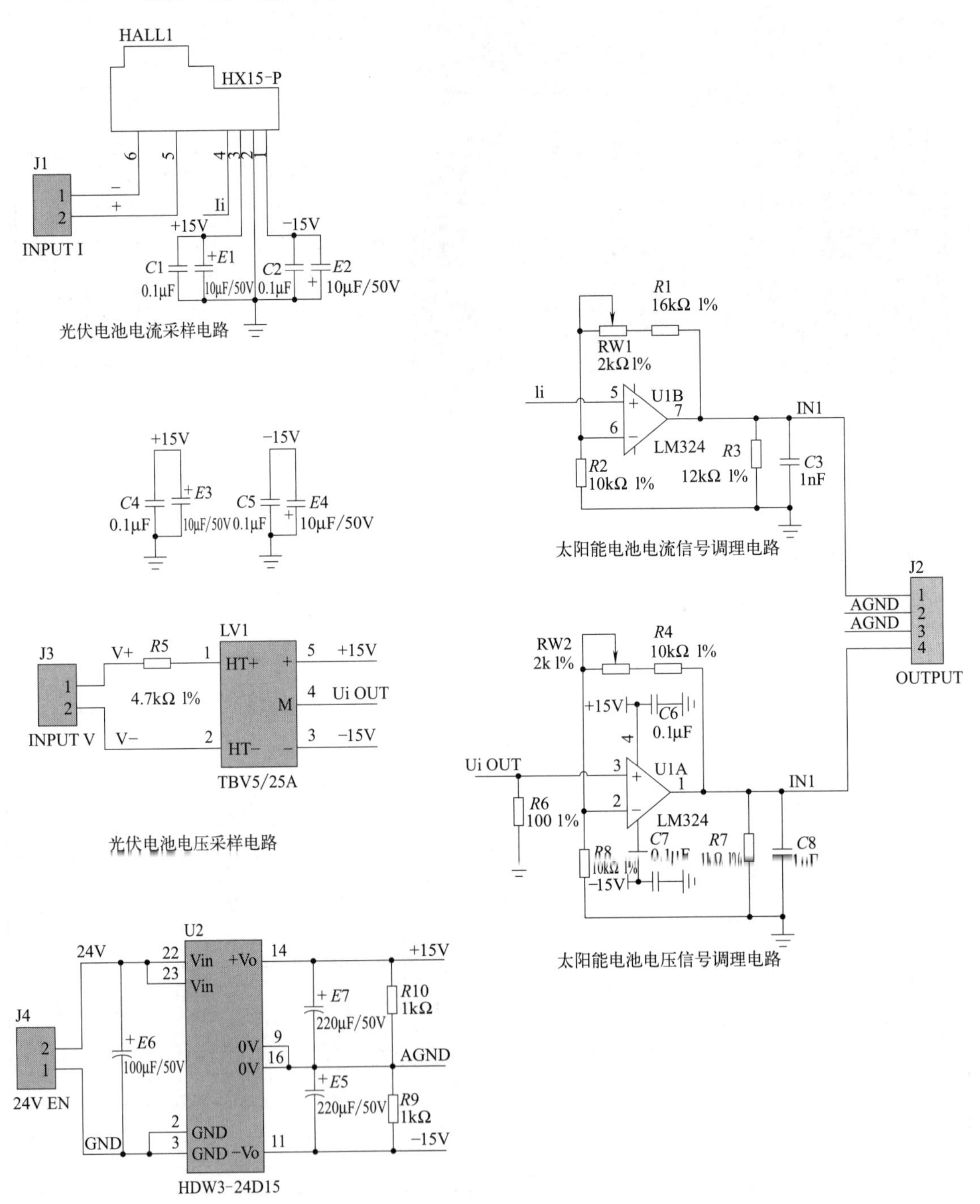

图 3.3.15 直流电压电流采样模块 1 原理图

2. 实物照片

直流电压电流采样模块 1 如图 3.3.16 所示。

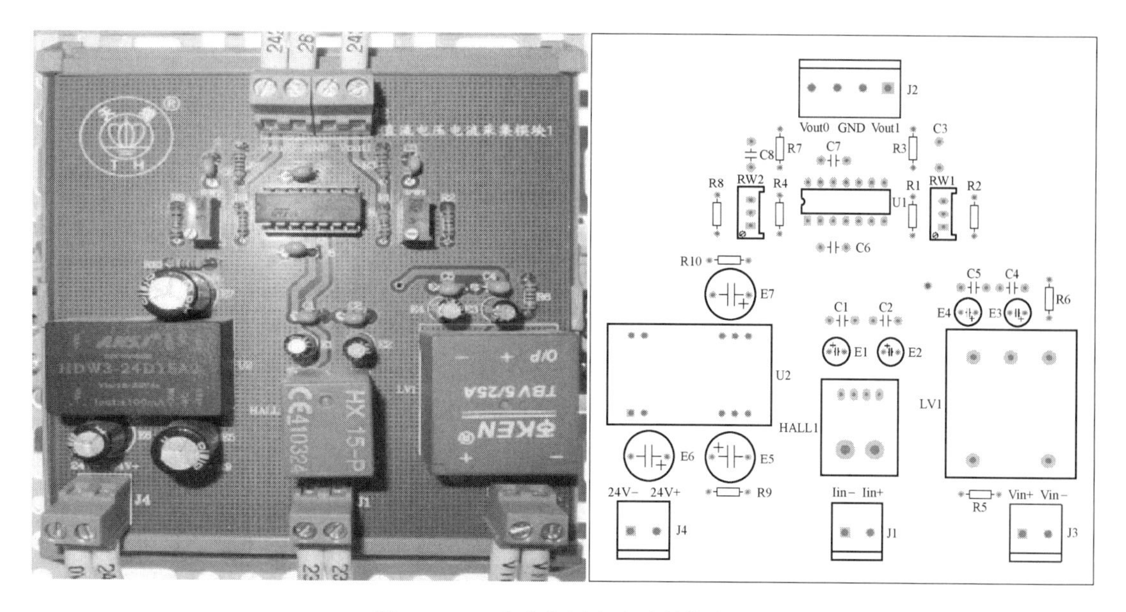

图 3.3.16 直流电压电流采样模块 1

3. 端口定义

直流电压电流采样模块 1 端口定义如表 3.3.11 所示。

表 3.3.11 端口定义表

序号	名称	说明
1	J4:24V+	隔离电源 DC/DC 24V 输入
2	J4:24V−	
3	J1:Iin−	电流采样输入
4	J1:Iin+	
5	J3:Vin−	电压采样输入
6	J3:Vin+	
7	J2:Vout0	电流采样调理输出
8	J2:GND	
9	J2:GND	电压采样调理输出
10	J2:Vout1	

九、直流电压电流采样模块 2

通过电压霍尔传感器和电流传感器将输出的电压和电流转换成满足单片机输入端要求的电压信号。

1. 原理图

直流电压电流采样模块 2 原理如图 3.3.17 所示。

2. 实物照片

直流电压电流采样模块 2 实物照片如图 3.3.18 所示。

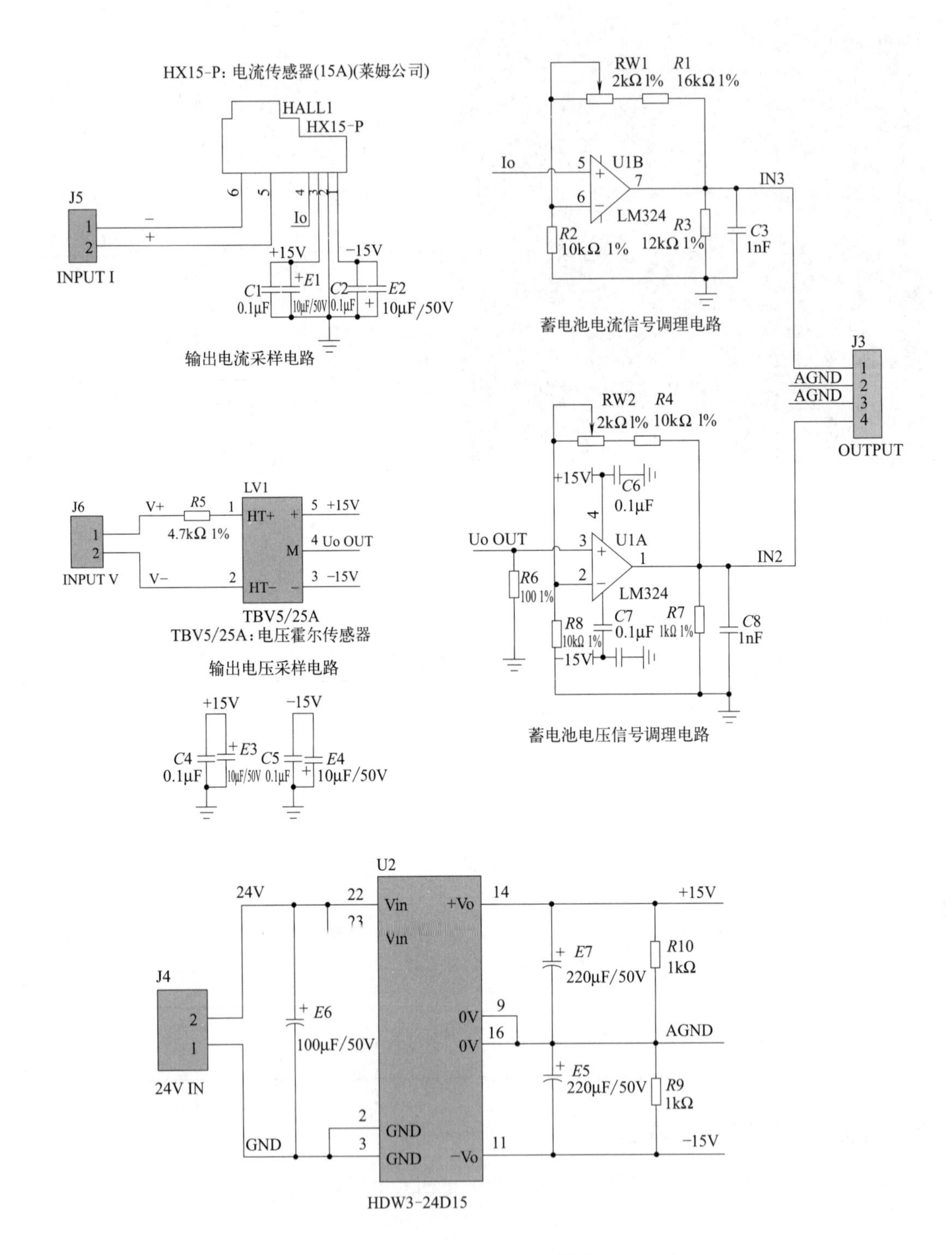

图 3.3.17　直流电压电流采样模块 2 原理图

3. 端口定义

直流电压电流采样模块 2 端口定义如表 3.3.12 所示。

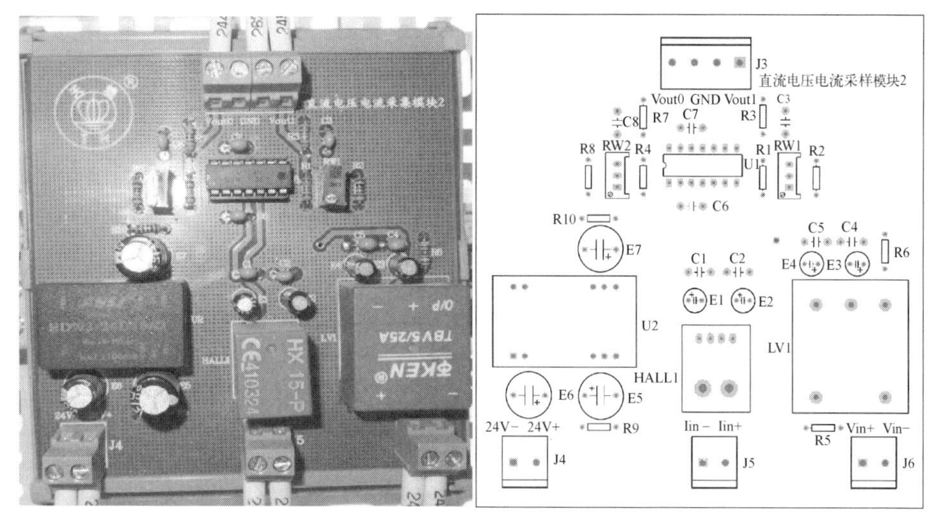

图 3.3.18　直流电压电流采样模块 2 实物图

表 3.3.12　端口定义表

序号	名称	说明
1	J4:24V+	隔离电源 DC/DC 24V 输入
2	J4:24V−	
3	J5:Iin−	电流采样输入
4	J5:Iin+	
5	J6:Vin−	电压采样输入
6	J6:Vin+	
7	J3:Vout0	电流采样调理输出
8	J3:GND	
9	J3:GND	电压采样调理输出
10	J3:Vout1	

十、主电路模块

通过实时采集风机阵列电池的功率来调整主电路的占空比，等效调节了负载的阻抗，使负载取用的功率得以改变，可始终跟随风机阵列电池输出的最大功率点，等效为一个阻抗变换器。

1. 原理图

主电路模块原理如图 3.3.19 所示。

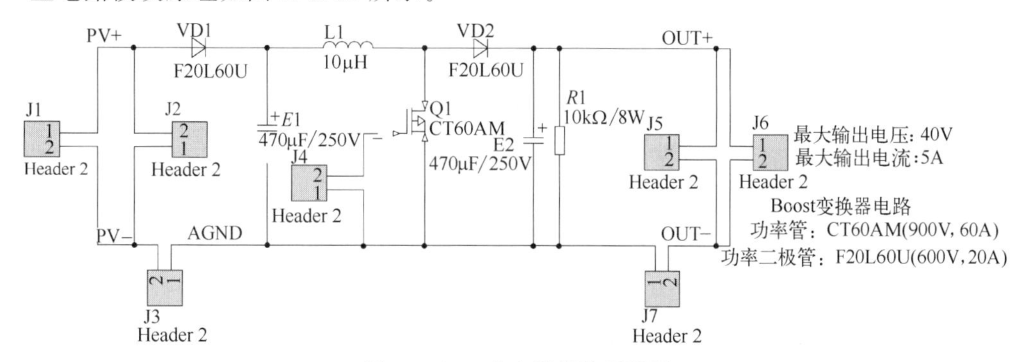

图 3.3.19　主电路模块原理图

2. 实物照片

主电路模块实物照片如图 3.3.20 所示。

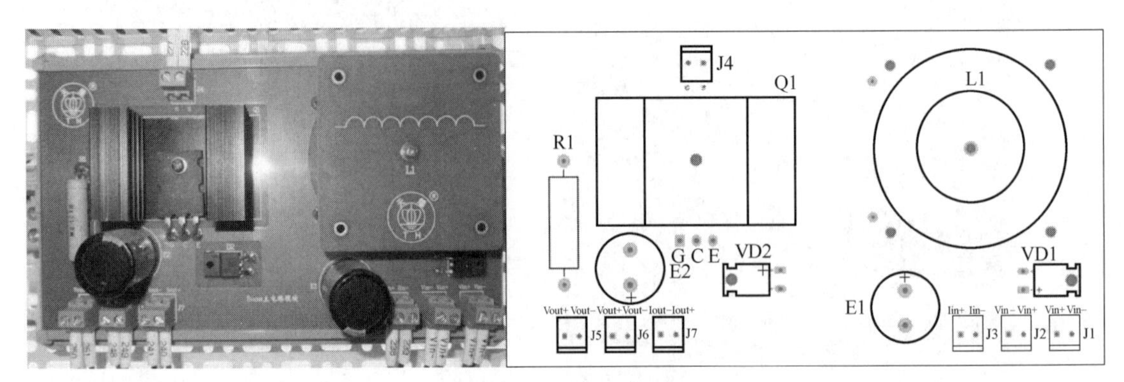

图 3.3.20 主电路模块

3. 端口定义

端口定义如表 3.3.13 所示。

表 3.3.13 端口定义表

序号	名称	说明
1	J1:Vin+	主电路输入电压
2	J1:Vin-	
3	J2:Vin+	输入电压采样输出
4	J2:Vin-	
5	J3:Iin-	输入电流采样输出
6	J3:Iin+	
7	J5:Vin+	主电路输出电压
8	J5:Vin-	
9	J6:Vin+	输出电压采样输出
10	J6:Vin-	
11	J7:Iin-	输出电流采样输出
12	J7:Iin+	
13	J4:G	隔离 PWM 驱动信号输入
14	J4:S	

十一、智能充放电控制器

根据蓄电池电压高低，调节充电状态和电流的大小，防止蓄电池过充或过放，延长蓄电池使用寿命。智能充放电控制器可控制风力发电机对蓄电池进行智能充电，设备外观大方，具有运行状态指示灯，充放电控制精确，液晶显示直观，操作方便。具有完善的保护功能。核心控制元件采用美国原装微控制器，控制软件采用德国先进的控制技术，限流恒压式充电，大大延长蓄电池寿命，设备充电效率高，空载损耗低。系统运行安全、稳定、可靠，使用寿命长。具有较高的性价比。

1. 主要功能

① 经常保持蓄电池处于饱满状态。

② 防止蓄电池过度充电。

③ 防止蓄电池过度放电。

④ 防止夜间蓄电池向太阳能板反向充电。

⑤ 蓄电池反接保护。

⑥ 蓄电池达到设定的停充电压后，断开充电回路，同时自动检测风力发电机输出电压，当风力发电机输出电压过高时，自动控制风力发电机刹车。

⑦ 电流过大时控制器自动保护并控制风力发电机刹车。

⑧ 风力发电机手动刹车保护。

⑨ 太阳能极板反接保护。

⑩ 负载电流超过控制器额定电流，控制器自动保护并锁死，显示“Load OverCurrent”。

⑪ 控制器可以进行防雷击保护。

⑫ 控制器长期累计并存储蓄电池充电安时数、发电机和太阳能发电总度数、放电安时数。

⑬ 控制器开机时会自动根据电池电压等级（12V/24V）自动设置停充电压、负载停机电压、负载恢复电压、刹车电流、刹车时间等参数。

⑭ 用户可以根据自己的需要，自行设置停充电压、负载停机电压、负载恢复电压、刹车电流、刹车时间五个参数。

⑮ 为防止电池过度充电，控制器自动控制最高停充电压，不得大于15V（对于12V电池）或30V（对于24V电池）（大于此值“+”键不起作用）。

2. 操作方法

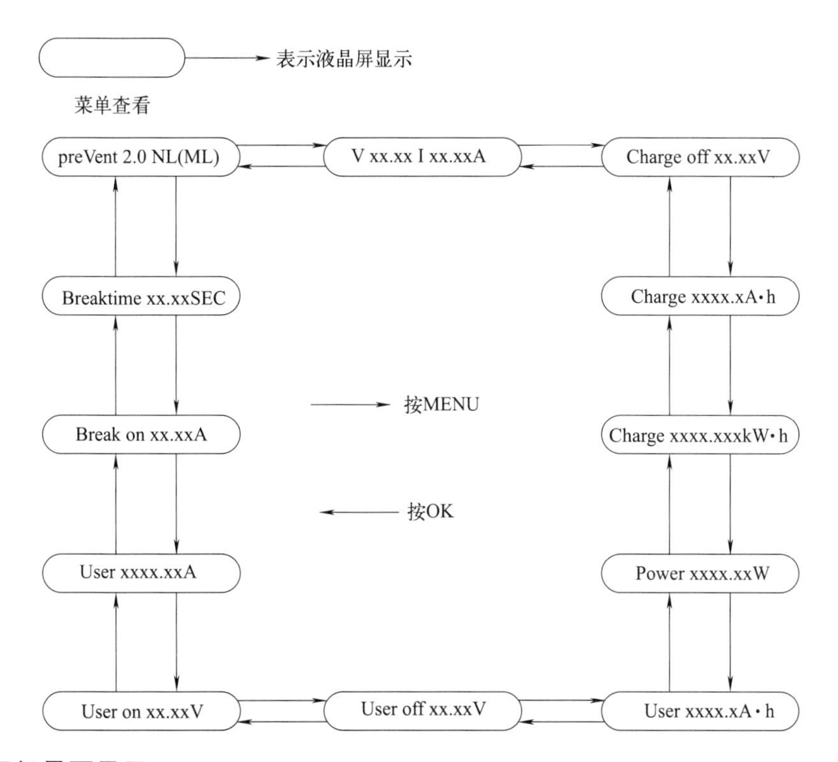

(1) 开机界面显示

preVent 2.0 NL 表示12V电池模式，preVent 2.0 ML 表示24V电池模式。

(2) 查看电池电压和充电电流

V xx.xx I xx.xxA V（电池电压，单位：V），I（充电电流，单位：A）。

(3) 设置电池停充电压

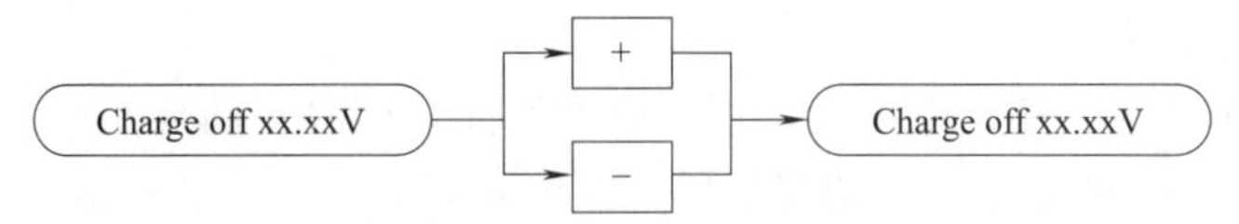

(4) 查看充电累计安时数

(Charge xxxx.xA·h) Charge（充电安时数，单位：A·h）。

(5) 查看发电机和太阳能发电累计度数

(Charge xxxx.xxxkW·h) Charge（发电度数，单位：kW·h）。

(6) 查看发电机和太阳能发电瞬时功率

(Power xxxx.xxW) Power（发电瞬时功率，单位：W）。

(7) 查看放电累计安时数

(User xxxx.xA·h) User（放电安时数，单位：A·h）。

(8) 设置负载停机电压

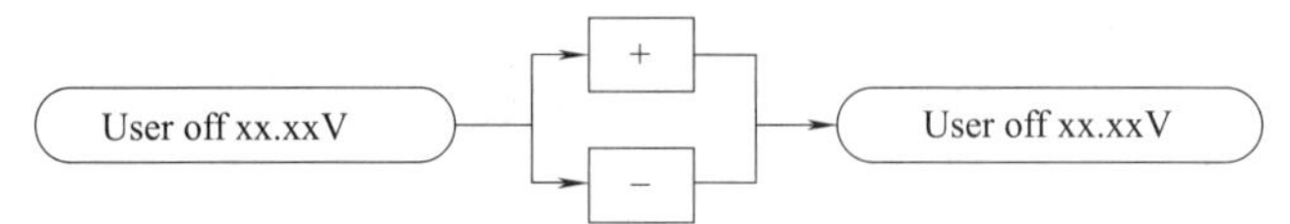

(9) 设置负载开机电压

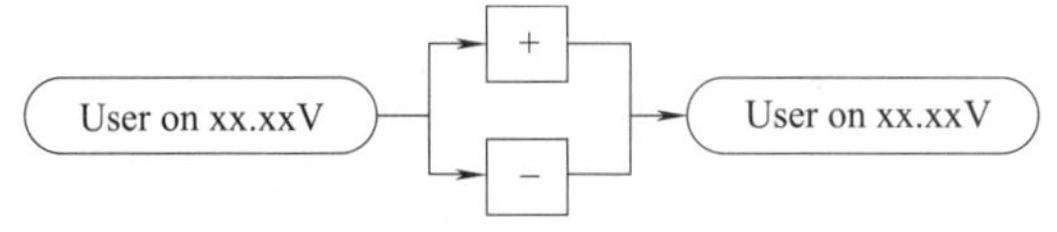

(10) 查看放电电流

(User xxxx.xxA) User（放电电流，单位：A）

(11) 设置刹车电流

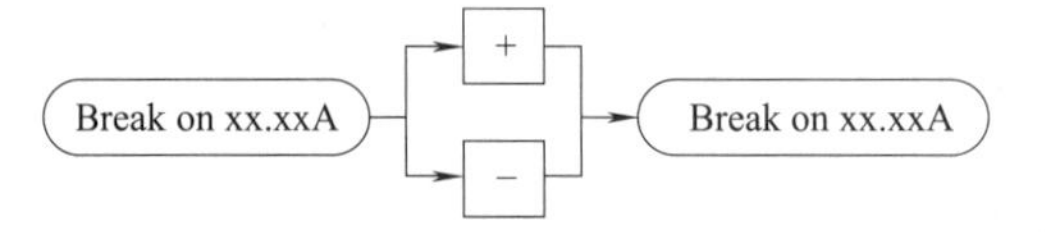

(12) 设置刹车时间

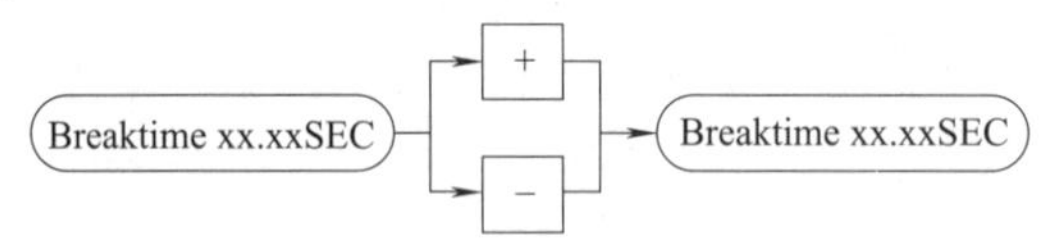

(13) 负载过流保护

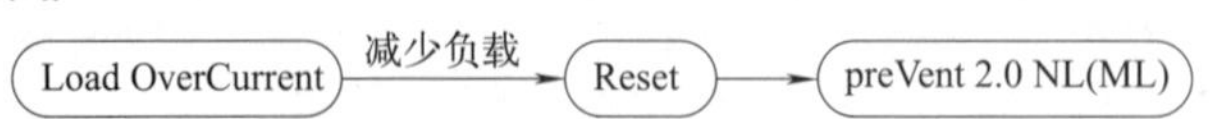

3. 技术参数

智能充放电控制器技术参数如表 3.3.14 所示。

表 3.3.14 技术参数表

额定电流	12V、24V 电压自动识别
最大负载电流	35.61A/17.80A
蓄电池充满断开电压	14.32V/28.64V(默认值,可设置,范围:0~15V/30V)
负载欠压断开电压	10.98V/21.97V(默认值,可设置)
负载开机恢复电压	12.44V/28.88V(默认值,可设置)

续表

刹车电流	15.79A/7.86A(默认值,可设置,范围:0～40A/20A)
刹车后自动恢复时间	10s(默认值,可设置)
自动刹车电压	17.5V/32.5V　　(风机整流后电压)
空载损耗	≤40mA
外形尺寸	228mm×133mm×75mm
质量	1.2kg
工作环境	环境温度－10～＋50℃,相对湿度 0～90%

4. 实物图

智能充放电控制器如图 3.3.21 所示。

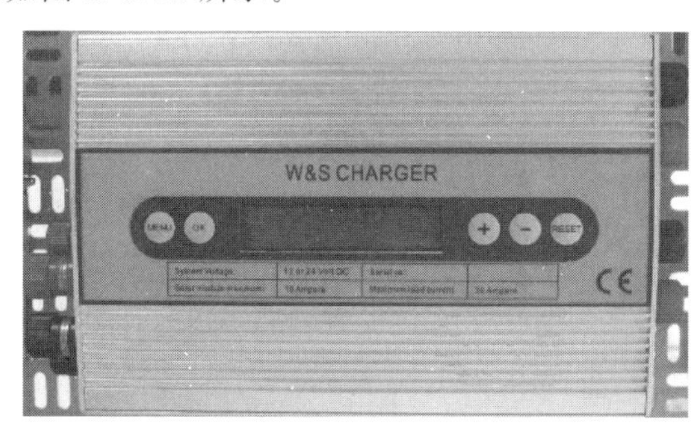

图 3.3.21　智能充放电控制器

项目实施

一、工具清单

① THWPWG-3B 型 大型风力发电系统实训平台风机对象；
② 剥线钳 1 把；
③ 压线钳 1 把；
④ 尖嘴钳 1 把；
⑤ 十字螺丝刀、一字螺丝刀各 1 把；
⑥ 多功能万用表一个。

二、安全注意事项

为了顺利完成实训项目，确保实训时设备的安全、可靠及长期的运行，实训人员要严格遵守如下安全规程：

1. 实训前的准备

① 实训前仔细阅读使用说明书，熟悉系统的相关部分；
② 实训前仔细阅读系统操作说明及实训的注意事项；
③ 实训前仔细阅读能源存储系统相关器件的使用说明；
④ 实训前确保系统总电源处于断开状态；
⑤ 实训前根据实训指导书中相关内容熟悉此次实训的操作步骤。

2. 实训中的注意事项

① 使用前先检查各电源是否正常。

② 接线前务必熟悉装置的各单元模块的功能及接线位置。
③ 实训接线前必须先断开总电源，严禁带电接线。
④ 接线完毕，检查无误后方可通电。
⑤ 熟练掌握能源存储系统各个模块的接线图及相关模块参数设置方法。
⑥ 控制柜中存在 AC 220V 的接入点，实训时要注意安全。
⑦ 实训始终，实训台要保持整洁，不可随意放置杂物，以免发生短路等故障。
⑧ 实训完毕，应及时关闭电源开关，并及时清理实训台。
⑨ 严格按照正确的操作步骤给系统上电和断电，以免误操作给系统带来损坏。
⑩ 在实训过程中，设备安装时注意防止高处跌落、挤压受伤。
⑪ 在实训过程中，有“危险”标志的地方为强电注意安全。

3. 实训的步骤

实训时要做到以下几点：

(1) 预习报告详细完整，熟悉设备

实训开始前，指导老师要对学生的预习报告做检查，要求学生了解本次实训的目的、内容和安全实训操作步骤，只有满足此要求后，方能允许开始实训。

指导老师要对实训装置做详细介绍，学生必须熟悉该次实训所用的各种设备，明确这些设备的功能与使用方法。

(2) 建立小组，合理分工

每次实训都以小组为单位进行，每组由 2～3 人组成。

(3) 试运行

在正式实训开始之前，先熟悉装置的操作，然后按一定安全操作规范接通电源，观察设备是否正常。如果设备出现异常，应立即切断电源，并排除故障；如果一切正常，即可正式开始实训。

(4) 认真负责，实训有始有终

实训完毕后，应请指导老师检查实训资料。经指导老师认可后，按照安全操作步骤关闭所有电源，并把实训中所用的物品整理好，放回原位。

4. 实训总结

这是实训的最后、最重要阶段，应分析实训现象并撰写实训报告。每位实训参与者要独立完成一份实训报告，实训报告的编写应持严肃认真、实事求是的态度。

实训报告是根据实训中观察发现的问题，经过自己分析研究或组员之间分析讨论后写出的实训总结和心得体会，应简明扼要、字迹清楚、结论明确。

实训报告应包括以下内容：

① 实训名称、专业、班级、学号、姓名、同组者姓名等。
② 实训目的、实训内容、实训步骤。
③ 实训设备的型号、规格。
④ 实训资料的整理。
⑤ 用理论知识对实训结果进行分析总结，得出正确的结论。
⑥ 对实训中出现的现象、遇到的问题进行分析讨论，写出心得体会，并提出自己的建议和改进措施。
⑦ 实训报告应写在一定规格的报告纸上，保持整洁。
⑧ 每次实训每人独立完成一份报告，按时送交指导老师批阅。

三、实训内容

1. 安装：器件布局

根据图 3.3.1 能源转换储存控制系统控制柜的器件布局，将铝导轨、线槽、各功能模块、端子、连接模块等固定到网孔板相应位置上，为连线做好准备。

2. 连接

(1) 端子排端口定义

端子排定义如图 3.3.22 所示。

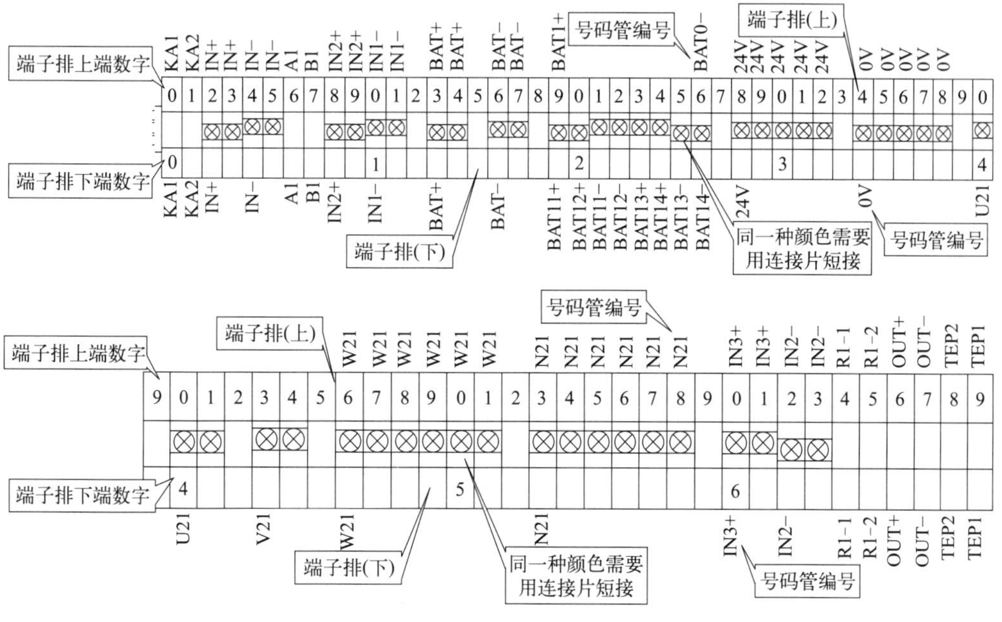

图 3.3.22　端子排定义

端子排编号定义：

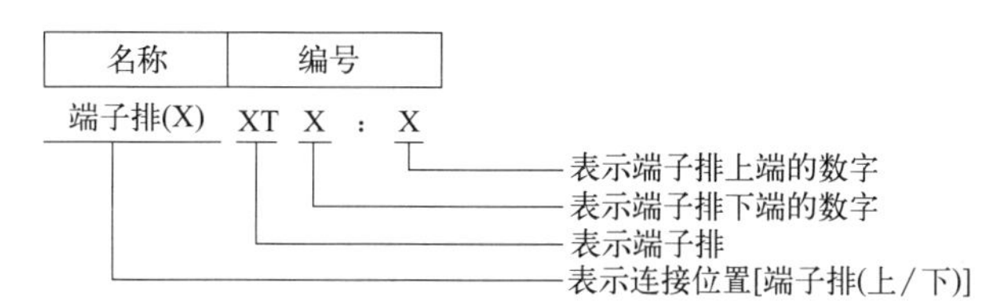

(2) 风机输出电压表接线

风机输出电压表接线如表 3.3.15 所示。

表 3.3.15　风机输出电压表接线表

序号	起始端位置	结束端位置		号码管编号	线型
	风机输出电压表	名称	编号		
1	INPUT＋	接线排（上）	1	IN＋	42 红
2	INPUT－		2	IN－	42 黑
3	POWER18		3	W21	23 红
4	POWER17		4	N21	23 黑
5	RS485A		5	A1	12 蓝
6	RS485B		6	B1	12 蓝

(3) 风机输出电压表接线排接线

风机输出电压表接线排接线如表 3.3.16 所示。

表 3.3.16 风机输出电压表接线排接线表

序号	起始端位置	结束端位置		号码管编号	线型
	风机输出电压表接线排(下)	名称	编号		
1	1	端子排(上)	XT0:2	IN+	42 红
2	2		XT0:4	IN-	42 黑
3	3	端子排(上)	XT4:8	W21	23 红
4	4		XT5:5	N21	23 黑
5	5	风机输出电流表端子排(上)	序号 5;XT0:6	A1	12 蓝
6	6		序号 6;XT0:7	B1	12 蓝

(4) 风机输出电流表接线

风机输出电流表接线如表 3.3.17 所示。

表 3.3.17 风机输出电流表接线表

序号	起始端位置	结束端位置		号码管编号	线型
	风机输出电流表	名称	编号		
1	INPUT+	接线排(上)	1	IN+	42 红
2	INPUT−		2	IN1+	42 红
3	POWER18		3	W21	23 红
4	POWER17		4	N21	23 黑
5	RS485A		5	A1	12 蓝
6	RS485B		6	B1	12 蓝

(5) 风机输出电流表接线排接线

风机输出电流表接线排接线如表 3.3.18 所示。

表 3.3.18 风机输出电流表接线排接线表

序号	起始端位置	结束端位置		号码管编号	线型
	风机输出电压表接线排(下)	名称	编号		
1	1	端子排(上)	XT0:3	IN+	42 红
2	2	风机输出断路器 K1	下左	IN1+	42 红
3	3	端子排(上)	XT4:9	W21	23 红
4	4		XT5:6	N21	23 黑
5	5	蓄电池电压表	序号 5	A1	12 蓝
6	6	风机输出电压表	序号 6	B1	12 蓝

(6) 风机输出断路器接线

风机输出断路器接线如表 3.3.19 所示。

表 3.3.19 风机输出断路器接线表

序号	起始端位置	结束端位置		号码管编号	线型
	风机输出断路器 K1	名称	编号		
1	上左	端子排(下)	XT0:8	IN2+	42 红
2	上右		XT1:0	IN1−	42 黑
3	下左	风机输出电流表	序号 2	IN1+	42 红
4	下右	端子排(上)	XT0:5	IN−	42 黑

(7) MPPT 断路器接线

MPPT 断路器接线如表 3.3.20 所示。

表 3.3.20　MPPT 断路器接线表

序号	起始端位置	结束端位置		号码管编号	线型
	MPPT 断路器 K2	名称	编号		
1	上左	端子排(上)	XT0:8	IN2+	42 红
2	上右		XT1:0	IN1−	42 黑
3	下左	主电路模块	J1:Vin+	Vin+	42 红
4	下右		J1:Vin−	Vin−	42 黑

(8) 充放电控制器断路器接线

充放电控制器断路器接线如表 3.3.21 所示。

表 3.3.21　充放电控制器断路器接线表

序号	起始端位置	结束端位置		号码管编号	线型
	充放电控制器断路器 K3	名称	编号		
1	上左	端子排(上)	XT0:9	IN2+	42 红
2	上右		XT1:1	IN1−	42 黑
3	下左	端子排(下)	XT6:0	IN3+	42 红
4	下右		XT6:2	IN2−	42 黑

(9) 蓄电池断路器接线

蓄电池断路器接线如表 3.3.22 所示。

表 3.3.22　蓄电池断路器接线表

序号	起始端位置	结束端位置		号码管编号	线型
	蓄电池断路器 K4	名称	编号		
1	上左	蓄电池电流表	序号 2	222	42 红
2	上右	端子排(下)	XT1:6	BAT−	42 黑
3	下左	熔断器	熔断器上端	BAT0+	42 红
4	下右	端子排(上)	XT2:6	BAT0−	42 黑

(10) 蓄电池电压表接线

蓄电池电压表接线如表 3.3.23 所示。

表 3.3.23　蓄电池电压表接线表

序号	起始端位置	结束端位置		号码管编号	线型
	蓄电池电压表	名称	编号		
1	INPUT+	接线排(上)	1	BAT+	42 红
2	INPUT−		2	BAT−	42 黑
3	POWER18		3	W21	23 红
4	POWER17		4	N21	23 黑
5	RS485A		5	A1	12 蓝
6	RS485B		6	B1	12 蓝

(11) 蓄电池电压表接线排接线

蓄电池电压表接线排接线如表 3.3.24 所示。

表 3.3.24　蓄电池电压表接线排接线

序号	起始端位置	结束端位置		号码管编号	线型
	蓄电池电压表接线排(下)	名称	编号		
1	1	接线排(上)	XT1:4	BAT+	42 红
2	2		XT1:7	BAT−	42 黑
3	3	接线排(上)	XT4:6	W21	23 红
4	4		XT5:3	N21	23 黑
5	5	蓄电池电流表	序号 5	A1	12 蓝
6	6	风机输出电流表	序号 6	B1	12 蓝

(12) 蓄电池电流表接线

蓄电池电流表接线如表 3.3.25 所示。

表 3.3.25　蓄电池电流表接线表

序号	起始端位置	结束端位置		号码管编号	线型
	蓄电池电流表	名称	编号		
1	INPUT+	接线排(上)	1	BAT+	42 红
2	INPUT−		2	222	42 红
3	POWER18		3	W21	23 红
4	POWER17		4	N21	23 黑
5	RS485A		5	A1	12 蓝
6	RS485B		6	B1	12 蓝

(13) 蓄电池电流表接线排接线

蓄电池电流表接线排接线如表 3.3.26 所示。

表 3.3.26　蓄电池电流表接线排接线表

序号	起始端位置	结束端位置		号码管编号	线型
	蓄电池电流表接线排(下)	名称	编号		
1	1	接线排(下)	XT1:3	BAT+	42 红
2	2	蓄电池断路器 K4	上左	222	42 红
3	3	端子排(上)	XT4:7	W21	23 红
4	4		XT5:4	N21	23 黑
5	5	蓄电池电压表	序号 5	A1	12 蓝
6	6		序号 6	B1	12 蓝

(14) 充放电控制器接线

充放电控制器接线如表 3.3.27 所示。

表 3.3.27　充放电控制器接线表

序号	起始端位置	结束端位置		号码管编号	线型
	充放电控制器	名称	编号		
1	WINDGENERATOR+	端子排(上)	XT6:0	IN3+	42 红
2	WINDGENERATOR−		XT6:2	IN2−	42 黑
3	BATTERY+	端子排(上)	XT1:3	BAT+	42 红
4	BATTERY−		XT1:6	BAT−	42 黑
5	OUTPUT+		XT6:6	OUT+	42 红
6	OUTPUT−		XT6:7	OUT−	42 黑
7	BRAKE+	温度告警模块	J3:IN1	BK1	12 蓝
8	BRAKE−		J3:GND	BK2	12 蓝

(15) 摇柄接线

摇柄接线如表 3.3.28 所示。

表 3.3.28 摇柄接线表

序号	起始端位置	结束端位置		号码管编号	线型
	摇柄(双联开关)	名称	编号		
1	下(与 COM2 通)	端子排(上)	XT6:1	IN3+	42 红
2	下(与 COM1 通)		XT6:3	IN2−	42 黑
3	上(与 COM1 通)		XT6:4	R1-1	42 红
4	上(与 COM2 通)		XT6:5	R1-2	42 黑
5	COM1	主电路模块	J5:Vout+	218	42 红
6	COM2		J5:Vout−	219	42 黑

(16) 温度告警模块接线

温度告警模块接线如表 3.3.29 所示。

表 3.3.29 温度告警模块接线表

序号	起始端位置	结束端位置		号码管编号	线型
	温度告警模块	名称	编号		
1	J2:KA1	端子排(上)	XT0:0	KA1	23 红
2	J2:KA2		XT0:1	KA2	23 红
3	J1:24V+		XT3:0	24V	23 红
4	J1:24V−		XT3:6	0V	23 黑
5	J3:IN1	充放电控制器	BRAKE+	BK1	12 蓝
6	J3:GND		BRAKE−	BK2	12 蓝
7	J4:GND	CPU 核心模块	J9:GND	GND	23 黑
8	J4:+5V		J9:+5V	+5V	23 红
9	J4:TM		J4:IN4	TM	12 蓝
10	J5:+5V	温度传感器	XT6:8	TEP1	12 蓝
11	J5:TM		XT6:9	TEP2	12 蓝

(17) 熔断器接线

熔断器接线如表 3.3.30 所示。

表 3.3.30 熔断器接线表

序号	起始端位置	结束端位置		号码管编号	线型
	熔断器	名称	编号		
1	熔断器(上)	蓄电池断路器 K4	下左	BAT0+	42 红
2	熔断器(下)	端子排(上)	XT1:9	BAT1+	42 红

(18) 开关电源接线

开关电源接线如表 3.3.31 所示。

表 3.3.31 开关电源接线表

序号	起始端位置	结束端位置		号码管编号	线型
	开关电源	名称	编号		
1	L	端子排(上)	XT5:0	W21	23 红
2	N		XT5:7	N21	23 黑
3	EARTH	机柜外壳		EARTH	23 黄绿
4	−V	端子排(下)	XT3:4	0V	42 黑
5	+V		XT2:8	24V	42 红

(19) 人机交互模块接线

人机交互模块接线如表 3.3.32 所示。

表 3.3.32 人机交互模块接线表

序号	起始端位置	结束端位置		号码管编号	线型
	人机交互模块	名称	编号		
1	J1	CPU 核心模块	J1	无	20P 双头排线

(20) CPU 核心模块接线

CPU 核心模块接线如表 3.3.33 所示。

表 3.3.33 CPU 核心模块接线表

序号	起始端位置	结束端位置		号码管编号	线型
	CPU 核心模块	名称	编号		
1	J2:IN0	直流电压电流采样模块 1	J2:Vout1	211	12 蓝
2	J2:GND		J2:GND	210	23 黑
3	J2:IN1		J2:Vout0	209	12 蓝
4	J3:IN2	直流电压电流采样模块 2	J3:Vout1	208	12 蓝
5	J3:GND		J3:GND	207	23 黑
6	J3:IN3		J3:Vout0	206	12 蓝
7	J4:IN4	温度告警模块	J4:TM	TM	12 蓝
9	J8:24V+	端子排(上)	XT3:1	24V	23 红
8	J8:24V−		XT3:7	0V	23 黑
10	J9:+5V	温度告警模块	J4:+5V	+5V	23 红
11	J9:GND		J4:GND	GND	23 黑
12	J11:GND	PWM 驱动模块	J1:GND	200	23 黑
13	J11:RXD		J1:TXD	201	12 蓝
14	J11:TXD		J1:RXD	202	12 蓝
15	J11:+5V		J1:+5V	203	23 红

(21) PWM 驱动模块接线

PWM 驱动模块接线如表 3.3.34 所示。

表 3.3.34 驱动模块接线表

序号	起始端位置	结束端位置		号码管编号	线型
	PWM 驱动模块	名称	编号		
1	J6:S2	主电路模块	J4:S	204	12 蓝
2	J6:G2		J4:G	205	12 蓝
3	J1: +5V	CPU 核心模块	J11:+5V	203	23 红
4	J1:GND		J11:GND	200	23 黑
5	J1:RXD		J11:TXD	202	12 蓝
6	J1:TXD		J11:RXD	201	12 蓝
7	J5:24V+	端子排(上)	XT3:2	24V	23 红
8	J5:24V−		XT3:8	0V	23 黑

(22) 直流电压电流采样模块 1 接线

直流电压电流采样模块 1 接线如表 3.3.35 所示。

(23) 直流电压电流采样模块 2 接线

直流电压电流采样模块 2 接线如表 3.3.36 所示。

(24) 主电路模块接线

主电路模块接线如表 3.3.37 所示。

表 3.3.35 直流电压电流采样模块 1 接线表

序号	起始端位置	结束端位置		号码管编号	线型
	直流电压电流采样模块 1	名称	编号		
1	J3:Vin+	主电路模块	J2:Vin+	223	42 红
2	J3:Vin−		J2:Vin−	224	42 黑
3	J1:Iin+		J3:Iin+	212	42 红
4	J1:Iin−		J3:Iin−	213	42 黑
5	J4:24+	端子排(上)	XT2:8	24V	23 红
6	J4:24−		XT3:4	0V	23 黑
7	J2:Vout0	CPU 核心模块	J2:IN1	209	12 蓝
8	J2:GND		J2:GND	210	23 黑
9	J2:Vout1		J2:IN0	211	12 蓝

表 3.3.36 直流电压电流采样模块 2 接线表

序号	起始端位置	结束端位置		号码管编号	线型
	直流电压电流采样模块 2	名称	编号		
1	J6:Vin+	主电路模块	J6:Vout+	216	42 红
2	J6:Vin−		J6:Vout−	217	42 黑
3	J5:Iin+		J7:Iout+	214	42 红
4	J5:Iin−		J7:Iout−	215	42 黑
5	J4:24+	端子排(上)	XT2:9	24V	23 红
6	J4:24−		XT3:5	0V	23 黑
7	J3:Vout0	CPU 核心模块	J3:IN3	206	12 蓝
8	J3:GND		J3:GND	207	23 黑
9	J3:Vout1		J3:IN2	208	12 蓝

表 3.3.37 主电路模块接线表

序号	起始端位置	结束端位置		号码管编号	线型
	主电路模块	名称	编号		
1	J1:Vin+	MPPT 断路器	下左	Vin+	42 红
2	J1:Vin−		下右	Vin−	42 黑
3	J2:Vin+	直流电压电流采样模块 1	J3:Vin+	223	42 红
4	J2:Vin−		J3:Vin−	224	42 黑
5	J3:Iin+		J1:Iin+	212	42 红
6	J3:Iin−		J1:Iin−	213	42 黑
7	J7:Iout+	直流电压电流采样模块 2	J5:Iin+	214	42 红
8	J7:Iout−		J5:Iin−	215	42 黑
9	J6:Vout+		J6:Vin+	216	42 红
10	J6:Vout−		J6 Vin−	217	42 黑
11	J5:Vout+	摇柄(双联开关)	COM1	218	42 红
12	J5:Vout−		COM2	219	42 黑
13	J4:S	PWM 驱动模块	J6:S2	204	12 蓝
14	J4:G		J6:G2	205	12 蓝

(25) 连接模块 XS4 接线

连接模块 XS4 接线如表 3.3.38 所示。

(26) 连接模块 XS5 接线

连接模块 XS5 接线如表 3.3.39 所示。

(27) 蓄电池组接线

蓄电池组接线如表 3.3.40 所示。

表 3.3.38 连接模块 XS4 接线表

序号	起始端位置	结束端位置		号码管编号	线型
	连接模块 XS4	名称	编号		
1	A1	端子排(下)	XT0:0	KA1	23 红
2	A2		XT0:1	KA2	23 黑
3	B1		XT0:2	IN+	23 红
4	B2		XT0:4	IN−	23 黑

表 3.3.39 连接模块 XS5 接线表

序号	起始端位置	结束端位置		号码管编号	线型
	连接模块 XS5	名称	编号		
1	A7	端子排(下)	XT0:6	A1	42 红
2	A8		XT0:7	B1	42 黑
3	A9		XT6:6	OUT+	42 红
4	A10		XT6:7	OUT−	42 黑

表 3.3.40 蓄电池组接线表

序号	起始端位置	结束端位置		号码管编号	线型
	蓄电池组	名称	编号		
1	BAT11+	端子排(下)	XT1:9	BAT11+	42 红
2	BAT11−		XT2:1	BAT11−	42 黑
3	BAT12+		XT2:0	BAT12+	42 红
4	BAT12−		XT2:2	BAT12−	42 黑
5	BAT13+		XT2:3	BAT13+	42 红
6	BAT13−		XT2:5	BAT13−	42 黑
7	BAT14+		XT2:4	BAT14+	42 红
8	BAT14−		XT2:6	BAT14−	42 黑

(28) 固定负载接线

固定负载接线如表 3.3.41 所示。

表 3.3.41 固定负载接线表

序号	起始端位置	结束端位置		号码管编号	线型
	固定负载	名称	编号		
1	1	端子排(下)	XT6:4	R1-1	42 红
2	2		XT6:5	R1-2	42 黑

(29) 温度传感器接线

温度传感器接线如表 3.3.42 所示。

表 3.3.42 温度传感器接线表

序号	起始端位置	结束端位置		号码管编号	线型
	温度传感器	名称	编号		
1	1	端子排(下)	XT6:8	TEP2	12 蓝
2	2		XT6:9	TEP1	12 蓝

四、操作说明

1. 工作模式开关设置

工作模式开关设置如表 3.3.43 所示。

表 3.3.43 工作模式开关设置表

序号	功能	MPPT(固定负载)	MPPT(蓄电池)	充放电控制器
1	风机输出(K1)	开	开	开
2	MPPT(K2)	开	开	关
3	充放电控制器(K3)	关	关	开
4	蓄电池(K4)	开	开	开
5	摇柄	上	下	中

2. 最大功率跟踪测试操作步骤

① 合上“能源储存控制单元”上的“总电源”开关，系统得电，三相电源指示灯亮。

② 将充放电控制器的刹车置于“RELEASE（自动刹车）”状态，合上“能源储存控制单元”的“蓄电池”断路器，蓄电池接入 MPPT 控制器，同时给充放电控制器供电，此时充放电控制器进行初始化，红色指示灯点亮（工作在刹车状态）。必须等红色指示灯熄灭（退出刹车状态）才能进行下一步操作。

③ 合上“能源储存控制单元”上的“风机输出”和“MPPT”断路器。

④ 按“CPU 核心模块”上的复位按钮 K1，系统复位。

⑤ 按“人机交互模块”上的“ENTER”键，进入手动功率跟踪界面，然后按“UP”和“DOWN”键手动调节占空比，测量多组风机输出电流、输出电压，记录于下表。

序号	电压/V	电流/A	功率/W
1			
2			
3			
4			
5			
6			
7			
8			
9			
10			

⑥ 通过记录的电压电流数据，计算每个电压、电流对应的功率。

⑦ 通过“人机交互模块”上“UP”“DOWN”，选择“功率跟踪 PQ”，按“ENTER”进入自动功率跟踪，一段时间后分别记录电压电流数据，并计算对应功率。

⑧ 依次关闭“能源储存控制单元”“风机输出”“MPPT”“蓄电池”断路器和“总电源”开关。

3. ISP 下载器的驱动安装

① 用 USB 线将 ISP 下载器连接到电脑上。第一次使用该下载器时，电脑会弹出发现新硬件的提示界面，如图 3.3.23 所示。

② 选择“从列表或指定位置安装（高级）(S)”，点击“下一步”。在出现的界面中选择 USB ISP 驱动所在路径。然后单击“下一步”，驱动开始安装，如图 3.3.24 所示。

图 3.3.23 新硬件向导

③ 安装完成后单击“完成”，如图 3.3.25 所示。

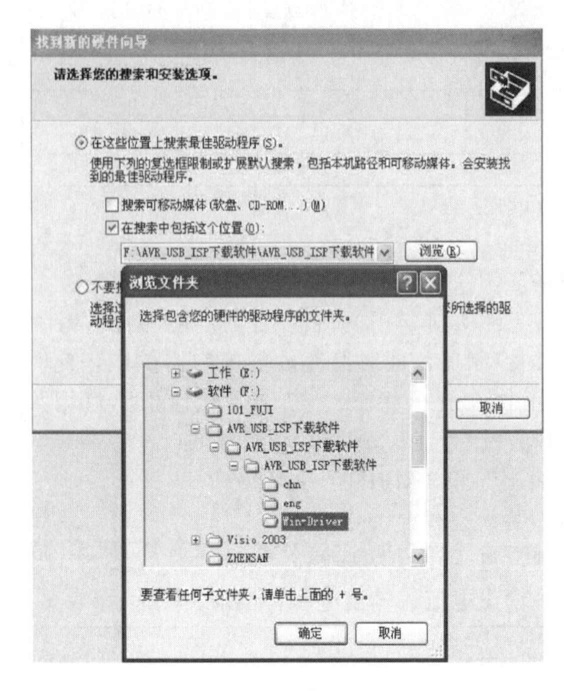

图 3.3.24 选择路径

图 3.3.25 完成安装

④ 右击“我的电脑”，打开“设备管理器”，将 ISP 下载器连接到电脑上后会出现如图 3.3.26 所示这一项，这表示驱动安装成功。

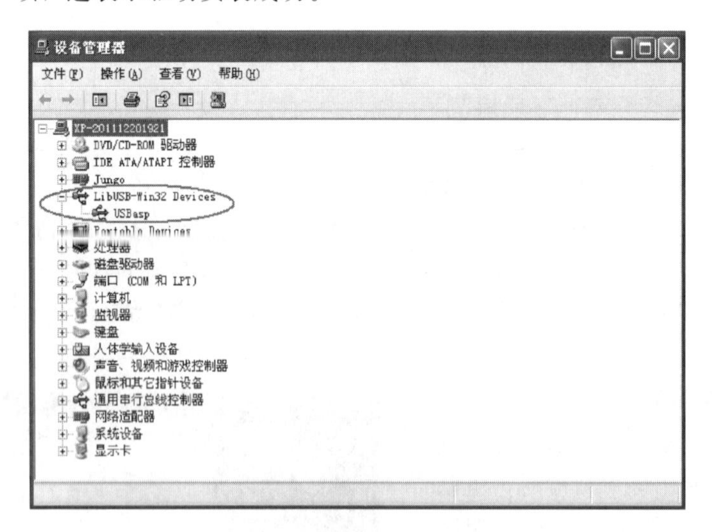

图 3.3.26 打开“设备管理器”

⑤ 把 ISP 下载器的一端连接到 THWPG020.PCB 的 J10 上，另一端通过 USB 线连接到电脑，电脑上打开“PROGISP”下载软件，按图 3.3.27 所示进行设置。

⑥ 单击“调入 Flash”按钮，弹出“打开”对话框。然后单击“自动”按钮，开始烧录程序。烧录成功后会有提示“1：Erase，Blank，Write Flash，Verify Flash，Successfully done”，断开电源。

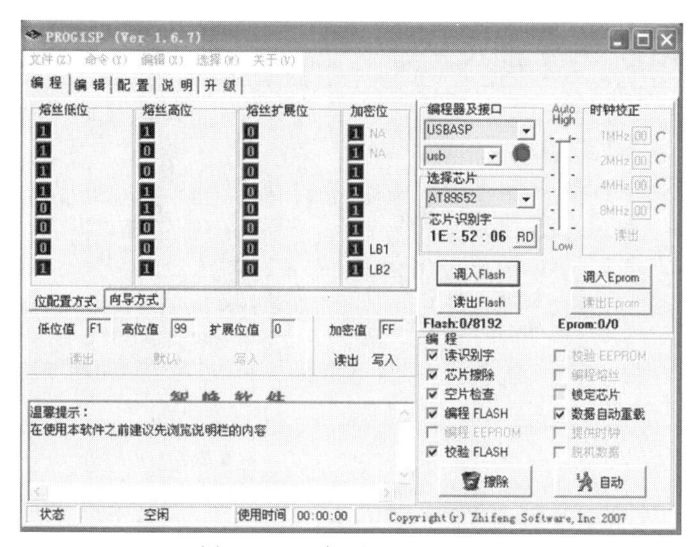

图 3.3.27　打开“PROGISP”

五、常见故障处理

常见故障分析及处理如表 3.3.44 所示。

表 3.3.44　常见故障分析及处理汇总表

故障现象	故障的可能原因	故障的排除方法和步骤
供电异常	1. 熔丝损坏 2. 存在短路或者漏电	1. 查看熔丝座内的熔丝是否烧毁，如烧毁则更换新熔丝 2. 仔细检查线路是否存在短路，并用万用表测量各相之间阻值和各相间对地阻值是否正常，如果阻值为零或阻值很小，说明存在短路现象，应采取逐次断开的方法检查
风机对象工作正常，但不能向蓄电池充电	1. 断路器损坏 2. 断路器未闭合 3. 蓄电池熔断芯损坏 4. 充放电控制器损坏 5. MPPT 控制器模块损坏 6. 线路开路或接触不良	1. 检查风机输出、MPPT、蓄电池断路器输入与输出是否导通。 2. 闭合风机输出、MPPT、蓄电池断路器。 3. 检查熔断器内 10A 熔断芯是否损坏，如已损坏则更换熔断芯。 4. 如线路正常且更换熔断芯后故障仍然存在，则再断开充放电控制器风机输入端连线，用万用表测量风机输出电压，如电压正常，则需更换或维修控制器；如没有电压，则需更换或维修发电机，并检查 MPPT 控制器各模块是否有损坏。 5. 检查 MPPT 控制器模块是否损坏。 6. 用万用表检查系统线路是否存在开路或者接触不良现象
充放电控制器绿灯熄，负载被切断	蓄电池电压不足	给蓄电池充电或者更换蓄电池
充放电控制器绿灯熄，负载被切断，液晶屏显示 Load Over Current	负载超载或者负载短路	减少负载或消除短路的故障，按 Reset 键重新开机
充放电控制器红灯熄	未充电	如果是夜间，属正常现象，若长时间不充电应检查风机连线是否正确，有无松动和开路
风机最大功率跟踪不能进入稳定状态	1. 电压电流采样模块 1 异常 2. ADC0809 损坏	1. 用万用表检测电压电流采样模块 1 供电及线路连接情况。 2. 更换 ADC0809

续表

故障现象	故障的可能原因	故障的排除方法和步骤
不能进行风机最大功率跟踪(风机输出电压正常,电流无)	1. 断路器损坏 2. 断路器未闭合 3. CPU 核心模块异常 4. 通信接口模块异常 5. PWM 驱动模块异常 6. 主电路模块异常 7. MPPT 断路器线路接触不良	1. 检查风机输出、MPPT、蓄电池断路器输入与输出是否导通。 2. 闭合风机输出、MPPT、蓄电池断路器。 3. 检查 CPU 核心模块供电是否正常,如供电正常,用万用表检测线路连接情况。 4. 检查通信接口模块供电是否正常,如供电正常,用万用表检测线路连接情况。 5. 检查 PWM 驱动模块供电是否正常,如供电正常,用万用表检测线路连接情况。 6. 用万用表检测主电路模块线路连接情况。 7. 用万用表检测 MPPT 断路器线路连接情况
开启风机后立即进入保护状态	1. 断路器损坏 2. 断路器未闭合 3. 充放电控制器刹车保护 4. 蓄电池馈电	1. 检查蓄电池断路器输入与输出是否导通。 2. 闭合蓄电池断路器。 3. 查看充放电控制器手动刹车按钮是否为"BRAKE"刹车状态,如果为"BRAKE"则手动打到"RELEASE"处。充放电控制器在刚刚开启时,系统初始化需要一段时间,此时充放电控制器处于刹车保护阶段,需等待系统初始化结束后才能进行充电。 4. 观察蓄电池电压是否低于 18V,如果低于 18V,充放电控制器将自动进入 12V 模式,只要输入电压高于 15V,充放电控制器就会发出刹车保护信号,由于风力发电机为三相 12V 输出,所以充放电控制器会一直处于保护状态,需对蓄电池进行充电,恢复正常电压后才能正常开启风力发电机

项目作业

(1) 能源转换储存控制系统主要由哪些模块组成?各有什么功能?

(2) 用文字或流程图描述最大功率跟踪测试项目的操作流程。

项目四　偏航功能的实现

项目描述

了解偏航控制系统工作原理,实施手动偏航控制和主控计算机自动偏航控制;完成程序编辑,实现给定角度的偏航功能。

能力目标:

① 掌握绝对值编码器和格雷码工作原理;

② 理解偏航控制过程并能用 PLC 编程实现。

项目环境

偏航控制系统主要由从 PLC、变频器、绝对值编码器、偏航电机(交流减速电机)、控制按钮等器件组成。可完成偏航控制系统的安装、手动与自动偏航 PLC 控制程序的编写与调试,以及从(slave)PLC 与监控管理主(master)PLC 的通信。偏航变桨控制系统控制柜如图 3.4.1 所示。

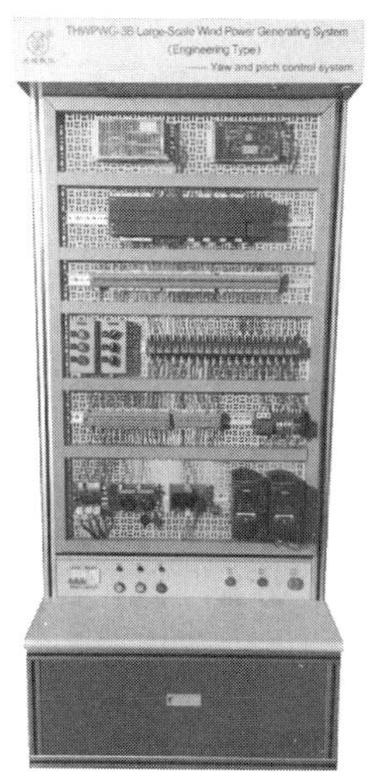

图 3.4.1　偏航变桨控制系统控制柜

一、偏航控制系统主要设备

偏航控制系统主要设备清单见表 3.4.1。

表 3.4.1　偏航控制系统主要设备清单

序号	名称	规　格	数量
1	偏航变桨控制系统控制柜	880mm×620mm×2120mm	1 台
2	变频器	MM420、MM440 各一台；三相输入；功率：0.75kW	2 台
3	按钮模块	24V/6A 开关电源 1 个；急停按钮 1 个；复位、启动、停止按钮黄、绿、红各 1 个；自锁按钮黄、绿、红各 1 个；转换开关 2 个；24V 指示灯黄、绿、红各 2 个	1 组
4	PLC	西门子 CPU ST40 主机	1 台
		EM DR32 数字量输入/输出模块，16×24V DC 输入/16 点继电器输出	1 台
		EM AQ02 模块，2 路模拟量输出(2AI)	1 台
		EM AR02 模块，2 路热电阻输入模块	3 台
		EM DP01 通信模块	1 台

二、MM420 变频器主要参数设置

MM420 变频器主要参数表见表 3.4.2。

表 3.4.2　MM420 变频器主要参数表

序号	变频器参数	设定值	功能说明
1	P0010	30	
2	P0970	1	恢复出厂设置
3	P0010	1	快速调试
4	P0003	2	允许访问扩展参数
5	P0304	380	电动机的额定电压
6	P0305	3	电动机的额定电流
7	P0307	0.3	电动机的额定功率
8	P0310	50.00	电动机的额定频率
9	P0311	1300	电动机的额定转速
10	P1000	2	模拟输入
11	P1080	0	电动机的最小频率(0Hz)
12	P1082	50.00	电动机的最大频率(50Hz)

续表

序号	变频器参数	设定值	功能说明
13	P1120	5.0	斜坡上升时间
14	P1121	0.1	斜坡下降时间
15	P0003	3	
16	P0010	0	
17	P0700	2	选择命令源（由端子排输入）
18	P0701	1	ON/OFF(接通正转/停车命令 1)
19	P0702	2	接通反转

三、主要控制原理与接线图

电气原理图如图 3.4.2 所示，变频器接线图如图 3.4.3 所示。

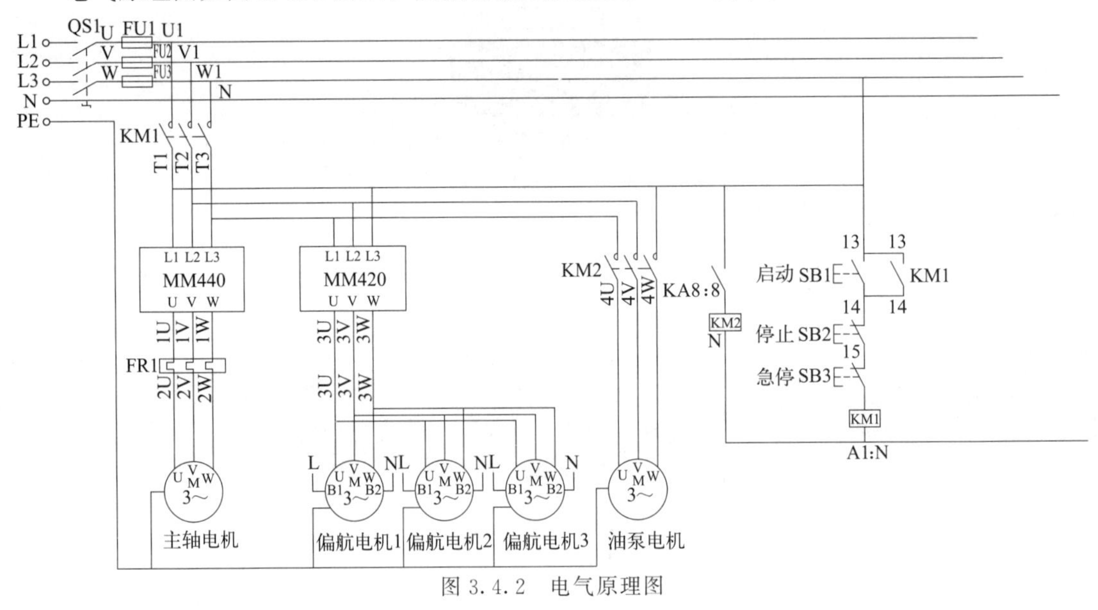

图 3.4.2　电气原理图

图 3.4.3　420 变频器（偏航变频器）接线图

偏航变桨控制系统控制柜分解图如图 3.4.4 所示。

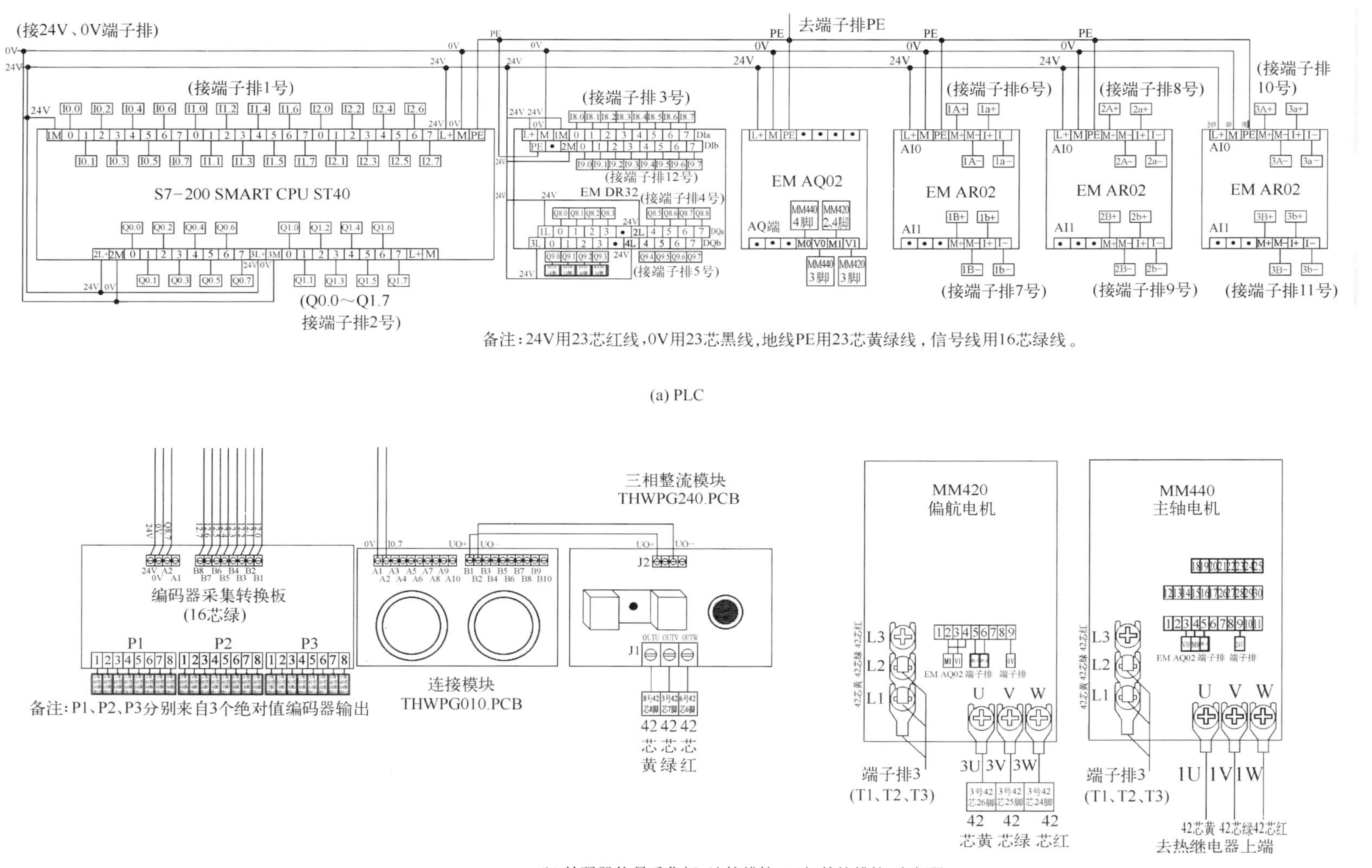

(a) PLC

(b) 编码器信号采集板、连接模块、三相整流模块、变频器

图 3. 4. 4

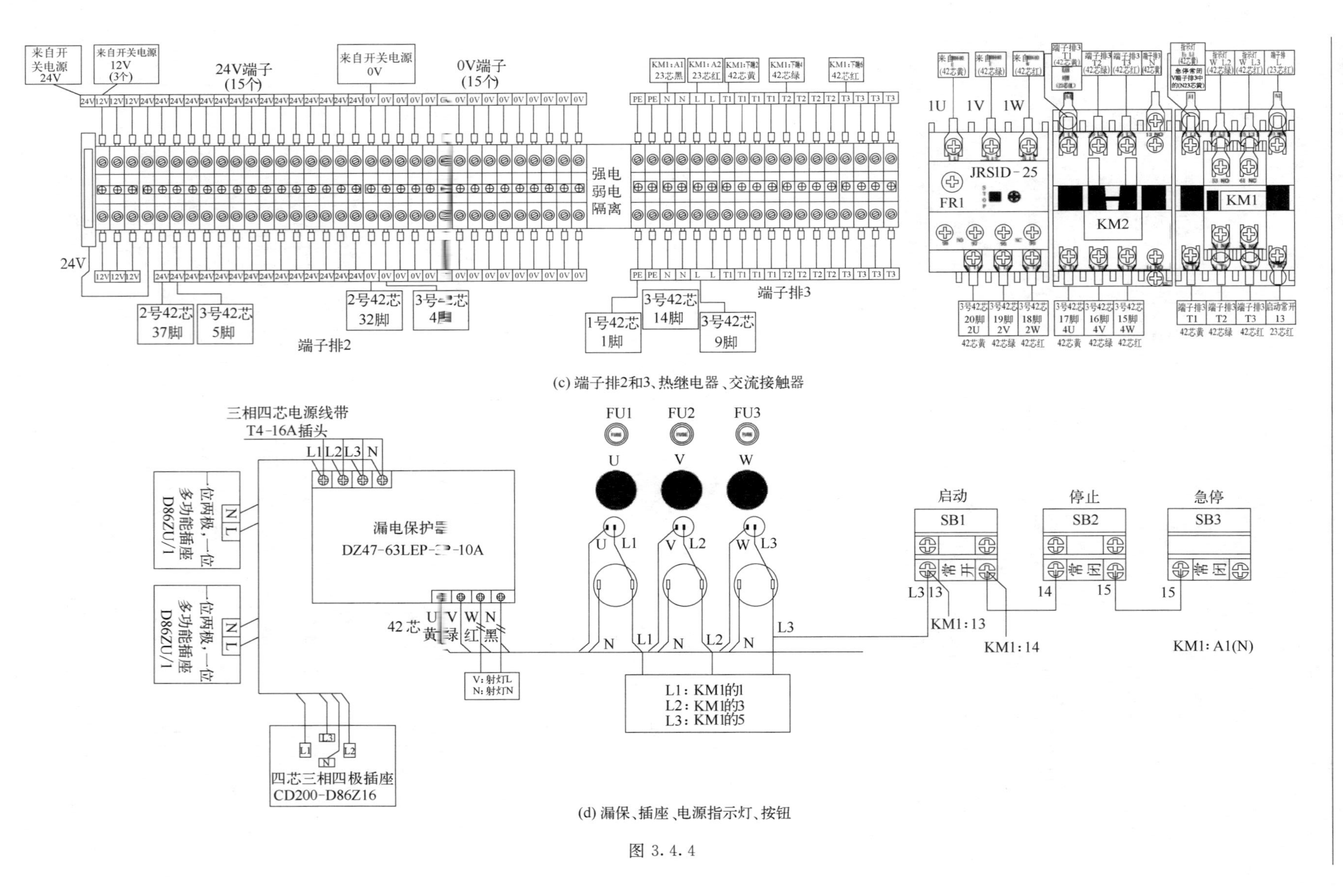

(c) 端子排2和3、热继电器、交流接触器

(d) 漏保、插座、电源指示灯、按钮

图 3.4.4

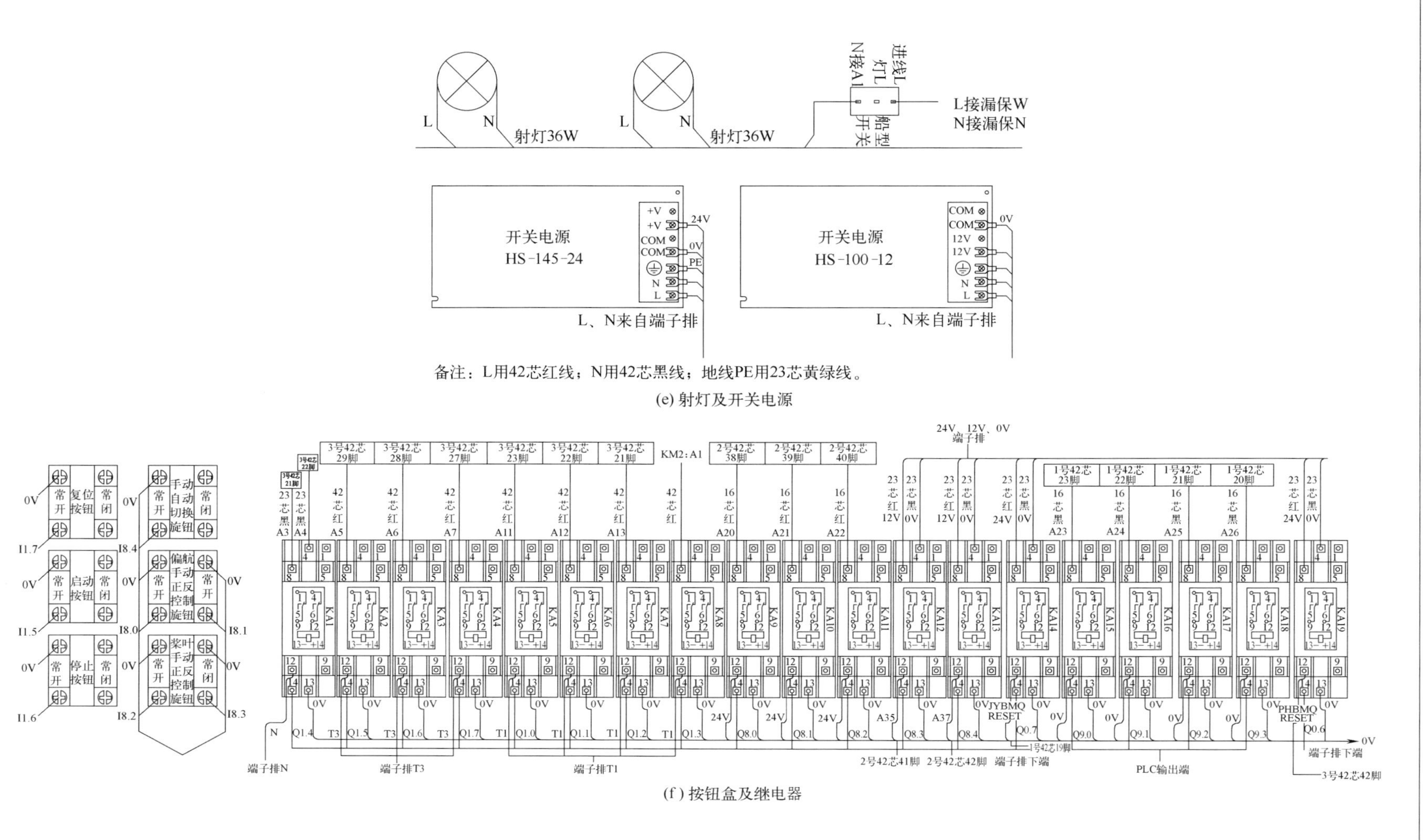

备注：L用42芯红线；N用42芯黑线；地线PE用23芯黄绿线。

(e) 射灯及开关电源

(f) 按钮盒及继电器

图 3.4.4

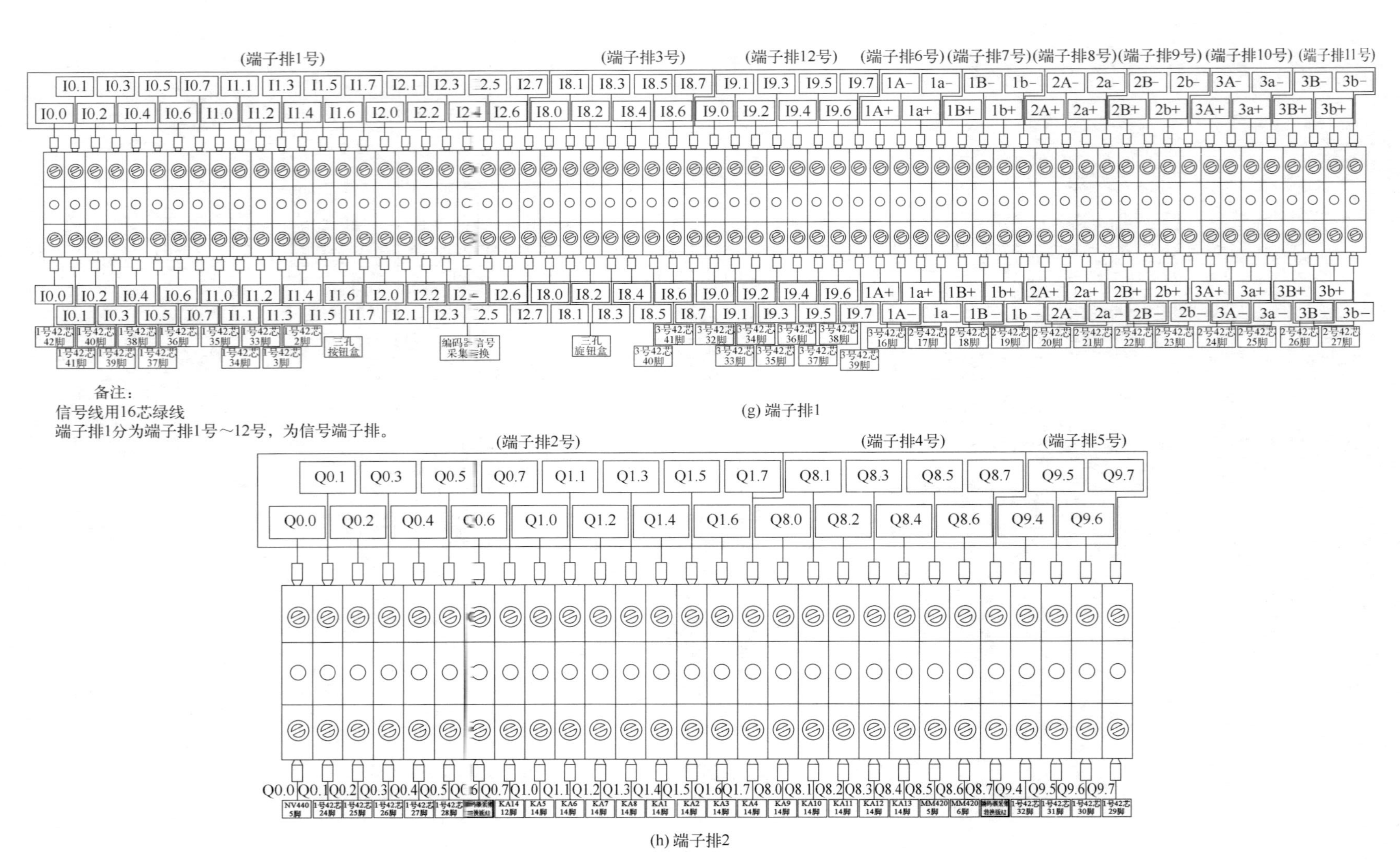

图 3.4.4 偏航变桨控制系统控制柜分解图

对接插帽定义如图 3.4.5 所示。

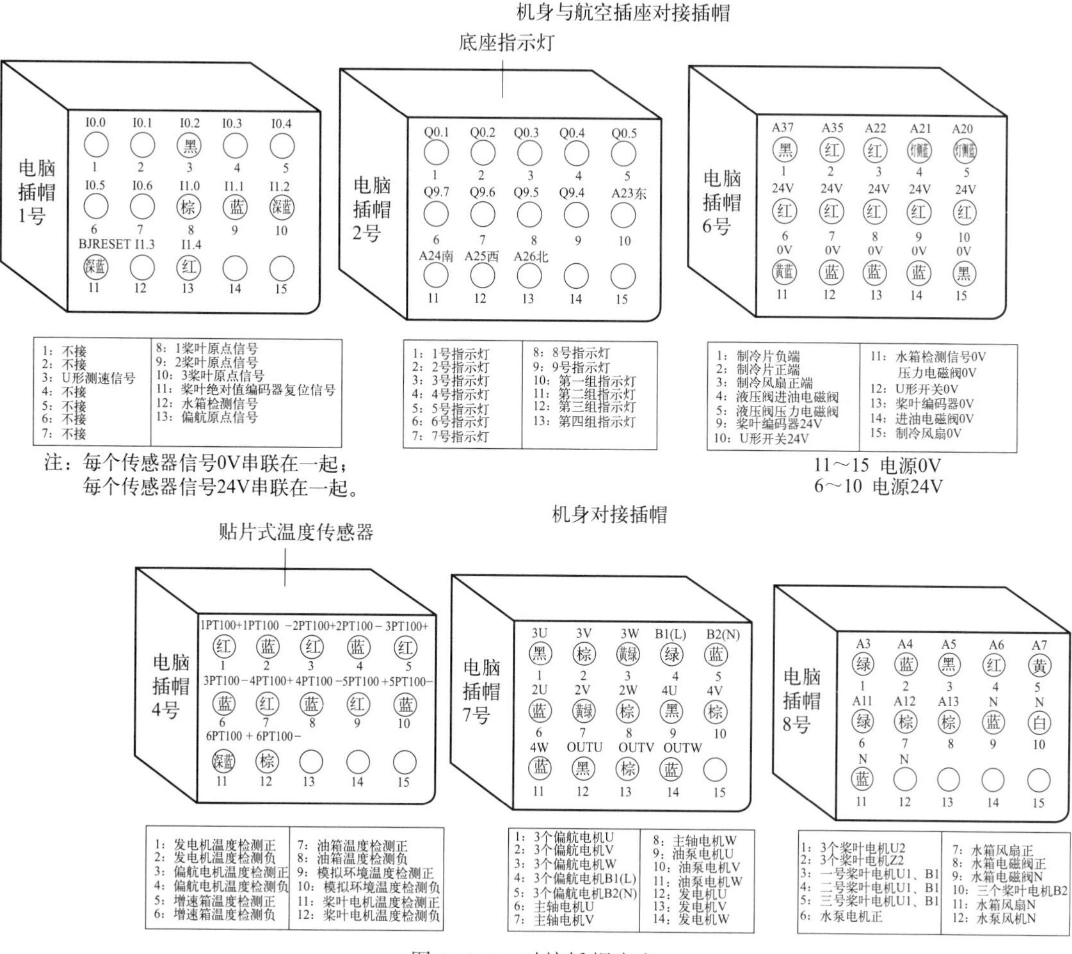

图 3.4.5　对接插帽定义

偏航编码器信号如图 3.4.6 所示，偏航信号指示灯接线图如图 3.4.7 所示。

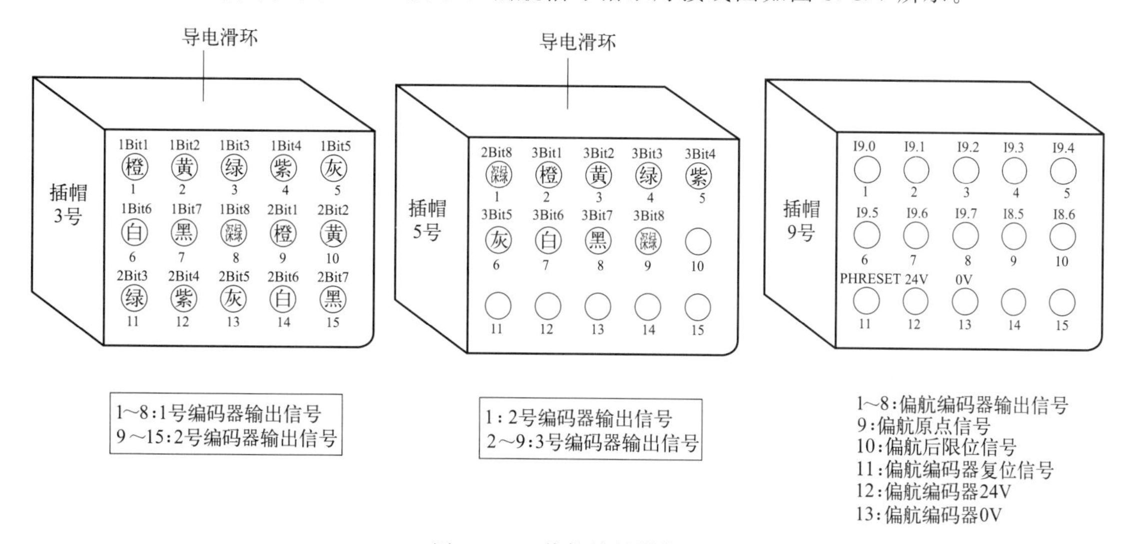

图 3.4.6　偏航编码器信号

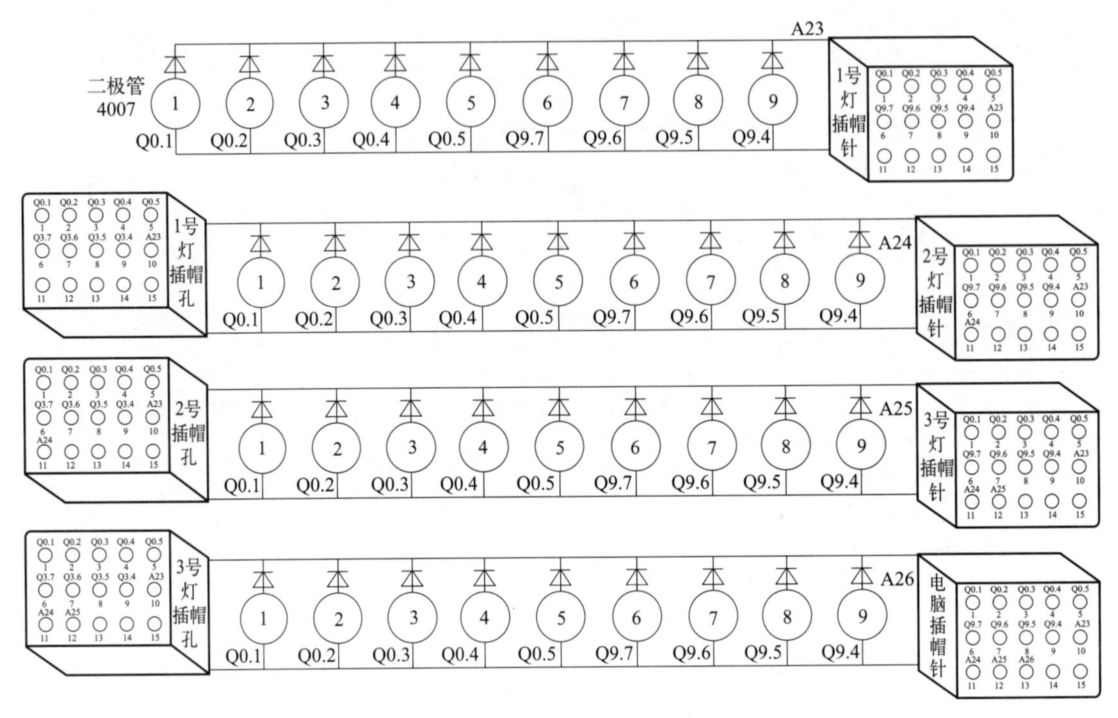

图 3.4.7　偏航信号指示灯接线图

四、偏航控制

1. 偏航控制涉及的 I/O 分配

偏航控制相关 I/O 分配见表 3.4.3。

表 3.4.3　偏航控制相关 I/O 分配表

输入信号/设备	输入端子	输出信号/设备	输出端子
手动偏航正转	I8.0	偏航位置 1 号指示灯	Q0.1
手动偏航反转	I8.1	偏航位置 2 号指示灯	Q0.2
偏航原点	I8.5	偏航位置 3 号指示灯	Q0.3
偏航后限位	I8.6	偏航位置 4 号指示灯	Q0.4
偏航 bit1	I9.0	偏航位置 5 号指示灯	Q0.5
偏航 bit2	I9.1	偏航编码器 RESET	Q0.6
偏航 bit3	I9.2	偏航变频器正转	Q8.6
偏航 bit4	I9.3	偏航变频器反转	Q8.5
偏航 bit5	I9.4	偏航位置指示灯第一组(东)	Q9.0
偏航 bit6	I9.5	偏航位置指示灯第二组(南)	Q9.1
偏航 bit7	I9.6	偏航位置指示灯第三组(西)	Q9.2
偏航 bit8	I9.7	偏航位置指示灯第四组(北)	Q9.3
读取偏航电机温度	AIW50	偏航位置 9 号指示灯	Q9.4
		偏航位置 8 号指示灯	Q9.5
		偏航位置 7 号指示灯	Q9.6
		偏航位置 6 号指示灯	Q9.7
		偏航变频器模拟值	AQW34

2. 偏航控制涉及的主要变量

偏航控制相关变量见表 3.4.4。

表 3.4.4　偏航控制相关变量表

变量名称	地址	变量名称	地址
偏航角度输入	VW102	当前偏航角度显示	VW112
		偏航电机温度显示	VW120

3. 偏航控制程序逻辑

将“手自动切换”按钮旋转到“自动”状态下，按下控制柜上的“复位”按钮，偏航系统复位至原点。

复位完成后将“手自动切换”按钮旋转到“手动”状态下，按下控制柜上的“启动”按钮进入运行状态，风机根据给定的转速开始运行；运行一段时间后，偏航系统根据第一次给定的角度值进行偏转，达到指定角度后停止；再运行一段时间后，偏航系统根据第二次给定的角度值进行偏转，达到指定角度后停止；继续运行一段时间后刹车系统工作，风机停止运行，偏航系统重新回到原点，到达原点后重复上述过程。

复位完成后将“手自动切换”按钮旋转到“自动”状态下，按下控制柜上的“启动”按钮或者上位机控制界面的“启动”键，在上位机控制界面上设置偏航的角度，系统会根据设定的角度值进行偏航。

启动偏航和即将达到偏航给定角度时，偏航速度较小；在偏航过程的中间段，偏航速度较大。

在风机对象底座上有 9×4 共 36 个偏航角度指示灯，当偏航运行到给定的角度值后停止，相对应位置的指示灯点亮，LED 点阵屏界面显示当前偏航角度。

项目原理及基础知识

一、编码器

绝对值编码器光码盘上有许多道刻线，每道刻线依次以 2 线、4 线、8 线、16 线……编排，编码器通过读取每一个位置每道刻线的通、暗，获得一组从 2^0 到 2^{n-1} 次方的唯一的二进制编码，这就称为 n 位绝对值编码器。因不受停电、干扰的影响，无需记忆，无需找参考点，而且不用一直计数，编码器的抗干扰特性、数据的可靠性大大提高了。绝对值编码器多用于角度的计量。

二、格雷码

绝对式编码器是直接输出数字量的传感器，利用自然二进制或循环二进制（格雷码）方式进行光电转换，编码的设计一般是采用自然二进制码、循环二进制码、二进制补码等。抗干扰能力强，没有累积误差；电源切断后位置信息不会丢失，但分辨率是由二进制的位数决定的，根据不同的精度要求，可以选择不同的分辨率即位数。采用循环二进制编码的绝对式编码器，输出信号是一种数字排序，不是权重码，每一位没有确定的大小，不能直接进行比较大小和算术运算，也不能直接转换成其他信号，要经过一次码变换，变成自然二进制码，再由上位机读取以实现相应的控制，见表 3.4.5。

1. 自然二进制码转换成二进制格雷码

自然二进制码转换成二进制格雷码，法则是保留自然二进制码的最高位作为格雷码的最高位，而次高位格雷码为二进制码的高位与次高位相异或，而格雷码其余各位与次高位的求法相类似。

表 3.4.5　几种自然二进制码与格雷码的对照表

十进制	二进制	格雷码	十进制	二进制	格雷码
0	0000	0000	8	1000	1100
1	0001	0001	9	1001	1101
2	0010	0011	10	1010	1111
3	0011	0010	11	1011	1110
4	0100	0110	12	1100	1010
5	0101	0111	13	1101	1011
6	0110	0101	14	1110	1001
7	0111	0100	15	1111	1000

2. 二进制格雷码转换成自然二进制码

二进制格雷码转换成自然二进制码，其法则是保留格雷码的最高位作为自然二进制码的最高位，而次高位自然二进制码为高位自然二进制码与次高位格雷码相异或，而自然二进制码的其余各位与次高位自然二进制码的求法相类似。

西门子 S7-200 SMART 没有格雷码转换库，需要在官网上下载格雷码库并将库添加进来，在编程的时候只需要调用即可完成格雷码到二进制的转换，从而直观地看出绝对值编码器的值。

三、偏航原理

本套设备上绝对值编码器采用八线制接线，可转换数值范围为 0～255，由于绝对值编码器通过绝对位置来反映转动角度，因此，编码器轴每转动一周，编码器输出值就执行一个轮回，所以，偏航系统所能反映的偏航角度范围就与偏航电机所在的小齿轮数与大齿轮数之比息息相关。

本套设备偏航小齿轮齿数为 21，偏航大齿轮齿数为 94，故偏航电机轴每转动一圈(360°)，风机对象转动角度值为 21/94×360°＝80.43°，即绝对值编码器转换后的数值 0～255 所对应的角度范围为 0～80.43°。

上位机给定的偏航角度存储在 PLC 的 VW102 中，是一个角度值。角度值与绝对值编码器输出值之间的对应关系与偏航小齿轮齿数/偏航大齿轮齿数有关，计算方法如下：

编码器输出值 255 对应变桨角度　21/94 ×360°≈80°

1 个编码器输出值对应角度　80°/255＝0.3137 (°)/个

给定偏航角度对应的编码值＝给定角度值/0.3137

因此，本实训项目主要工作为将绝对值编码器发出的数值经过整理后转换为直观的数值，进行比较，从而确定偏航方向及角度值。

项目实施

一、仪表、设备、工具清单

THWPWG-3B 型 大型风力发电系统实训平台风机对象模型、偏航变桨控制系统控制柜、能源控制·监控管理·气象站控制柜。

二、安全操作规范

① 熟练掌握偏航电机控制原理及程序控制方法；

② 实训前仔细阅读使用说明书，熟悉与偏航相关的部分，仔细阅读相关操作说明，确保各系统控制柜电源处于断开状态，根据实训指导书中相关内容熟悉此次实训的操作步骤；

③ 严格按照正确的操作步骤给系统上电和断电，以免误操作给系统带来损坏；

④ 先检查各电源和设备是否正常，确保正常后再正式开始实训任务；

⑤ 在实训过程中，始终保持实训台整洁，不可随意放置杂物，以免发生短路等故障，有“危险”标志的地方为强电注意安全；

⑥ 实训完毕，应及时关闭电源开关，并及时清理实训台。

三、实训步骤

1. 偏航回原点

① 打开控制系统总电源，系统上电，电源指示灯亮。

② 将三位旋钮“M/A”旋钮打到“A”状态。

③ 按下“RESET”按钮。

④ 风机对象偏航回原点。

⑤ 当风机对象偏航回到原点后，原点传感器指示灯亮。

⑥ KA19 继电器动作，偏航绝对值编码器复位，桨叶采集信号 I9.0～I9.7 信号清零，PLC 上相应指示灯熄灭，偏航复位完成。

2. 偏航手动正反转控制

① 打开控制系统总电源，系统上电，电源指示灯亮。

② 将三位旋钮“M/A”旋钮打到“M”状态。

③ 将三位旋钮中“Y：FWD/REV”旋钮打到“FWD”，偏航顺时针旋转。

④ 将三位旋钮中“Y：FWD/REV”旋钮打到“REV”，偏航逆时针旋转。

3. 主控计算机正反转控制

① 将三位旋钮“M/A”旋钮打到“A”状态。

② 打开主控计算机控制软件，按步骤连接。

③ 首先将三个桨叶复位到原点。

④ 在主控计算机中输入偏航角度，按下“confirm”按钮，再按下“Start”按钮。

⑤ 偏航旋转到设置角度。

⑥ 指示灯根据偏转角度亮起或熄灭。

⑦ 当下一次偏航时，先复位，再进行偏航操作。

4. 程序阅读——偏航控制

① 掌握风力发电机组的偏航控制逻辑。

② 列出与偏航控制相关的主要 I/O 分配及主要中间继电器、变量的含义。

③ 用流程图或自然语言描述网络 5、HSC_INIT、网络 61、63、30、31、32、19 的偏航控制逻辑。

5. 程序编写——偏航控制

① 编写风力发电机组的风机偏航控制程序。

② 用户通过上位机软件给定偏航角度，根据偏航的不同阶段调节输入变频器 MM420 的模拟量。

在同等风速下，当偏航角度距离原点±6°范围内时，模拟量输出值减去 500；当桨叶角度在±3°范围内时，模拟量输出值减去 1000。

按下启动按钮，启动主轴变频器；按下停止按钮或风速超过安全风速，停止主轴变频器。

a. 测量主轴转速;

b. 主轴变频器模拟量给定;

c. 偏航变频器模拟量给定;

6. 撰写实训报告

项目作业

用文字或流程图，描述偏航功能实现流程。

项目五　变桨功能的实现

项目描述

了解变桨控制系统工作原理，实施手动变桨控制和上位机自动变桨控制；完成程序编辑，实现给定角度的变桨功能。

能力目标：

① 掌握绝对值编码器和格雷码工作原理；

② 理解变桨控制过程并能用 PLC 编程实现。

项目环境

变桨控制系统主要由从 PLC、变频器、绝对值编码器、变桨电机（交流减速电机）、控制按钮等器件组成。可完成变桨控制系统的安装、手动与自动变桨 PLC 控制程序的编写与调试，以及从（slave）PLC 与监控管理主（master）PLC 的通信。

一、变桨控制系统主要设备

变桨控制系统主要设备清单见表 3.5.1。

表 3.5.1　变桨控制系统主要设备清单

序号	名称	规　　格	数量
1	偏航变桨控制系统控制柜	880mm×600mm×2100mm	1台
2	按钮模块	24V/6A 开关电源 1 个;急停按钮 1 个;复位、启动、停止按钮黄、绿、红各 1 个;自锁按钮黄、绿、红各 1 个;转换开关 1 个;24V 指示灯黄、绿、红各 2 个	1组
3	PLC	西门子 CPU ST40 主机	1台
		EM DR32 数字量输入/输出模块,16×24V DC 输入/16 点继电器输出	1台
		EM AQ02 模块,2 路模拟量输出(2AI)	1台
		EM AR02 模块,2 路热电阻输入模块(PT100)	3台
		EM DP01 通信模块	1台
4	开关电源	HS-145-24,24V 输出	1个
		HS-100-12,12V 输出	1个
5	交流接触器	LC1-D0610M5N　220V	2个

二、主要控制原理与接线图

变桨电机接线图见图 3.5.1，导电滑环接线见图 3.5.2，每个桨叶电机 U2、Z2 之间接运行电容。

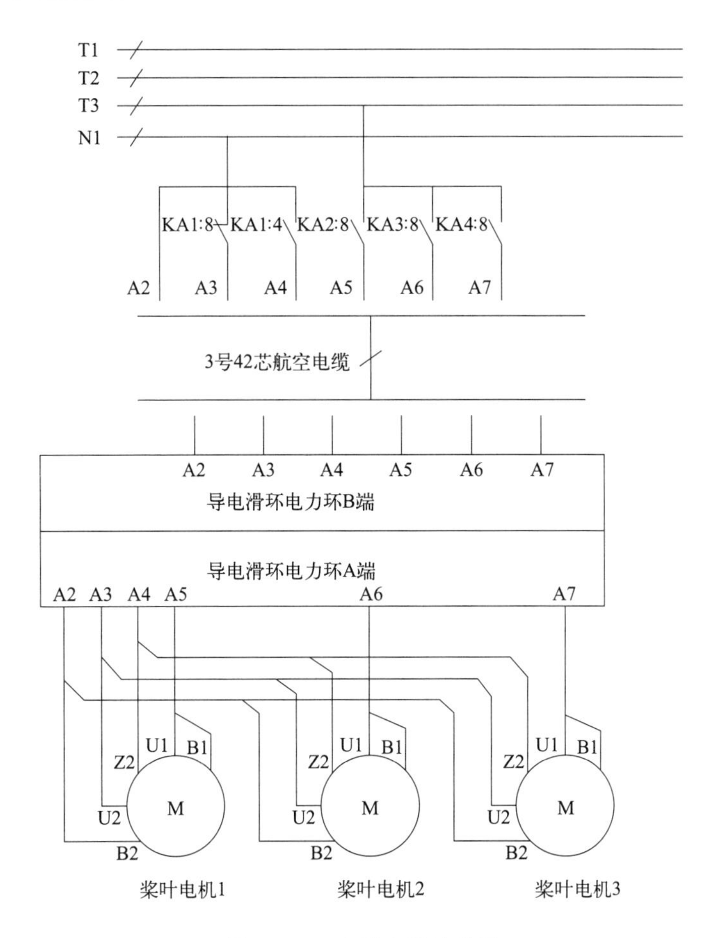

图 3.5.1　变桨电机接线图

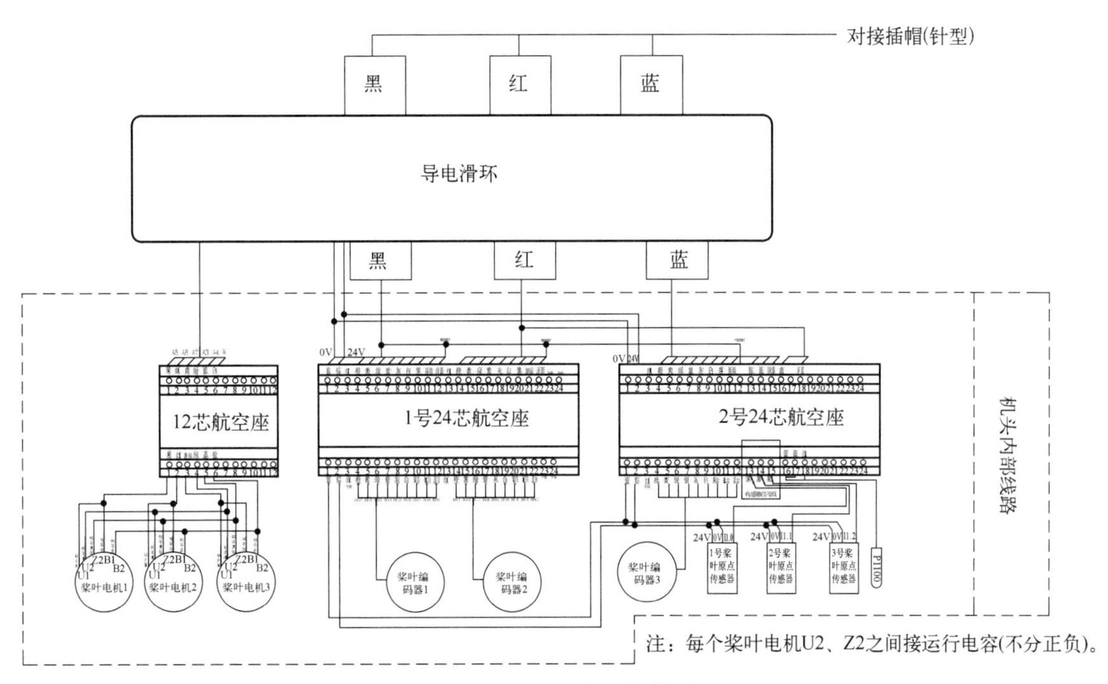

图 3.5.2　导电滑环接线图

偏航变桨控制系统控制柜装连及分解图、对接插帽定义见图 3.4.4～图 3.4.6。

三、变桨控制

1. 变桨控制涉及的 I/O 分配

变桨控制相关 I/O 分配见表 3.5.2。

表 3.5.2 变桨控制相关 I/O 分配表

输入信号/设备	输入端子	输出信号/设备	输出端子
桨叶 1 原点	I1.0	桨叶编码器 RESET	Q0.7
桨叶 2 原点	I1.1	桨叶电机正转	Q1.4
桨叶 3 原点	I1.2	桨叶 1 启动	Q1.5
桨叶 Bit1	I2.0	桨叶 2 启动	Q1.6
桨叶 Bit2	I2.1	桨叶 3 启动	Q1.7
桨叶 Bit3	I2.2		
桨叶 Bit4	I2.3		
桨叶 Bit5	I2.4		
桨叶 Bit6	I2.5		
桨叶 Bit7	I2.6		
桨叶 Bit8	I2.7		
手动桨叶反转	I8.2		
手动桨叶正转	I8.3		
读取桨叶电机温度	AIW82		

2. 变桨控制涉及的主要变量

变桨控制相关变量见表 3.5.3。

表 3.5.3 变桨控制相关变量表

变量名称	地址	变量名称	地址
变桨角度输入	VW104	当前桨叶角度显示	VW114
		桨叶电机温度显示	VW128

3. 变桨控制程序逻辑

将“手自动切换”按钮旋转到“自动”状态下，按下控制柜上的“复位”按钮，桨叶系统复位至原点。

复位完成后将“手自动切换”按钮旋转到“手动”状态下，按下控制柜上的“启动”按钮进入运行状态，风机根据给定的转速开始运行；运行一段时间并完成第一次偏航后，桨叶系统根据第一次给定的角度进行旋转，到达指定角度后停止；再运行一段时间并完成第二次偏航后，桨叶系统根据第二次给定的角度进行旋转，到达指定角度后停止；继续运行一段时间后刹车系统工作，风机停止运行，桨叶系统重新回到原点，到达原点后重复上述过程。

复位完成后将“手自动切换”按钮旋转到“自动”状态下，按下控制柜上的“启动”按钮或者上位机控制界面的“启动”键，在上位机控制界面上设置桨叶角度。系统会根据设定的角度值进行变桨。

变桨控制系统根据上位机软件内给定的变桨角度值来进行变桨，同时在上位机软件界面内显示当前变桨角度。桨叶运行到给定的角度值后停止，LED 点阵屏界面显示当前偏航角度。

项目原理及基础知识

一、编码器（见项目四）

二、格雷码（见项目四）

三、变桨原理

本套设备上绝对值编码器采用八线制接线，可转换数值范围为0～255，由于绝对值编码器通过绝对位置来反映转动角度，因此，编码器轴每转动一周，编码器输出值就执行一个轮回，所以，变桨系统所能反映的变桨角度范围就与变桨电机所在的小齿轮齿数与变桨大齿轮齿数之比息息相关。

本套设备变桨小齿轮齿数为20，变桨轴承内齿轮齿数为37，故变桨电机轴每转动一圈（360°），桨叶转动角度值为20/37×360°＝194.594595°，即绝对值编码器转换后的数值0～255所对应的角度范围为0～194.594595°。

因此，本实训项目主要工作为将绝对值编码器发出的数值经过整理后转换为直观的数值，进行比较，从而确定变桨电机的转动方向及角度值。

项目实施

一、仪表、设备、工具清单

THWPWG-3B型大型风力发电系统实训平台风机对象模型、偏航变桨控制系统控制柜、能源控制·监控管理·气象站控制柜。

二、安全操作规范

① 熟练掌握变桨电机控制原理及程序控制方法；

② 实训前仔细阅读使用说明书，熟悉与变桨相关的部分，仔细阅读相关操作说明，确保各系统控制柜电源处于断开状态，根据实训指导书中相关内容熟悉此次实训的操作步骤；

③ 严格按照正确的操作步骤给系统上电和断电，以免误操作给系统带来损坏；

④ 先检查各电源和设备是否正常，确保正常后再正式开始实训任务；

⑤ 在实训过程中，始终保持实训台整洁，不可随意放置杂物，以免发生短路等故障，有“危险”标志的地方为强电注意安全；

⑥ 实训完毕，应及时关闭电源开关，并及时清理实训台。

三、实训步骤

1. 程序阅读理解

变桨绝对值编码器端子信息见表3.5.4。

如表3.5.4所示，绝对值编码器各种接线方法对应的各输出位权重已经给出，本套设备采256/180接线方法，橙～红/白线分别接入PLC输入点I2.0～I2.7。这些信号需要经过处理后才变成能够直观感受到的数据，具体方法就是先将这些数据当作二进制数来处理，通过不同的权重变换为十进制数，然后调用格雷码转换库，变为可以直接用的数据。

变桨绝对值编码器复位程序：

表 3.5.4 变桨绝对值编码器端子信息表

引脚编号、定义	线颜色	分辨率		
		512	256/180	128/90
1	蓝	0V	←	←
2	棕	10.8～26.4V	←	←
3	黑	不接	←	←
4	红	Bit1(2^0)	不接	←
5	橙	Bit2(2^1)	Bit1(2^0)	不接
6	黄	Bit3(2^2)	Bit2(2^1)	Bit1(2^0)
7	绿	Bit4(2^3)	Bit3(2^2)	Bit2(2^1)
8	紫	Bit5(2^4)	Bit4(2^3)	Bit3(2^2)
9	灰	Bit6(2^5)	Bit5(2^4)	Bit4(2^3)
10	白	Bit7(2^6)	Bit6(2^5)	Bit5(2^4)
11	黑/白	Bit8(2^7)	Bit7(2^6)	Bit6(2^5)
12	红/白	Bit9(2^8)	Bit8(2^7)	Bit7(2^6)
13	蓝/白	RESET	←	←

当电压在 0～24V DC（高电平）之间时，当前轴的位置被设置为编码器的“0”位置。正常工作时，编码器需保持 0～0.8V DC 的电压（低电平）。

Q0.7 用于驱动继电器 KA14 接通 24V，1s 后再接通 0V，编码器复位，程序如图 3.5.3 所示。

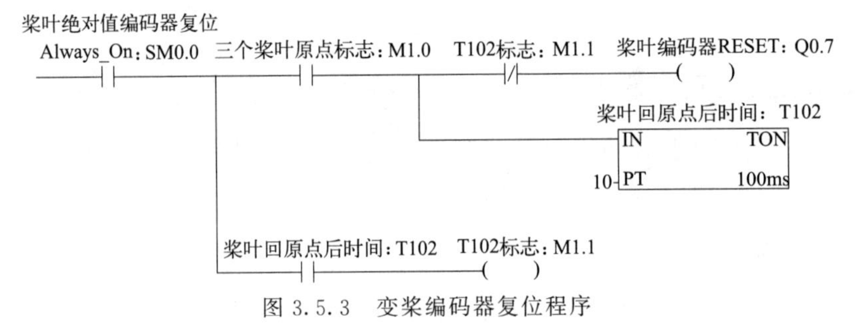

图 3.5.3 变桨编码器复位程序

变桨编码器信号采集程序如图 3.5.4 所示。

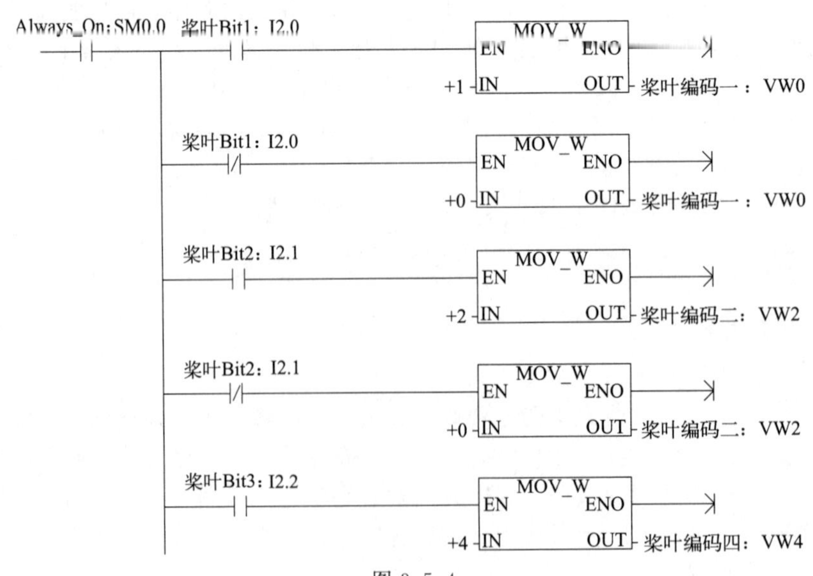

图 3.5.4

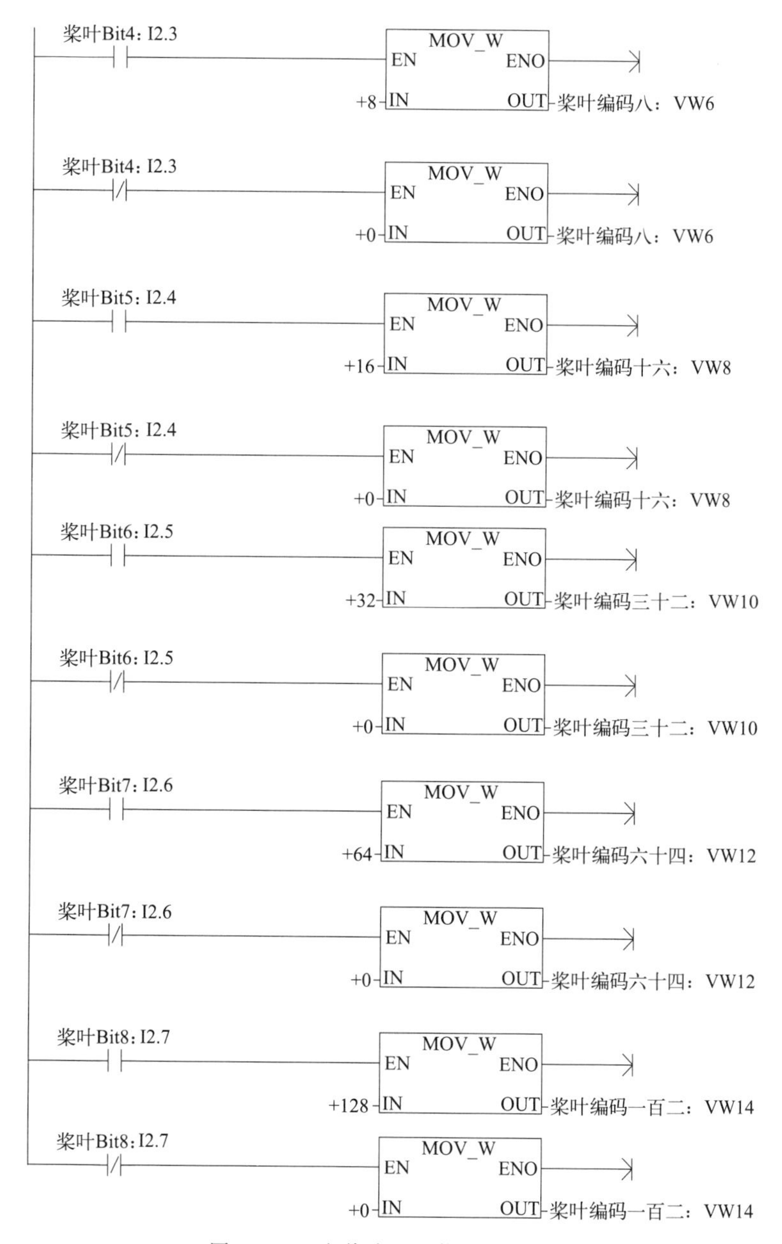

图 3.5.4　变桨编码器信号采集程序

图 3.5.4 所示程序将绝对值编码器传出的数据当作二进制处理，即当编码器 8 位信号都没有信号时，对应的信号存储区全为 0。

当 I2.0=1 时，将 $2^0=1$ 传送到 VW0。

当 I2.1=1 时，将 $2^1=2$ 传送到 VW2。

当 I2.2=1 时，将 $2^2=4$ 传送到 VW4。

当 I2.3=1 时，将 $2^3=8$ 传送到 VW6。

当 I2.4=1 时，将 $2^4=16$ 传送到 VW8。

当 I2.5=1 时，将 $2^5=32$ 传送到 VW10。

当 I2.6=1 时，将 $2^6=64$ 传送到 VW12。

当 I2.7=1 时，将 $2^7=128$ 传送到 VW14。

变桨编码器数值计算如图 3.5.5 所示。

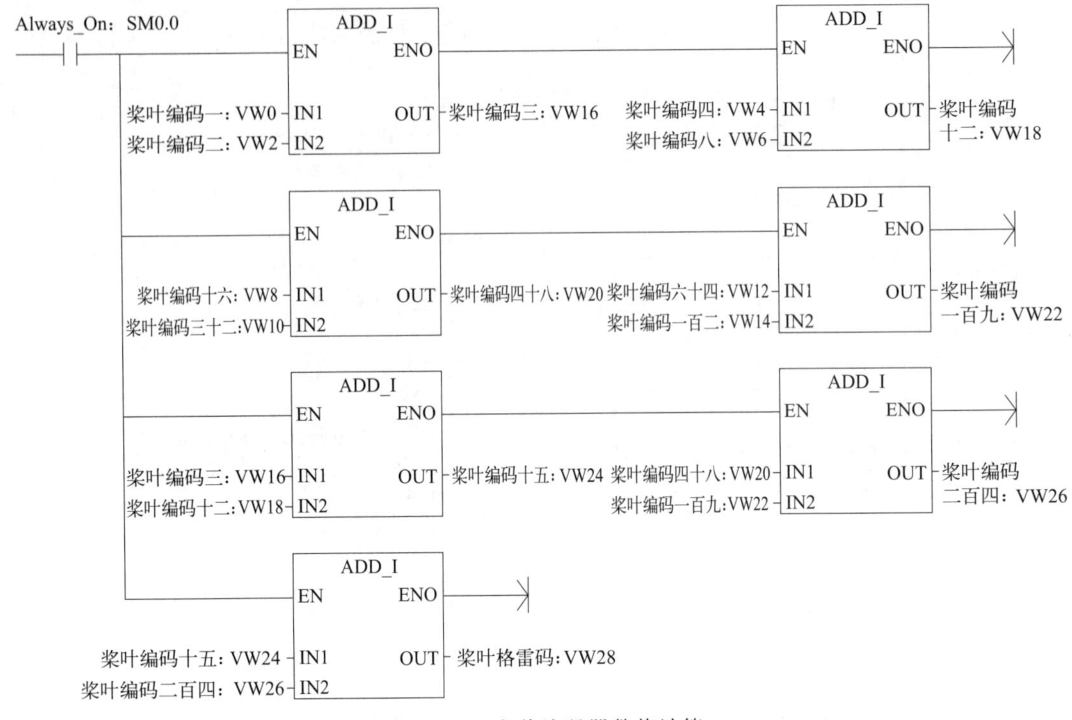

图 3.5.5　变桨编码器数值计算

图 3.5.5 所示程序将所得的二进制数转化为十进制数。将 VW0 和 VW2 的值相加，放在 VW16 中，将 VW4 和 VW6 的值相加，放在 VW18 中；将 VW8 和 VW10 的值相加，放在 VW20 中；将 VW12 和 VW14 的值相加，放在 VW22 中；将 VW16 和 VW18 的值相加，放在 VW24 中；将 VW20 和 VW22 的值相加，放在 VW26 中；将 VW24 和 VW26 的值相加，放在 VW28 中。

VW28 中得到的数值就是编码器产生的格雷码数值，如图 3.5.6 所示。

格雷码转换：

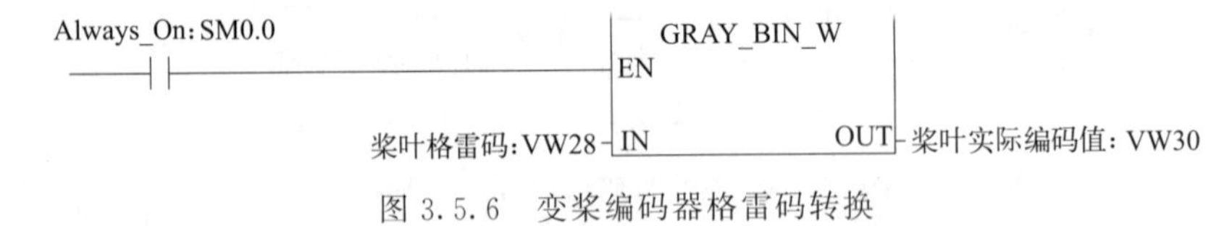

图 3.5.6　变桨编码器格雷码转换

图 3.5.6 所示程序通过调用格雷码转换库，将计算后的格雷码十进制数转换为可以直接用的十进制数，存储在 VW30 中。

变桨角度与绝对值编码器间数值的转换：

主控计算机给定的变桨角度存储在 PLC 的 VW104 中，是一个角度值。角度值与绝对值编码器输出值之间的对应关系与变桨小齿轮齿数（20）/变桨轴承内圈齿轮齿数（37）有关，计算方法如下：

编码器输出值 255 对应变桨角度：$\frac{20}{37}\times360°\approx194°$；

1 个编码器输出值对应角度：$\frac{194°}{255}=0.763116(°)/个$；

给定变桨角度对应的编码值$=\dfrac{\text{给定编码值}}{0.763116}$，实现程序如图 3.5.7 所示。

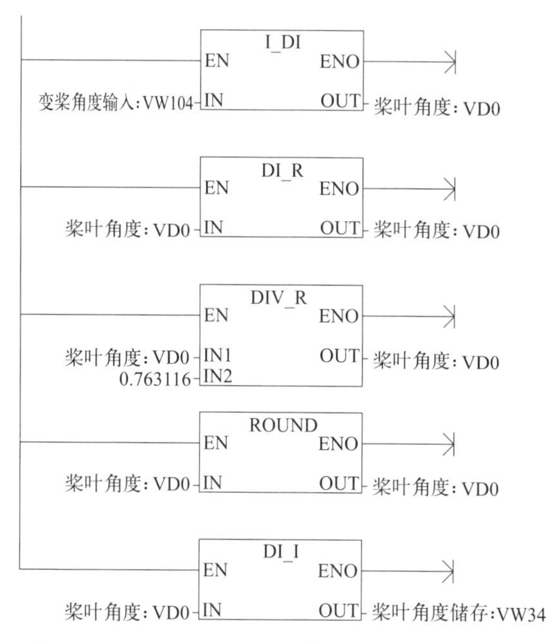

图 3.5.7 变桨角度与绝对值编码器间数值的转换

将 VW104 中输入的角度从整数转换成双精度整数，存放到 VD0 中；将 VD0 中数值转换成实数，方便进行后面的运算；

将转换成实数的 VD0 中的数值除以 0.763116，得到实数类型的桨叶角度重新存放到 VD0 中；

对 VD0 中的数值进行取整；

将取整后的 VD0 中的数值由双精度整数转换成整数类型，存放到 VW34 中；

这样，将得到的编码值 VW34 与实际编码值 VW30 相比较，应用到程序当中。

2. 桨叶控制

(1) 桨叶回原点

打开控制系统总电源，系统上电，电源指示灯亮。三个桨叶若未回到原点，即原点传感器指示灯未亮，则进行其他操作前需要对桨叶进行原点复位操作。将三位旋钮“M/A”旋钮，打到“M”状态。按下“RESET”按钮。桨叶一、二、三依次回原点。若三个桨叶均在原点，按下“RESET”按钮后桨叶不动作。若其中一个或两个桨叶不在原点，桨叶按一定顺序回原点。在运行过程中按下“STOP”按钮，桨叶停止复位运行。当三个桨叶均回到原点后，三个桨叶传感器指示灯亮，KA14 继电器动作 1s，桨叶绝对值编码器复位，桨叶采集信号 I2.0～I2.7 信号清零，PLC 上相应指示灯熄灭。桨叶复位完成。

(2) 桨叶手动正反转控制

将三位旋钮“M/A”旋钮，打到“M”状态。将三位旋钮中“P：FWD/REV”旋钮打到“FWD”，桨叶顺时针旋转。将三位旋钮中“P：FWD/REV”旋钮打到“REV”，桨叶逆时针旋转。

(3) 桨叶自动正反转控制

将三位旋钮“M/A”旋钮，打到“A”状态。当输入角度在 0～180°时，首先按下

“RESET”按钮，三个桨叶复位到原点。在主控计算机或者PLC编程软件中输入角度后，按下“confirm”按钮，再按下“Start”按钮。桨叶正转到设置角度。当下一次的输入数值小于当前角度时，按下上位机“Start”按钮，桨叶反向到达输入角度。

输入角度在180°～360°时，与前述操作相同，但应该特别注意：完成第一次变桨后，若想要下一次输入角度值大于当前值（如当前值为270°，输入值为300°），此时应先按“复位”按钮，桨叶回到原点后，再输入设置角度（如上所述300°），这里与0～180°变桨有所区别。

3. 撰写实训报告

项目作业

用文字或流程图，描述变桨功能的实现（分该实训设备的变桨和工程实际机组的变桨两种情况）。

项目六　并网型逆变器工作原理实训

项目描述

查阅风力发电机组并网逆变器工作原理、结构组成，运用所学的知识和相关安装手册，完成1.5MW风力发电机组模拟并网逆变器工作原理实训。遵循安全用电规范和注意事项。

能力目标：

① 了解并网逆变器的工作原理。

② 掌握并网逆变器的使用及其作用。

项目环境

本实训任务主要涉及此实训平台的并网型逆变器，完成该任务需要参考THWPWG-3B型大型风力发电系统实训平台设备说明手册，认识并了解并网逆变器原理、结构，了解相关工具的使用规范和注意事项。

项目原理及基础知识

通常，把将交流电能变换成直流电能的过程称为整流，把完成整流功能的电路称为整流电路，把实现整流过程的装置称为整流设备或整流器。与之相对应，把将直流电能变换成交流电能的过程称为逆变，把完成逆变功能的电路称为逆变电路，把实现逆变过程的装置称为逆变设备或逆变器。

一、逆变器的控制方式

并网逆变器按控制方式分类，可分为电压源电压控制、电压源电流控制、电流源电压控制和电流源电流控制四种方法，但由于逆变回路中大电感往往会导致系统动态响应差，当前世界范围内大部分逆变器均采用以电压源输入为主。

逆变器与市电并联运行的输出控制可分为电压控制和电流控制，市电系统可视为容量无穷大的定值交流电压源，如果并网逆变器的输出采用电压控制，则实际上就是一个电压源与电压源并联运行的系统，这种情况下很难保证系统稳定运行；如果并网逆变器的输出采用电流控制，则只需控制并网逆变器的输出电流以跟踪市电电压，即可达到并联运行的目的。

并网逆变器一般都采用电压源输入、电流源输出的控制方式。

二、逆变器的基本结构

并网逆变器功能框图如图 3.6.1 所示，主电路拓扑结构由 DC/DC（Boost 升压电路）+DC/AC（单相并网逆变器）+滤波器组成，控制回路由母线电压检测+输出电流检测+电网电压检测+隔离驱动+DSP 控制电路+键盘+液晶屏组成。

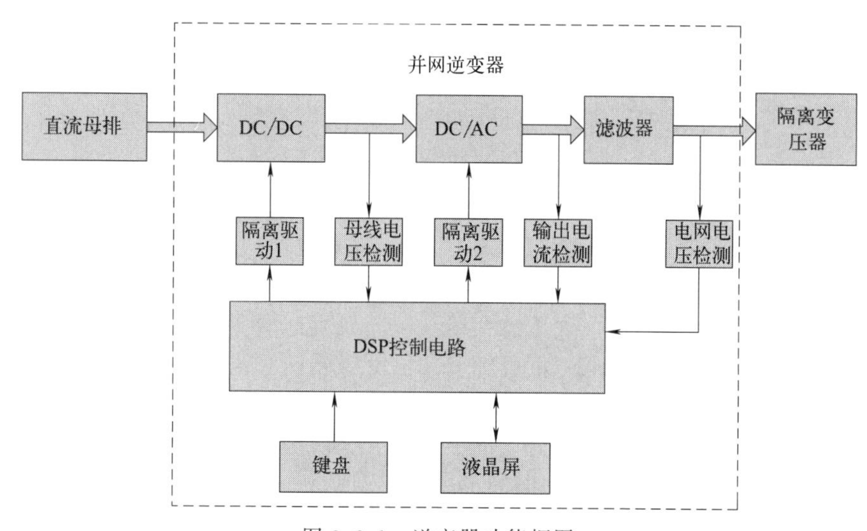

图 3.6.1 逆变器功能框图

并网逆变器模块说明：

① DC/DC 模块（Boost 升压电路）：Boost 升压电路主要将直流母排输出直流电压 U_{bus} 变换成能满足并网要求的直流母线电压 U。

② DC/AC 模块：DC/AC 将该直流母线电压经过 DC/AC 逆变成与电网电压同频、同相、同幅的正弦交流电以实现与电网的并网连接。

③ 滤波器：滤除逆变器输出高频 PWM 谐波电流，减小进网电流中的高频环流作用，又能在逆变器与电网间进行能量的传递，使并网逆变器获得一定的阻尼特性，减小冲击电流，有利于系统的稳定运行。

④ 母线电压检测：完成电压闭环及保护作用。

⑤ 输出电流检测：完成电流闭环及保护作用。

⑥ 电网电压检测：完成电网锁相、电压前馈及保护作用。

⑦ 隔离驱动 1、2：完成对开关管的隔离驱动作用。

⑧ DSP 控制电路：执行并网逆变器的软件算法功能。

⑨ 键盘、液晶屏：显示并网参数及设置影响并网电流质量的参数。

三、逆变器的工作原理

单相电压全控型 PWM 逆变器工作原理如图 3.6.2 所示，为通常使用的单相输出的全桥逆变主电路；其中，交流元件采用 IGBT 管 Q11、Q12、Q13、Q14，并由 PWM 脉宽调制控制 IGBT 管的导通或截止。

当逆变器电路接上直流电源后，先由 Q11、Q14 导通，Q12、Q13 截止，则电流由直流

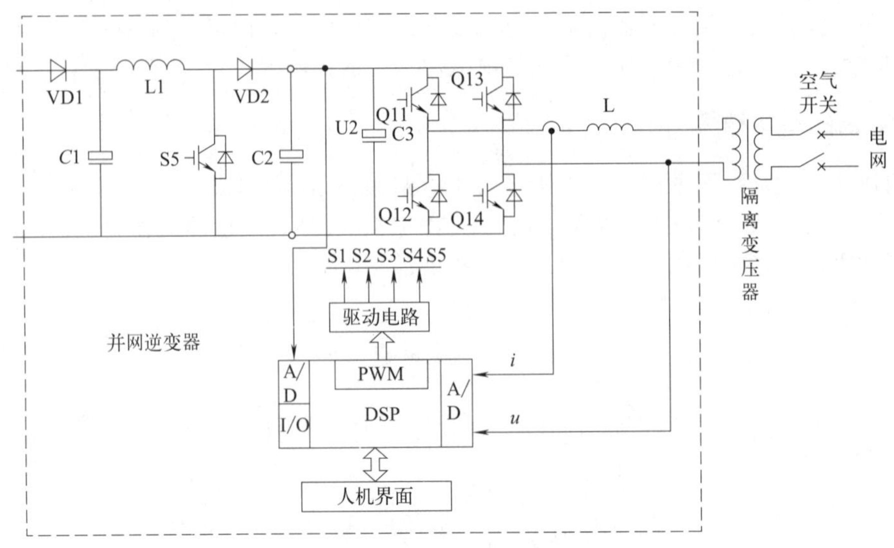

图 3.6.2 逆变器主电路

电源正极输出，经Q11、电感L、变压器初级线圈，到Q14回到电源负极。当Q11、Q14截止后，Q12、Q13导通，电流从电源正极经Q13、变压器初级线圈、电感到Q12回到电源负极。此时，在变压器初级线圈上，已形成正负交变方波，利用高频PWM控制，两对IGBT管交替重复，在变压器上产生交流电压。由于LC交流滤波器作用，使输出端形成正弦波交流电压。

四、并网逆变器参数设置操作说明

控制器上电，系统初始化，接口模块的工作指示灯和故障指示灯全亮，液晶屏显示“初始化…”，3～5s后，指示灯灭，液晶屏显示如图 3.6.3 所示界面。

1. 参数设定步骤

① 点击键盘上“确定”按键，出现参数设定界面，如图 3.6.4 所示。

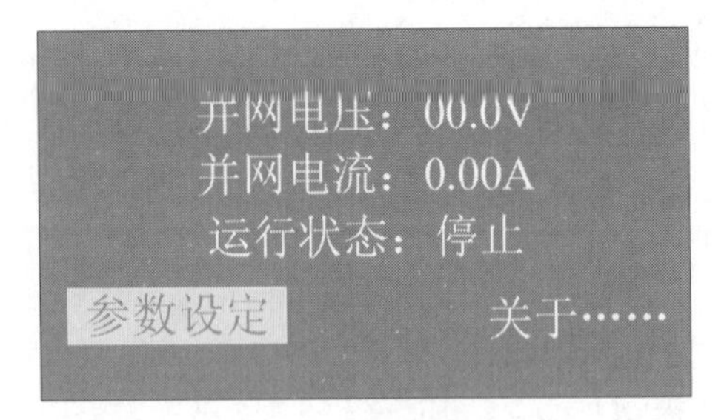

图 3.6.3 系统默认界面

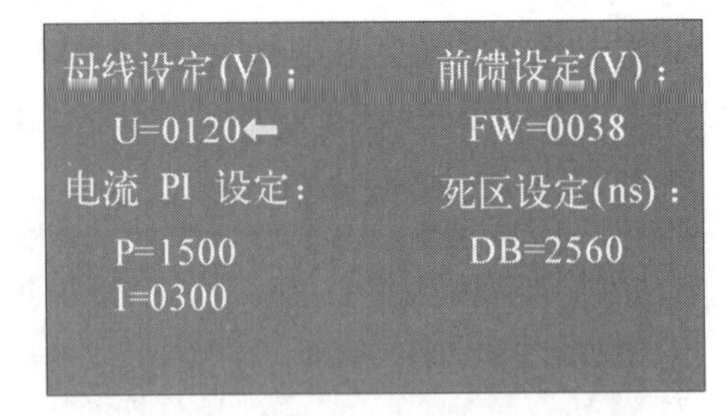

图 3.6.4 参数设定界面（1）

② 通过键盘上的“▼”或“▲”按键，液晶屏上光标移动，选择要设置的参数，默认选择母线设定，点击“设置”键，出现如图 3.6.5 所示界面。

③ 通过键盘上的数字键设定需要的值，例如输入“1”“0”“0”，显示 $U=100$，点击“确认”，完成参数设定，如图 3.6.6 所示。

④ 其他参数设定类似于母线设定，操作功能示意如图 3.6.7 所示。

2. 参数设置说明

为了保证系统能够安全运行，参数设置有效范围做如下限定：

母线设定(V)：　前馈设定(V)：
U=0000　FW=0038
电流 PI 设定：　死区设定(ns)：
P=1500　DB=2560
I=0300

图 3.6.5 参数设定界面（2）

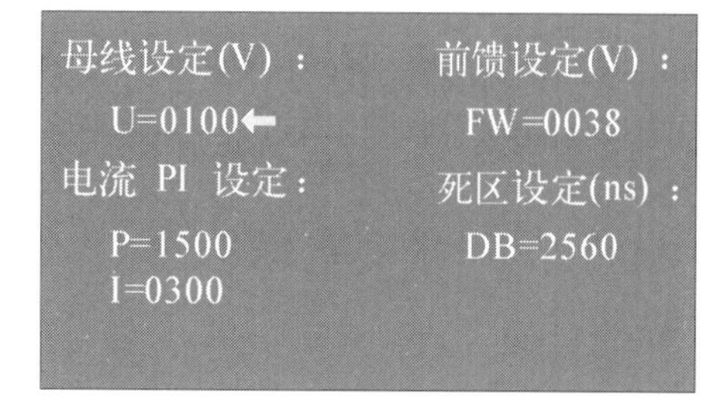

图 3.6.6 参数设定界面（3）

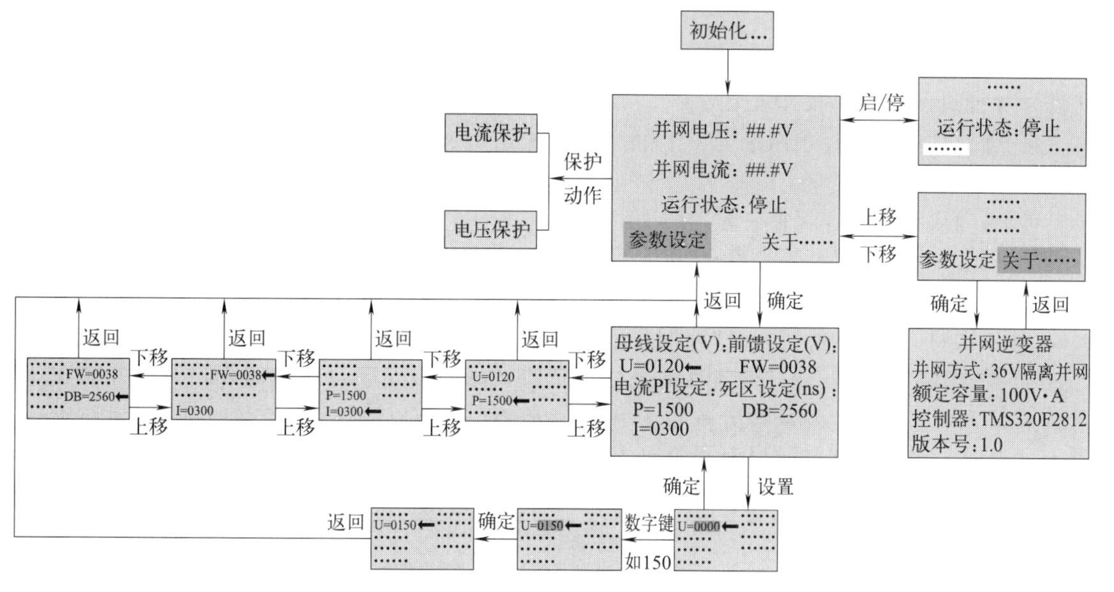

图 3.6.7 操作面板功能示意图

① 母线电压有效设置范围：60～120V；

② 电流环比例系数 P 有效设置范围：150～1500；

③ 电流环积分系数 I 有效设置范围：30～300；

④ 前馈电压有效设置范围：0～50V；

⑤ 死区设置有效值：2560ns、2780ns、2990ns、3200ns。

项目实施

一、仪表、设备、工具清单

① THWPWG-3B 型 大型风力发电系统实训平台。

② 万用表。

二、安全操作规范

① 使用前先检查各电源是否正常。

② 控制柜中存在 AC 220V 的接入点，实训时要注意安全。

③ 实训始终，实训台要保持整洁，不可随意放置杂物，以免发生短路等故障。

④ 实训完毕，应及时关闭电源开关，并及时清理实训台。

⑤ 严格按照正确的操作步骤给系统上电和断电，以免误操作给系统带来损坏。

⑥ 在操作系统的过程中，能源转换储存控制系统蓄电池开关打开之后有一个等待智能充放电控制器自检初始化的过程；必须等到智能充放电控制器的“红灯”灭掉后才能进行下一步操作。

⑦ 在实训过程中，有“危险”标志的地方为强电，须注意安全。

三、实训步骤

① 打开各控制系统总电源开关；电源指示灯有显示。

② 将充放电控制器的刹车置于“RELEASE（自动刹车）”状态，合上“能源转换储存控制系统”的“蓄电池”空气开关，接入蓄电池，同时给充放电控制器供电，此时充放电控制器进行初始化，红色指示灯点亮（工作在刹车状态）。必须等红色指示灯熄灭（退出刹车状态）才能进行下一步操作。

③ 合上“能源储存控制单元”上的“风机输出”和“充放电控制器”断路器，“摇柄”开关居中。

④ 启动风机，转速保持一定。

⑤ 打开“并网逆变控制系统”的“控制器”开关，并网逆变控制器上电，液晶屏初始化。

⑥ 依次打开“蓄电池”和“并网发电”开关，逆变器输入和输出电压部分有数值。

⑦ 操作人机界面，通过键盘移动光标来选择“参数设定”母线电压 $U=120$V，电流环比例系数 $P=1500$，电流环积分系数 $I=300$，前馈电压 FW=38V，死区时间 DB=2560ns；点击“返回”键，返回初始界面，再点击“启/停”键，启动逆变器；记录逆变器工作并网前后各电表的值：

序号	项目	逆变器输入		逆变器输出	
		U/V	I/A	U/V	I/A
1	并网前				
2	并网后				

⑧ 逆变器正常工作后，分别记录空载、直流负载、交流负载时逆变器输出电量表的参数：

序号	项目	逆变器输出电量表				
		U/V	I/A	P/kW	Q/kV·A	PF
1	并网后(空载)					
2	并网后(电机)					
3	并网后(LED灯)					

⑨ 实训结束后，点击“启/停”键使逆变器停止工作，风机停止。

⑩ 依次关闭“并网逆变控制系统”的“并网发电”“直流负载”“交流负载”“蓄电池”“控制器”空气开关和“能源转换储存控制系统”的“风机输出”“充放电控制器”“蓄电池”空气开关，最后再关闭各控制系统的“总电源”开关。

项目作业

逆变器正常工作后，改变母线电压 U 的值来观测各电表值。

项目七 并网逆变器参数设置及电能质量分析

项目描述

完成该实训任务，基于 THWPWG-3B 型大型风力发电系统实训平台，学习并网逆变器参数设置与电能质量分析。

能力目标：

① 了解并网逆变器的工作原理。

② 掌握并网逆变器的使用方法和作用。

③ 了解电能质量及谐波治理。

项目环境

随着新能源的发展应用，并网逆变器的研究越来越受到关注，它是将发电系统产生的直流电能转换成适合于电网使用的交流电能的装置；随着投放使用的并网逆变装置增多，其输出的进网电流谐波对电网的污染不容忽视，通常采用进网电流总谐波畸变率（THD）来描述电能质量，进网电流中谐波除了会给电网造成额外的损耗外，还可能损坏电网中的用电设备。

THWPWG-3B 型大型风力发电系统实训平台的并网逆变器控制采用电压型 PWM 技术，完成对电网的锁相、直流母线调节和电流调节实现电能并网和带负载运行，具有正弦波电流输出和单位功率因数特点，有效地解决了并网装置的谐波污染、功率因数低等问题。THWPWG-3B 型大型风力发电系统实训平台的并网逆变器框图如图 3.7.1 所示，主要由并网逆变器、逆变输出电量表、隔离变压器、直流负载、交流负载等组成。

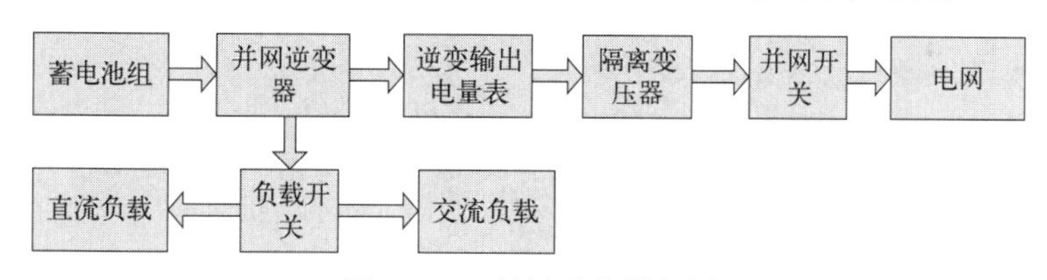

图 3.7.1 并网逆变器框图

在并网逆变控制单元中并网逆变器的功能为将蓄电池组过来的直流电转换为交流电输出，逆变输出电量表具有监测电量参数，如电压、电流、有功功率、无功功率、功率因数、电压/电流谐波畸变率等功能，隔离变压器可以将逆变出来的交流电升压为 220V 合并到电网中去，给电网供电，同时起保护作用。

逆变输出电量表是 THWPWG-3B 型大型风力发电系统实训平台中与电能质量分析相关的元件的主要元件。

一、逆变输出电量表介绍

ACR 系列谐波表具有全面的交流电量测量、复费率电能计算、谐波分析、遥信输入、遥控输出以及网络通信等功能；同时还具有电网波形实时跟踪显示和 SOE 事件记录功能，主要用于对电网供电质量的综合监控诊断及电能管理。

逆变输出电量表（图 3.7.2）主要应用于监测并网电流、电压、频率、功率、功率因数、电压及电流 THD 值、电压及电流的谐波波形等。

图 3.7.2 逆变输出电量表

二、逆变输出电量表型号及性能指标

在该平台中选用 ACR230ELH 型逆变输出电量表，该逆变输出电量表包含单相电流、电压输入、LCD 点阵式显示和一个 RS485 通信口，其性能指标如下：

① 输入网络。单相 2 线。

② 输入频率。45～65Hz。

③ 输入电压。AC 100V；过负荷 1.2 额定值（连续），2 倍额定值 1s；功耗小于 0.2V · A。

④ 输入电流。AC 5A；过负荷 1.2 额定值（连续），10 倍额定值 1s；功耗小于 0.2V · A。

⑤ 输出电能。集电极开路的光耦脉冲。

⑥ 通信。RS485 接口、Modbus-RTU 协议。

⑦ 开关量输入。接点输入，内置电源。

⑧ 开关量输出。继电器常开触点输出；触点容量：AC：250V 3A，DC：30V 3A。

⑨ 测量精度。频率 0.05Hz、无功电度 1 级、其他 0.5 级。

⑩ 电源。AC/DC 85～270V；功耗小于 6V · A。

⑪ 安全性。工频耐压：电源、电压电流输入回路之间，各输入端子（并联）和外壳之间 AC 2kV/1min；电源与开关量输入回路、通信回路、变送输出回路，各输入端子与各输出端子之间 AC 1.5kV/1min。绝缘：输入、输出端对机壳大于 100MΩ。

根据功能的应用不同分为直流负载和交流负载。

三、逆变输出电量表接线及操作

逆变输出电量表的接线图如图 3.7.3 所示：

按 SET 键：返回上一级菜单；按左键：向上翻页；按右键：向下翻页。按回车键：确认项目的选择和参数的修改。其端口定义如表 3.7.1 所示。

表 3.7.1 逆变输出电量表端口说明

序号	定义	说明
1	U+	被测电压输入
2	U−	
3	I+	被测电流输入
4	I−	

续表

序号	定义	说明
8	A	RS485
9	B	
17	L	交流 220V 电源输入
18	N	

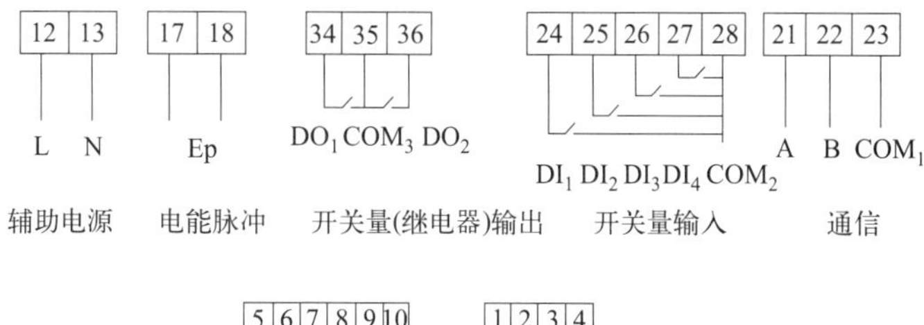

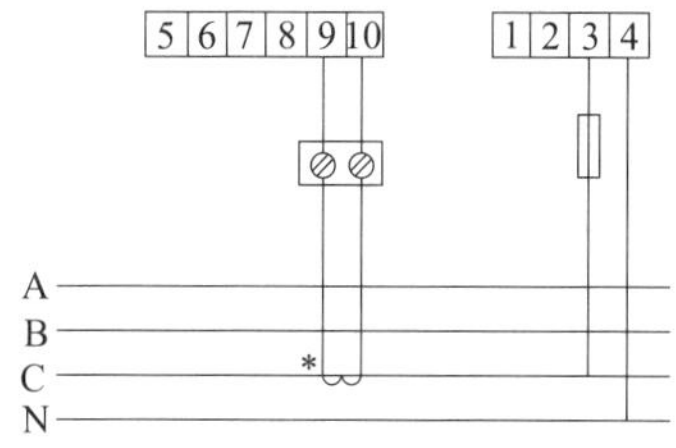

图 3.7.3　逆变输出电量表接线图

项目原理及基础知识

一、电流控制策略

采用电压外环、电流内环控制，首先建立两相同步旋转坐标系下逆变器数学模型，在此基础上给出了基于空间矢量调制的电流闭环控制策略，实现了并网电流有功分量和无功分量的独立控制。电流内环完成并网电流相位及幅值的控制，即跟踪并网电流指令 i_{ref}，电流内环一般采用加入并网电压前馈的 PI 控制器，将电流反馈采样值和电流指令值比较，其误差通过 PI 控制器输出，输出指令类型为电压，与电网前馈电压相加后，得到所需的并网电压指令。加入并网电压前馈，实际上是抵消了电网电压，使得电流 PI 环输出值即为电感电压，进而微调控制并网电流，是一种超前控制。在电压前馈 PI 控制中，由于 PI 控制器在跟踪正弦信号时不可避免会出现稳态误差，使得实际输出电流无法与输出电流指令相同。

电流内环控制一般有电流滞环跟踪控制和恒定开关频率电流控制两种，其中恒定开关频率电流控制原理框图如图 3.7.4 所示，正弦电流基准值 i_{ref} 和输出瞬时电流 i_{o} 比较得到误差量进入控制器调节后加前馈电压送到比较器，与三角波比较得到 SPWM 信号去控制主电路功率管的导通与截止。平台选择单极性倍频正弦脉宽调制（SPWM）的方式产生 SPWM，在不提高开关频率的前提下，提高了 SPWM 波形的谐波频率，从而使输出电压的谐波分量可以得到有效控制。

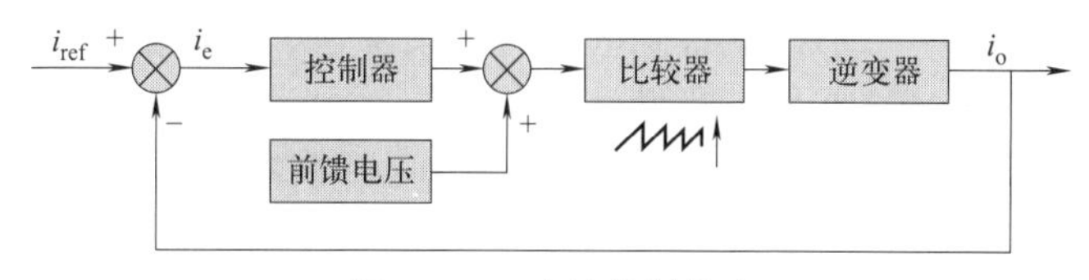

图 3.7.4　电流控制策略

电压外环的作用主要是通过 PI 控制器实现对直流电压指令 U_{dc} 的跟踪，同时给出电流内环指令 i_{ref}，其中直流电压指令 U_{dc} 是固定常数，通过前端 Boost 电路提供的母线电压 U 与其比较调节产生电流给定指令 i_{ref}。

二、影响电流并网质量的其他参数

并网逆变器的供电电能质量除与控制策略有关外，还受并网开关闭合时产生的进网电流环流、开关死区时间影响，直流电压影响，电网扰动及隔离变压器铁芯的饱和非线性特性影响等。

并网开关闭合前，逆变器输出电压与电网电压不一致，包括幅值、相位、频率、直流分量等不一致，即使此时进网电流给定值为 0，都会在逆变器与电网之间有电流流过，这里称其为环流。环流包括直流环流、基波环流、谐波环流，环流在轻载时占的比例较大，重载时占的比例减小。由于逆变器的等效输出阻抗和线路阻抗很小，所以逆变器输出电压与电网电压之间微小的电压差都会带来很大的环流，并且在逆变器与电网之间有电感存在，使环流滞后电压 90°，所以该环流的存在严重影响了网侧电流质量。逆变器的控制采用 PI 调节，而 PI 调节并不能实现无静差的跟踪，所以逆变器输出电压与电网电压之间存在幅值差和相位差，所以减小环流的问题就转化为减小 PI 调节稳态误差的问题。当逆变器控制开关频率较高时，死区时间对输出电流波形影响不可忽略。

电网电压扰动是一个暂态过程，其影响不是持续的，而变压器受到制造及运行条件等因素影响，实际运行时铁芯非线性饱和常常会产生较强的谐波电流扰动，且扰动是持续的，对输出电流波形质量产生较大的影响。

锁相精度采用数字锁相方法，正弦表格由 4000 个点构成，则锁相精度为 360/4000＝0.09，cos0.09＝0.99999，所以该锁相方法能获得很好的锁相效果，从而有较高的 PF 值。锁相采用在旋转坐标下跟踪电网电压合成矢量来实现，亦即求出电网电压合成矢量的空间位置角，作为控制策略中所用坐标变换的给定角度，便可实现电网相位与相序的跟踪。

项目实施

一、仪表、设备、工具清单

① THWPWG-3B 型 大型风力发电系统实训平台。

② 万用表。

二、安全操作规范

① 使用前先检查各电源是否正常。

② 控制柜中存在 AC 220V 的接入点，实训时要注意安全。

③ 实训始终，实训台要保持整洁，不可随意放置杂物，以免发生短路等故障。

④ 实训完毕，应及时关闭电源开关，并及时清理实训台。

⑤ 严格按照正确的操作步骤给系统上电和断电，以免误操作给系统带来损坏。

⑥ 在操作系统的过程中，能源转换储存控制系统蓄电池开关打开之后有一个等待智能充放电控制器自检初始化的过程；必须等到智能充放电控制器的“红灯”灭掉后才能进行下一步操作。

⑦ 在实训过程中，有“危险”标志的地方为强电注意安全。

三、实训步骤

① 该实训主要观察能源转换储存控制系统和并网逆变控制系统。
② 打开总电源，电源指示灯亮。
③ 将连接线一端插入能源转换储存控制系统控制柜连接模块的 A 端。
④ 另一端插入并网逆变控制系统控制柜连接模块的 A 端。
⑤ 将充放电控制器的刹车置于“自动刹车”状态。
⑥ 合上“蓄电池”空气开关。
⑦ 此时充放电控制器进行初始化。
⑧ 红色指示灯点亮。
⑨ 必须等红色指示灯熄灭才能进行下一步操作。
⑩ 合上“能源储存控制单元”上的“风机输出”和“充放电控制器”断路器。
⑪“摇柄”开关居中。
⑫ 在程序中输入主轴变频器启动信号 Q0.0 等于 1，给定模拟量 10000。
⑬ 启动风机，转速保持一定。
⑭ 打开“并网逆变控制系统”的“控制器”开关。
⑮ 并网逆变控制器上电，液晶屏初始化。
⑯ 依次打开“蓄电池”和“并网发电”开关。
⑰ 点击键盘上“确定”按键。
⑱ 进入参数设定人机界面。
⑲ 点击“返回”键，返回初始界面。
⑳ 点击“启、停”键，启动逆变器。
㉑ 改变电流环 PID 参数的设定。
㉒ 记录逆变器输出电量表的数值于下表。

序号	项目	逆变器输出电量表				
		U/V	I/A	PF	电压THD	电流THD
1	P=150 ;I=30					
2	P=500 ;I=100					
3	P=800 ;I=180					
4	P=1200;I=250					
5	P=1500;I=300					

㉓ 画出相应的谐波波形。
㉔ 点击“启、停”键使逆变器停止工作。
㉕ 点击键盘上“确定”按键。
㉖ 进入参数设定人机界面。
㉗ 通过键盘移动光标来选择“参数设定”。
㉘ 设定死区时间 DB=2780。
㉙ 点击“返回”键，返回初始界面。
㉚ 点击“启、停”键，启动逆变器。
㉛ 改变死区时间 DB 参数的设定。
㉜ 记录逆变器输出电量表的数值于下表。

序号	项目	逆变器输出电量表				
		U/V	I/A	PF	电压 THD	电流 THD
1	DB=2560ns					
2	DB=2780ns					
3	DB=2990ns					
4	DB=3200ns					

㉝ 实训结束后，点击“启、停”键使逆变器停止工作。

㉞ 在程序中输入 Q0.0 等于 0，给定模拟量 0，停止风机。

㉟ 依次关闭“并网逆变控制系统”的各断路器。

㊱ 依次关闭“能源转换储存控制系统”的各断路器，最后再关闭各控制系统的“总电源”。

项目作业

完成本项目任务需要完成实训前的准备，学习安全注意事项，完成“逆变器正常工作后，通过改变母线电压值观测电流谐波畸变率（THD）的值”的项目任务后，撰写项目报告，其中项目报告是根据项目实施过程中观察发现的问题，经过自己分析研究或组员之间分析讨论后写出的实训总结和心得体会，应简明扼要、字迹清楚、结论明确。

项目八　并网逆变控制系统安装与调试

项目描述

查阅风力发电机组并网逆变控制系统的安装手册，了解并网逆变控制的电路原理、结构组成，以及相关接线工艺，运用所学的知识和相关安装手册，完成风力发电机组模拟并网逆变控制系统安装与调试。

能力目标：

① 掌握并网逆变控制原理。

② 能够读懂并网逆变控制单元电路框图。

③ 能够根据给定的控制接线图和端口分配表，连接控制主电路。

项目环境

本实训任务主要涉及此实训平台的并网逆变控制部分，完成该任务需要参考 THWP-WG-3B 型 大型风力发电系统实训平台设备说明手册，认识并了解并网逆变控制系统原理、各部件及结构组成，了解相关工具的使用规范和注意事项。

一、并网逆变控制系统组成与功能

并网逆变控制系统框图如图 3.8.1 所示，主要由并网逆变器、逆变输出电量表、隔离变压器、直流负载、交流负载等组成。

并网逆变控制单元模块说明：

① 并网逆变器：将蓄电池组过来的直流电转换为交流电输出。

② 逆变输出电量表：监测电量参数，如电压、电流、有功功率、无功功率、功率因数、

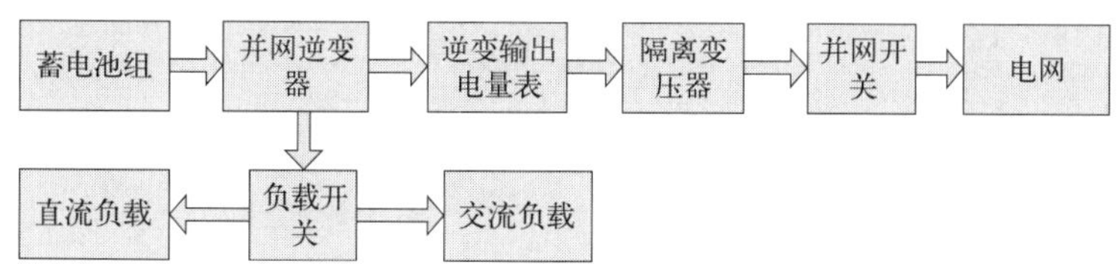

图 3.8.1　并网逆变器控制系统图

电压/电流谐波畸变率等。

③ 隔离变压器：将逆变出来的交流电升压为220V合并到电网中去，给电网供电，同时起保护作用。

④ 直流负载：24V直流电机作本地负载用。

⑤ 交流负载：36V交流指示灯作本地负载用。

功能：蓄电池出来的直流电压由并网逆变器逆变为与电网电压同频、同相、同幅的正弦交流电以实现与电网的并网连接；电量监测仪监测逆变输出端的电能参数；逆变输出的交流电压一部分经过隔离变压器升压为220V与电网并网连接，一部分供给本地负载。

二、并网逆变控制系统结构组成

并网逆变控制系统主要由核心模块、接口模块、液晶显示模块、键盘接口模块、驱动模块、直流电压升压模块、直流电压采样模块、交流电压采样模块、交流电流采样模块、温度告警模块、通信模块、开关电源、直流电机、方形指示灯、直流电压表、直流电流表、多功能数显表、变压器等组成，见表3.8.1。主要技术参数如下：

① 额定输入电压：DC 24V。

② 额定输出电压：220V±10%、50Hz±1Hz。

③ 额定功率：100V·A。

④ 输出功率因数：≥0.80（感性负载、容性负载）。

⑤ 逆变效率：≥80%。

表 3.8.1　并网逆变控制系统设备

序号	名　称	主要部件、器件及规格	数量	备注
1	并网逆变器	功率:100W 额定输入电压:DC 24V 额定输出电压:AC 36V	1台	
2	开关电源	额定输入电压:AC 220V 额定输出电压:DC 24V 额定功率:35W	1台	
3	直流电机	额定电压:DC 24V	1台	
4	方形指示灯	额定电压:AC 36V	1台	
5	直流电流表	输入电流范围:0～5A 精度:0.5%±5个字 通信:RS485通信接口	1个	
6	多功能数显表	测量电压范围:AC 0～250V 测量电流范围:AC 0～5A 通信:RS485通信接口	1个	
7	变压器	变比:36/220	1个	

三、并网逆变控制系统原理框图

并网逆变控制系统原理框图如图3.8.2所示。

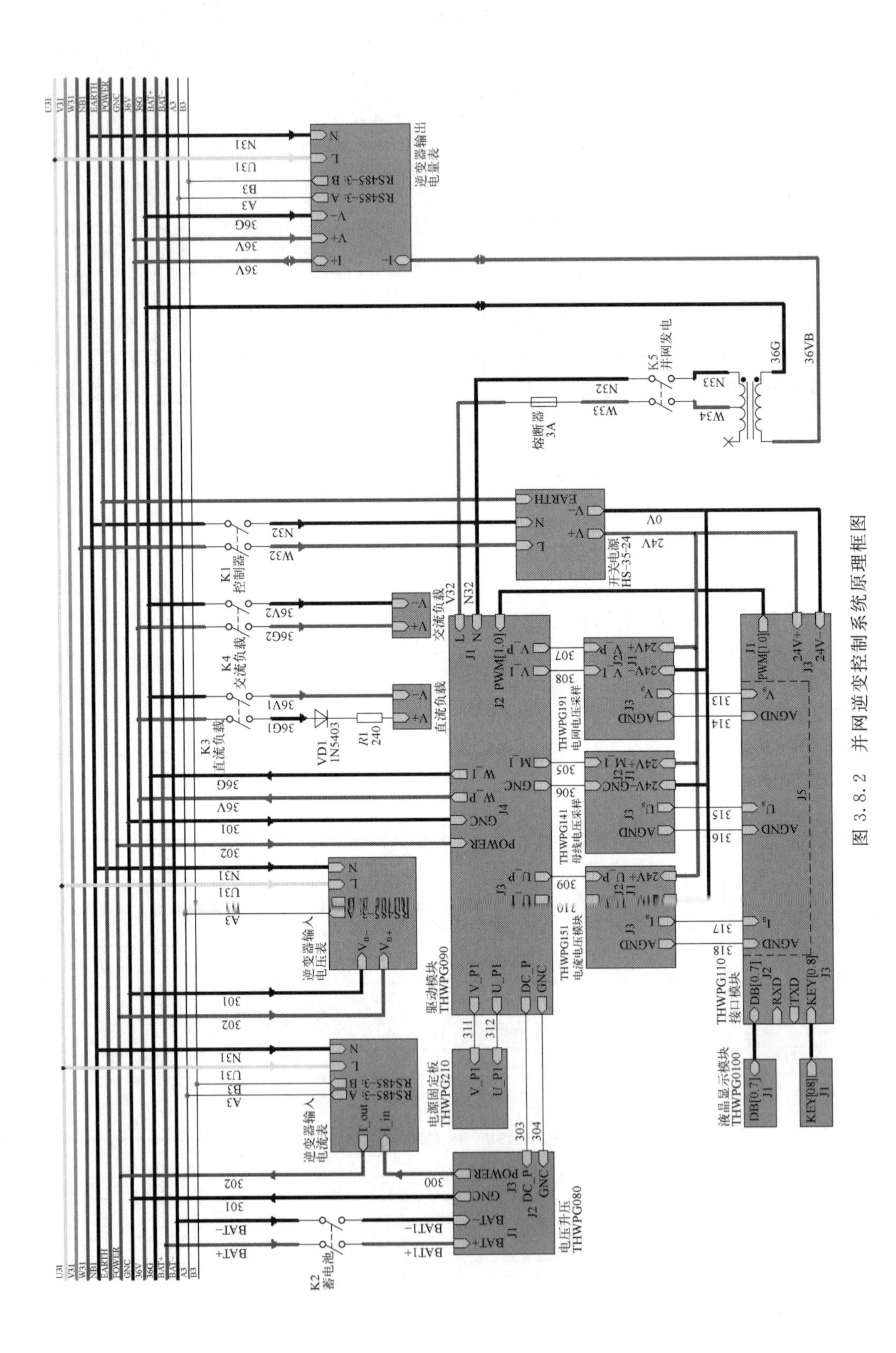

图 3.8.2 并网逆变控制系统原理框图

项目原理及基础知识

一、并网逆变器

并网逆变器功能框图如图 3.8.3 所示，主电路拓扑结构由 DC/DC(Boost 电路模块)+DC/AC(驱动电路模块)+滤波器(滤波板) 组成，控制回路由母线电压采样模块+电流采样模块+电网电压采样模块+温度告警模块+隔离驱动+DSP 控制电路+键盘接口模块+液晶显示模块组成。

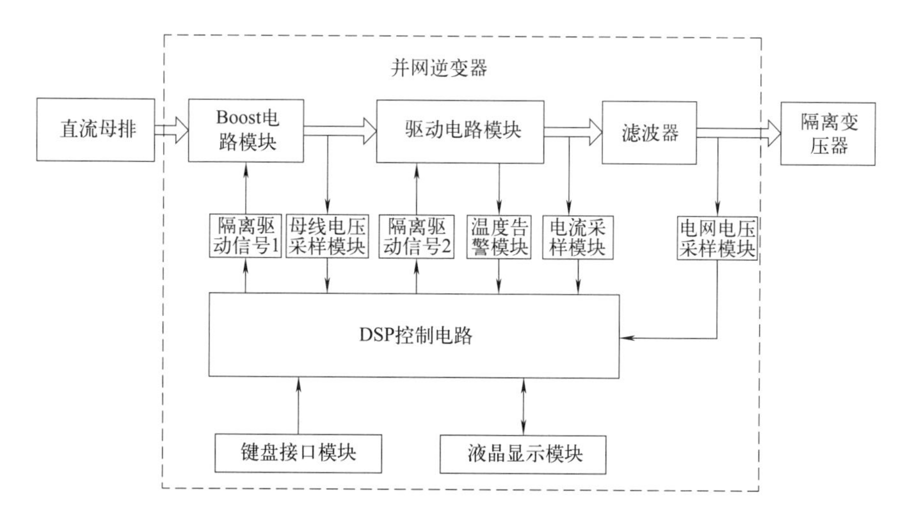

图 3.8.3 并网逆变器功能框图

① Boost 电路模块：Boost 升压电路主要将直流母排输出直流电压变换成能满足并网要求的母线电压。

② 驱动电路模块：驱动电路将该直流母线电压经过 DC/AC 逆变成与电网电压同频、同相、同幅的正弦交流电以实现与电网的并网连接。

③ 滤波器 (滤波板)：滤除逆变器输出的高频 PWM 谐波电流，减小进网电流中的高频环流，又能在逆变器与电网间进行能量的传递，使并网逆变器获得一定的阻尼特性，减小冲击电流，有利于系统的稳定运行。

④ 母线电压采样模块：母线电压检测，完成电压闭环及保护作用。

⑤ 电流采样模块：输出电流检测，完成电流闭环及保护作用。

⑥ 电网电压采样模块：电网电压检测，完成电网电压锁相、电压前馈及保护作用。

⑦ 隔离驱动信号 1：完成对 Boost 电路模块开关管的隔离驱动作用。

⑧ 隔离驱动信号 2：完成对驱动电路 IPM 智能模块的隔离驱动作用。

⑨ DSP 控制电路：执行并网逆变器的软件算法功能。

⑩ 键盘接口模块：设置影响并网电流质量的参数。

⑪ 液晶显示模块：显示并网参数。

二、Boost 电路模块

Boost 电路模块主要将蓄电池组输出的 24V 直流电压变换成能满足并网要求的直流母线电压，如图 3.8.4 所示。

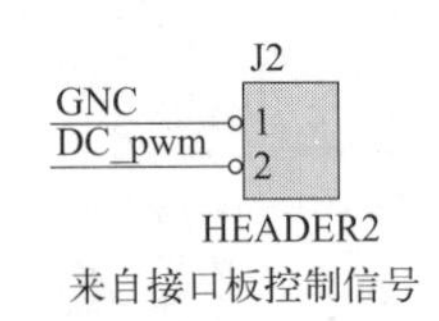

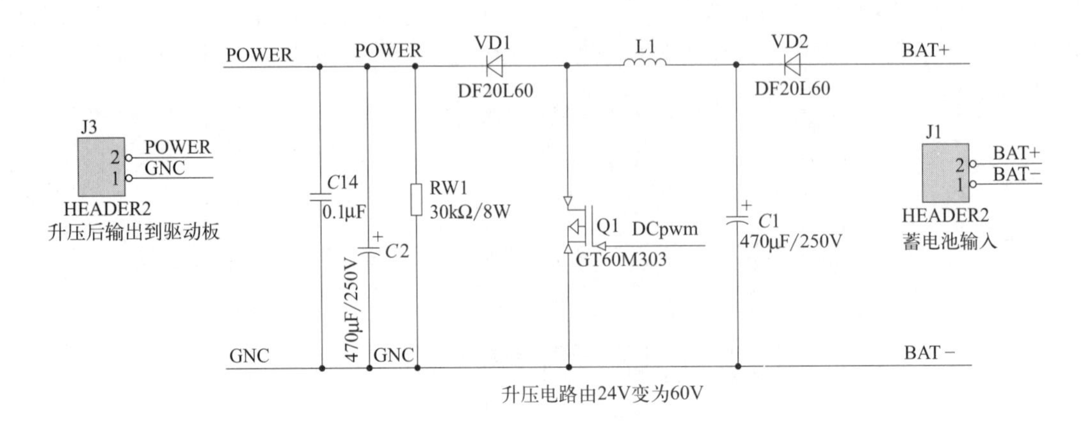

图 3.8.4　Boost 电路模块原理图

Boost 电路模块实物照片如图 3.8.5 所示。

图 3.8.5　Boost 电路模块

三、驱动电路模块

驱动电路模块将蓄电池电压通过 Boost 升压后，逆变成与电网电压同频、同相、同幅的正弦交流电以实现与电网的并网连接；主要是三菱 IPM 智能模块的应用。驱动电路模块原理图如图 3.8.6 所示。

实物照片如图 3.8.7 所示。

端子排端口定义如图 3.8.8 所示。

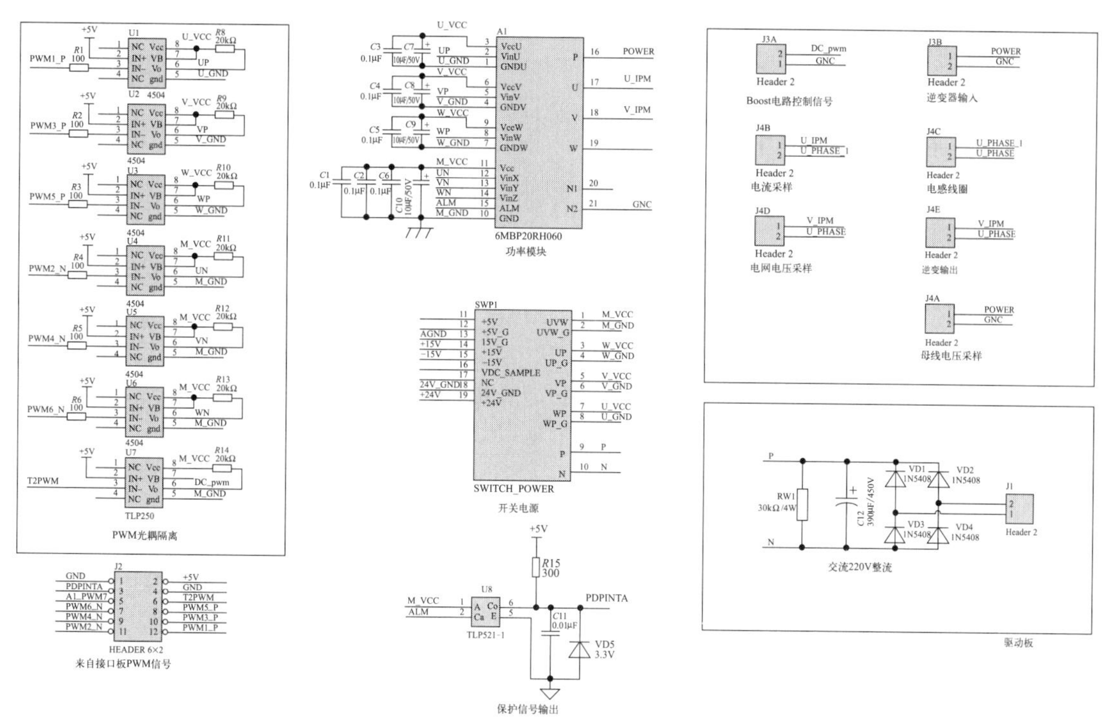

图 3.8.6　驱动电路模块原理图

图 3.8.7　驱动电路模块

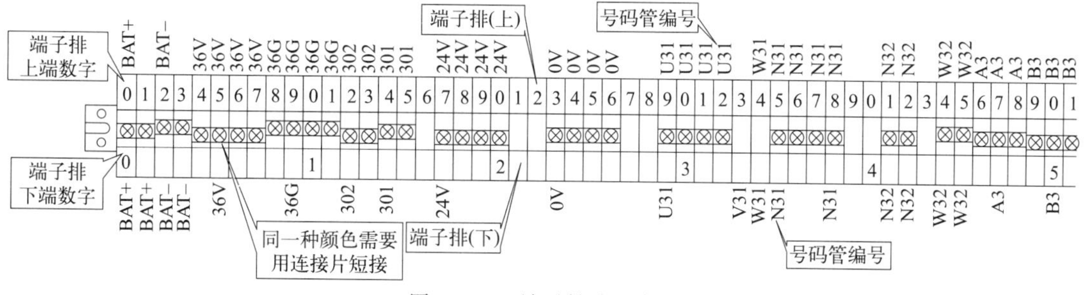

图 3.8.8　端子排编号定义

端子排编号定义

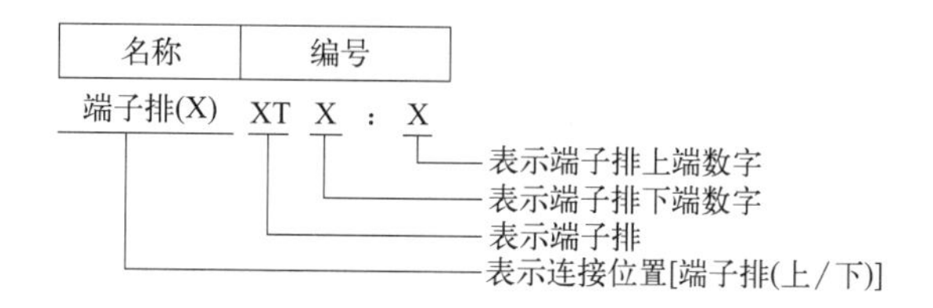

项目实施

一、仪表、设备、工具清单

① THWPWG-3B 型 大型风力发电系统实训平台风机对象模型；

② 剥线钳、压线钳、一字螺丝刀、十字螺丝刀。

二、安全操作规范

① 使用前先检查各电源是否正常。

② 接线前务必熟悉装置的各单元模块的功能及接线位置。

③ 实训接线前必须先断开总电源，严禁带电接线。

④ 接线完毕，检查无误后方可通电。

⑤ 熟练掌握偏航变桨电机控制原理及程序控制方法。

⑥ 控制柜中存在 AC 220V 的接入点，实训时要注意安全。

⑦ 实训始终，实训台要保持整洁，不可随意放置杂物，以免发生短路等故障。

⑧ 实训完毕，应及时关闭电源开关，并及时清理实训台。

⑨ 严格按照正确的操作步骤给系统上电和断电，以免误操作给系统带来损坏。

⑩ 在操作系统的过程中，能源转换储存控制系统蓄电池开关打开之后有一个等待智能充放电控制器自检初始化的过程，必须等到智能充放电控制器的“红灯”灭掉后才能进行下一步操作。

⑪ 在实训过程中，设备安装时注意防止高处跌落、挤压受伤。

⑫ 在实训过程中，有“危险”标志的地方为强电注意安全。

三、实训步骤

① 设备认识。并网逆变控制系统主要由核心模块、接口模块、液晶显示模块、键盘接口模块、驱动模块、直流电压升压模块、直流电压采样模块、交流电压采样模块、交流电流采样模块、温度告警模块、通信模块、开关电源、直流电机、方形指示灯、直流电压表、直流电流表、多功能数显表、变压器等组成。设备的布局如图 3.8.9 所示。

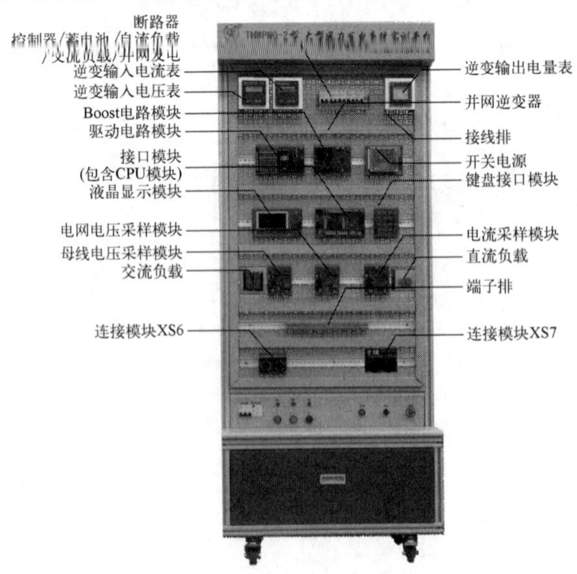

图 3.8.9　并网逆变控制单元器件布局图

请使用手机等设备，将对应设备拍照，照片插入到表 3.8.2 外观中。

表 3.8.2 并网控制系统设备

序号	名称	主要部件、器件及规格	数量	外观
1	并网逆变器	功率:100W 额定输入电压:DC 24V 额定输出电压:AC 36V	1台	
2	开关电源	额定输入电压:AC 220V 额定输出电压:DC 24V 额定功率:35W	1台	
3	直流电机	额定电压:DC 24V	1台	
4	方形指示灯	额定电压:AC 36V	1台	
5	直流电流表	输入电流范围:0～5A 精度:0.5%±5 个字 通信:RS485 通信接口	1个	
6	多功能数显表	测量电压范围:AC 0～250V 测量电流范围:AC 0～5A 通信:RS485 通信接口	1个	
7	变压器	变比:36/220	1个	

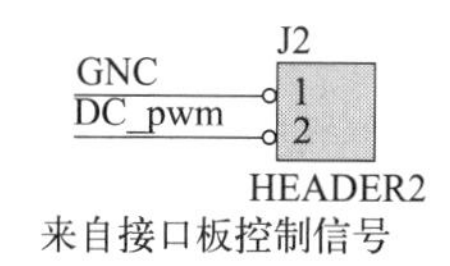

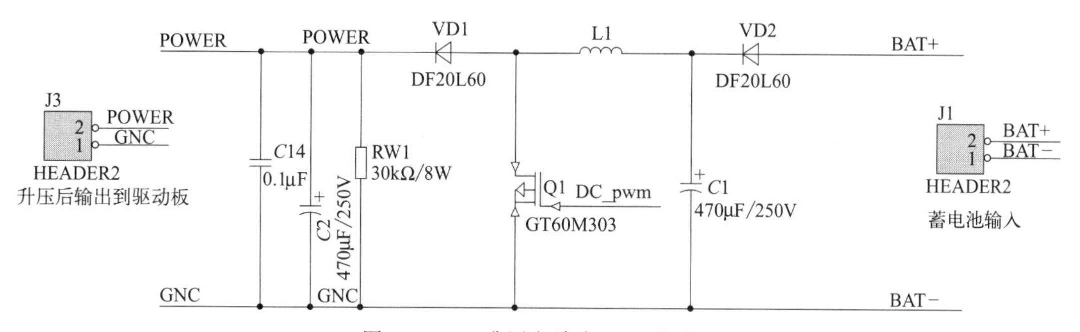

图 3.8.10 升压电路由 24V 变为 60V

② 绘制并仿真 Boost 升压电路，并描述现象。

③ 参考图 3.8.10 电路框图叙述并网逆变控制单元电路框图中各模块功能。

④ 安装驱动电路模块。根据图 3.8.11 所示的器件布局图，将铝导轨、驱动电路模块等固定到网孔板上，为连线做好准备。

根据驱动电路模块接线表完成接线，如表 3.8.3 所示。

⑤ 安装 Boost 电路模块。根据图 3.8.9 器件布局图，将铝导轨、Boost 电路模块（图 3.8.12）等固定到网孔板上，为连线做好准备。

图 3.8.11 驱动电路模块

表 3.8.3 驱动电路模块接线表

序号	起始端位置	结束端位置		号码管编号	线型
	驱动电路模块	名称	编号		
1	J1:AC-220V	端子排(上)	XT4:4	W32	23 红
2	J1:AC-220V		XT4:1	N32	23 黑
3	J2	接口模块	J1		14P 排线
4	J3: DC_P	Boost 电路模块	J2: DC_P	303	12 蓝
5	J3: GNC		J2: GNC	304	12 蓝
6	J3: GNC	母线电压采样模块	J2:GNC	306	42 黑
7	J3: M_I		J2:M_I	305	42 红
8	J3: W_I	端子排(下)	XT0:9	36G	42 黑
9	J3: W_P		XT0:5	36V	42 红
10	J4: V_I	电网电压采样模块	J2:V_I	308	42 黑
11	J4: V_P		J2:V_P	307	42 红
12	J4: V_P1	滤波板	J1:1	311	42 红
13	J4: U_P1		J1:2	312	42 红
14	J4: U_P	电流采样模块	J2:U_P	309	42 红
15	J4: U_I		J2:U_I	310	42 红
16	J4: GNC	端子排(上)	XT1:5	301	42 黑
17	J4: POWER		XT1:3	302	42 红

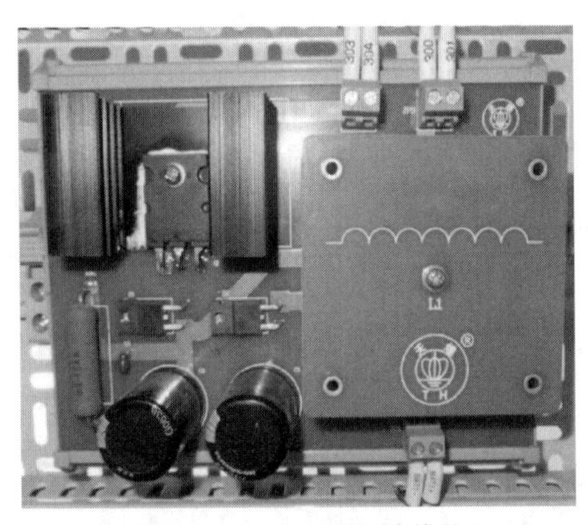

图 3.8.12 Boost 电路模块

根据 Boost 电路模块接线表完成接线，见表 3.8.4。

表 3.8.4 Boost 电路模块接线表

序号	起始端位置	结束端位置		号码管编号	线型
	Boost 电路模块	名称	编号		
1	J2: DC_P	驱动电路模块	J3: DC_P	303	12 蓝
2	J2: GNC		J3: GNC	304	12 蓝
3	J1:BAT+	蓄电池断路器	下左	BAT1+	42 红
4	J1:BAT−		下右	BAT1−	42 黑
5	J3: POWER	逆变输入电流表接线排(下)	I+	300	42 红
6	J3: GNC	端子排(下)	XT1:4	301	42 黑

项目作业

(1) 描述驱动模块接线中遇到的问题及解决方法

(2) 描述驱动模块、升压模块的主要功能。

项目九 发电系统整体运行

项目描述

学习风力发电技术的关键内容，了解风力发电的基本原理，通过实训平台，完成风力发电系统的运行与调试。

能力目标：

① 掌握风力发电系统的原理。

② 掌握风力发电系统各组成部分的作用。

项目环境

能源转换储存控制系统主要由直流电压电流采样模块、温度告警模块、PWM 驱动模块、CPU 核心模块、人机交互模块、通信模块、防雷器、智能型充放电控制器、蓄电池组、开关电源、直流电压表、直流电流表等组成，如图 3.3.1 所示。

项目原理及基础知识

风力发电系统主要由风力发电机组、整流器、控制器、蓄电池及逆变器、直流负载、交流负载组成。

① 风力发电机组。风力发电机组主要将风能转换为电能，输出电能经整流后送蓄电池中存储起来，也可以经并网逆变器送给电网。

② 蓄电池。蓄电池的作用主要是存储风力发电机组发出的电能，并可随时向负载供电。风力发电系统对蓄电池的基本要求：自放电率低、使用寿命长、充电效率高、深放电能力强、工作温度范围宽、少维护或免维护以及价格低廉。目前为风力发电系统配套使用的主要是免维护铅酸电池，在小型、微型系统中，也可用镍氢电池、镍镉电池、锂电池或超级电容。当需要大容量电能存储时，就需要将多只蓄电池串、并联起来构成蓄电池组。

③ 控制器。控制器的作用是控制整个系统的工作状态，其功能主要有：防止蓄电池过充电保护、防止蓄电池过放电保护、系统短路保护、夜间防反充保护等。

④ 负载。负载包括直流负载和交流负载，风力发电系统的直流负载电源直接来自控制器，而交流负载的电源来自逆变器，交流逆变器是把蓄电池输出的直流电转换成交流电供应给电网或者交流负载使用的设备。逆变器按运行方式可分为独立运行逆变器和并网逆变器。独立运行逆变器用于独立运行的风力发电系统，为独立负载供电。并网逆变器用于并网运行的风力发电系统。

风力发电系统的工作原理如下。

风力发电系统从大类上可分为独立（离网）风力发电系统和并网风力发电系统两大类。图 3.9.1 是独立型风力发电系统的工作原理示意图。风力发电的核心部件是风力发电机组，它将风能直接转换成电能，存储于蓄电池中。当负载用电时，蓄电池中的电能通过控制器合理地分配到各个负载上。风力发电机组所产生的电能为变压变频交流电能，经

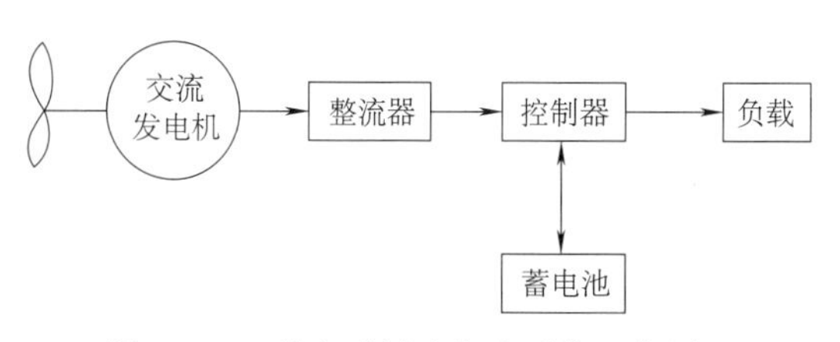

图 3.9.1 独立型风力发电系统工作原理

控制器转为直流电并对蓄电池充电，通过蓄电池将电能存储起来，在需要时使用。

图 3.9.2 是并网型风力发电系统工作原理示意图。并网型风力发电系统由风能直接转换成电能，并经配线箱进入并网逆变器，有些类型的并网型风力发电系统还要配置蓄电池组存储直流电能。并网逆变器由充放电控制、功率调节、交流逆变、并网保护切换等部分构成。经逆变器输出的交流电供负载使用，多余的电能通过电力变压器等设备馈入公共电网（可称为卖电）。当风力发电系统因天气原因发电不足或自身用电量偏大时，可由公共电网向交流负载供电（称为买电）。

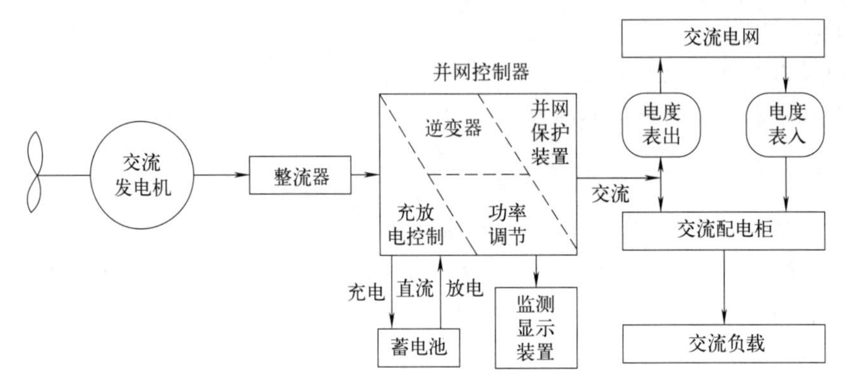

图 3.9.2　并网型风力发电系统工作原理

项目实施

一、仪表、设备、工具清单

THWPWG-3B 型 大型风力发电系统实训平台风机对象模型和能源控制 · 监控管理 · 气象站控制柜。

二、安全操作规范

① 熟练掌握风力发电机组发电原理及程序控制方法；

② 实训前仔细阅读使用说明书，熟悉与发电运行相关的部分，仔细阅读相关操作说明，确保各系统控制柜电源处于断开状态，根据实训指导书中相关内容熟悉此次实训的操作步骤；

③ 严格按照正确的操作步骤给系统上电和断电，以免误操作给系统带来损坏；

④ 先检查各电源和设备是否正常，确保正常后再正式开始实训任务；

⑤ 在实训过程中，始终保持实训台整洁，不可随意放置杂物，以免发生短路等故障，有“危险”标志的地方为强电注意安全；

⑥ 实训完毕，应及时关闭电源开关，并及时清理实训台。

三、实训步骤

① 合上“能源储存控制单元”上的“总电源”开关，系统得电，三相电源指示灯亮。

② 将充放电控制器的刹车置于“RELEASE（自动刹车）”状态，合上“能源储存控制单元”的“蓄电池”断路器，接入蓄电池，同时给充放电控制器供电，此时充放电控制器进行初始化，红色指示灯点亮（工作在刹车状态）。必须等红色指示灯熄灭（退出刹车状态）才能进行下一步操作。

③ 合上“能源储存控制单元”上的“风机输出”和“MPPT”断路器，“摇柄”开关向下。

④ 按“CPU核心模块”上的复位按钮K1，系统复位。

⑤ 启动风机，转速保持一定。

⑥ 按“人机交互模块”上的“ENTER”键，进入手动调节界面，然后按“UP”和“DOWN”键手动调节占空比，测量多组风机输出电流、电压及蓄电池输入电流、电压，记录到下表中。

序号	占空比(液晶上的数字)	风机输出			蓄电池		
		电压/V	电流/A	功率/W	电压/V	电流/A	功率/W
1							
2							
3							
4							
5							
6							
7							
8							
9							
10							

⑦ 通过记录的电压电流数据，计算每个电压、电流对应的功率。

⑧ 结束后，依次关闭“能源储存控制单元”的“风机输出”“MPPT”“蓄电池”断路器和“总电源”开关。

项目作业

画出实验平台的风力发电系统工作原理框图，并写出能量传递过程。

参考文献

［1］ 中国大唐集团公司赤峰风电培训基地．风力发电技术基础［M］．武汉：中国电力出版社，2020．

［2］ 任清晨，刘胜军，王维征．风力发电机组安装．运行．维护［M］．2版．北京：机械工业出版社，2019．

［3］ 张军利．双馈风力发电机组控制技术［M］．西安：西北工业大学出版社，2018．

［4］ 姚兴佳，宋俊．风力发电机组原理与应用［M］．4版．北京：机械工业出版社，2020．

Bilingual Textbooks of Vocational Education
职业教育双语教材

Installation and Commissioning of Large-scale Wind Power Generation System

大型风力发电系统安装与调试

Edited by Wang Xin Li Liangjun
Reviewed by Li Yunmei

王　欣　李良君　主编
李云梅　主审

Chemical Industry Press
化学工业出版社
· Beijing ·
· 北 京 ·

Foreword

In order to expand the vocational education cooperation with countries along "the Belt and Road" and implement the requirements of the plan launched by Tianjin to export excellent vocational education achievements to countries and share them with the world, vocational education, as a form of education most closely connected with manufacturing industry, is playing an important role. In order to cooperate with the theoretical and practical teaching of "Luban Workshop", carry out exchange and cooperation, improve the international influence of Chinese vocational education, innovate the international cooperation mode of vocational colleges and export excellent resources of vocational education in China, we compiled *Installation and Commissioning of Large-scale Wind Power Generation System*.

This textbook adopts the concept of project-oriented and task-driven to construct the course content combining basic knowledge of wind power generation technology with practical simulation system training. This textbook consists of three parts: The first part is the wind power generation technology, mainly introducing the professional basic knowledge of wind power generation; the second part is the large-scale wind power generation system and structure, focusing on the basic parameters of the training platform of the large-scale wind power generation system; the third part is the operational training project, which is the main content of the whole book. Among them, Project Ⅰ focuses on mastering the theoretical basis of wind power generation and the basic structure of wind power generating unit; Project Ⅱ uses common tools to disassemble and assemble the blades and pitch bearings to master the structure composition of pitch system and pitch principle; Project Ⅲ mainly introduce installation and commissioning of energy storage system; Project Ⅳ and Project Ⅴ mainly introduce the working process and function control of yaw and pitch; Project Ⅵ to Project Ⅷ: master the working principle and parameter setting of grid-connected inverter, and overall installation and commissioning of grid-connected inverter control system; Project Ⅸ realizes the overall operation of power generation system on the basis of mastering the contents of the preceding eight projects.

This textbook is compiled by Wang Xin and Li Liangjun, with Li Yunmei as the chief reviewer, Zhang Runhua, Yuan Jinfeng, Ma Sining, Yao Song, Shen Jie, Li Na and Cui Lipeng participated in the compilation. Wang Xin designed the framework, Zhang Runhua was responsible for the overall draft, and Li Liangjun reviewed the overall framework and all contents. The first and second parts are compiled by Wang Xin and Li Liangjun; in the third part, Yao Song was responsible for Project Ⅰ, Li Liangjun and Yuan

Jinfeng for Project Ⅱ, Yuan Jinfeng and Shen Jie for Project Ⅲ, Li Liangjun for Project Ⅳ, Shen jie and Cui Lipeng for Project Ⅴ, Li Na for Project Ⅵ, Ma Sining for Project Ⅶ, Zhang Runhua for Project Ⅷ and Wang Xin for Project Ⅸ.

Limited to the level of editor, there are certainly many omissions in the book, and we sincerely invite readers to criticize and correct them.

Editor

Contents

Part Ⅰ

Wind Power Generation Technology

Ⅰ. Understanding Wind Power Generation

1. Basic Principle of Wind Power Generation

Wind power generation is a process in which the captured wind energy is converted by wind power generating unit (fan) into mechanical energy, and the mechanical energy is transmitted to the generator through transmission mechanisms such as spindle and gear box, and then the mechanical energy is converted into electric energy by the generator, as shown in Figure 1.1.1. The electricity generated is boosted up to the power grid, so people can use it.

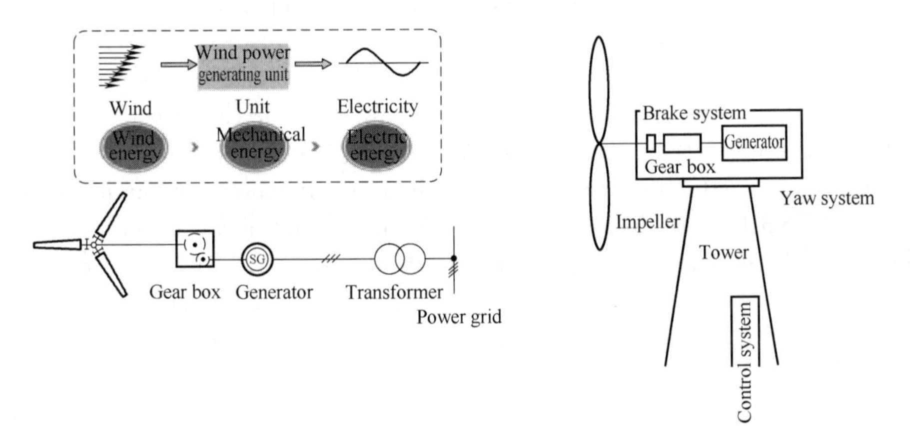

Figure 1.1.1 Schematic Diagram of Basic Principle of Wind Power Generation

According to current wind power generation technology, wind speed of about 3m/s can start generating electricity.

Due to the high damage rate of gear box of MW-level wind power generating unit, direct-driven wind power generating unit (without gear box) is adopted.

The rotor shaft of direct-driven wind power generating unit is directly connected with the generator, and multi-pole low-speed synchronous generator is adopted. And the double-fed wind power generating unit adopts a high-speed induction generator.

2. Classification of Wind Power Generating Unit

Classification by structural form: Horizontal axis, vertical axis;

Classification by power size: Micro (below 1kW), small (1-10kW), medium (10-100kW), large (above 100kW);

Classification by tower position: Windward (face the wind), downwind (follow the wind);

Classification by working principle of blade: Lift type, drag type;

Classification by number of blade: Single blade, double blade, triple blade, multiple

blade;

Classification by power regulation mode: Fixed blade type (stall regulation), variable pitch type (variable pitch regulation);

Classification by rotor speed: Constant speed, variable speed;

Classification by transmission mechanism: Gear box speed-up type, direct-driven type;

Classification by generator: Asynchronous type, synchronous type;

Classification by grid connection mode: Grid-connected, off-grid, mixed.

Common types of large wind power generating unit: Grid-connected horizontal axis windward triple blade variable speed variable pitch wind power generating unit, in which the double-fed type, i. e. speed-up type (induction generator) and direct-driven type unit (synchronous generator) are more.

3. Wind Power Generation Mode

The main technical direction of onshore wind power generation is low wind speed power generation, and the main machine type is 2-5MW large wind power generating unit. The key of this mode is to transmit electricity to power grid.

For offshore wind power generation, large-scale wind farms are mainly arranged on tidal flats and offshore waters, and large wind power generating units of more than 5MW are generally installed. The major constraint to this mode is the planning and construction costs of wind farms. However, the advantage of offshore wind power generation is obvious, that is, it does not occupy land and the offshore wind power resources are better.

4. Development History and Trend of Wind Power Generation

The installation of the first wind turbine with automatic operation and power generation was completed from 1887 to the winter of 1888. So far, wind power generation has roughly experienced the development course of experimental research, demonstration first, commercial development, energy accumulation, competitive development, scale leading and leaping development. At present, the development trend of wind power generation is as follows:

① The variable pitch regulation mode rapidly replaces the stall regulation mode;

② The variable speed operation mode rapidly replaces the constant speed operation mode;

③ The unit scale develops towards large scale;

④ Two forms of direct-driven permanent magnet and asynchronous double-fed develops together, and semi-direct- driven units come to the front;

⑤ The grid-connected control is becoming more and more friendly and automatic;

⑥ Rapid growth of offshore wind power generation;

⑦ The global wind power industry market is highly concentrated, and the emerging market will develop rapidly in the future;

⑧ Wind power generation costs are becoming more competitive.

The technology of wind power generation has made rapid progress, and the wind power generating unit are highly technological and reliable. Although the cost of wind power generation is still relatively high at present, the cost will continue to decrease with the increase of production batch and further technical improvement.

The outstanding advantage of wind power generation is that it has good environmental benefits and does not emit any harmful gases and wastes. Although the onshore wind farm occupies a large area of land, the basic area of wind power generating unit is very small,

which does not affect the normal production of farmland and pasture. Since wind-rich places are often barren beaches or mountainous region, the construction of wind farms will also develop tourism resources.

Ⅱ. Understanding Wind Power Generation System

1. Wind Power Generating Unit

(1) Double-Fed Unit

Double-fed asynchronous wind power generator is a kind of wound-rotor induction generator. The stator winding of the double-fed induction generator is directly connected to the power grid, and the rotor winding is connected to the power grid through frequency converter. The frequency, voltage, amplitude and phase of the rotor winding power supply are automatically adjusted by the frequency converter according to the operation requirements. The unit can realize constant frequency power generation at different rotational speeds to meet the requirements of electric load and grid connection. Because of the adoption of AC excitation, the generator and the power system form a "flexible connection", that is, the excitation current can be adjusted according to the voltage, current of the power grid and rotation speed of the generator, and the output voltage of the generator can be adjusted accurately to meet the requirements. The impeller of double-fed wind power generating unit drives the generator through multi-stage gear speed-increasing box, and its main structure includes wind wheel, transmission device, generator, current transformer system, control system, tower, etc, as shown in Figure 1. 1. 2.

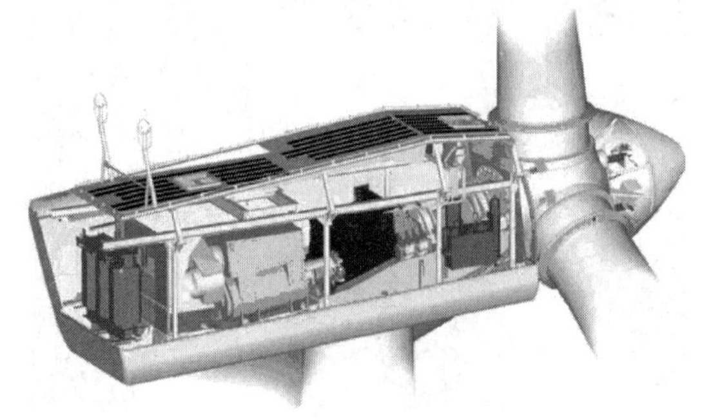

Figure 1. 1. 2 Structural Schematic Diagram of Double-Fed Wind Power Generating Unit

The double-fed wind power generating unit converts the mechanical energy transmitted from the gear box to the rotor shaft of the generator into electric energy, which is transmitted to the power grid through the stator and rotor of the generator, as shown in Figure 1. 1. 3. The generator stator winding is directly connected to the power grid, and the rotor winding can be connected to the current transformer whose frequency, amplitude and phase can be adjusted as required. The current transformer controls the generator to maintain power generation at both sub-synchronous and super-synchronous speeds. During super-synchronous power generation, energy is fed to the power grid through two channels of stator and rotor at the same time, and the current transformer feeds the DC side energy back to the power grid. During sub-synchronous power generation, the stator feeds energy to the power grid, and the rotor absorbs the energy to generate braking torque to make the generator work. The current converter system is called double-fed technology because it can be fed in two ways.

Double-fed wind power generating units have the following advantages:

① The reactive power can be controlled and the active power and reactive power control are decoupled by independently controlling the rotor excitation current.

② Double-fed induction generators do not need to be excited from the power grid, but are excited from the rotor circuit.

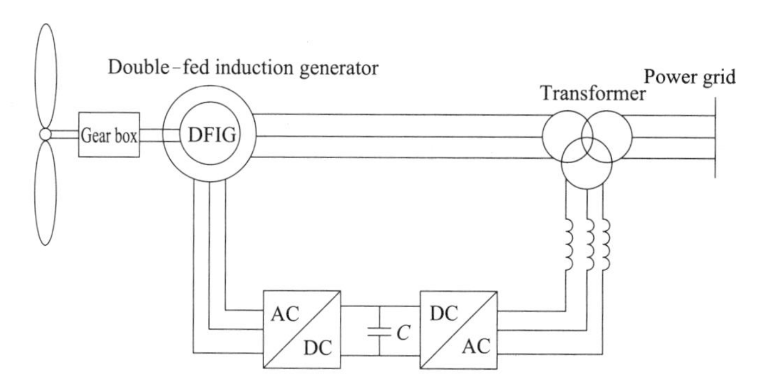

Figure 1. 1. 3 Schematic Diagram of Variable Speed Constant Frequency Unit of Double-Fed Wind Power Generation

③ It also generates reactive power which can be transmitted to the stator via the grid side current transformer.

(2) Direct-Driven Unit

Direct-driven wind power generating unit is directly driven by wind power, also called gearless wind power generating unit. It is driven by direct connection of multi-pole motor and impeller without the traditional gear box. It mainly consists of wind wheel, transmission device, generator, current transformer, control system, tower and so on, as shown in Figure 1. 1. 4. In order to improve the efficiency of low-speed generator, the direct-driven wind power generating unit adopts the way of increasing number of pole pairs greatly to improve the utilization rate of wind energy, and adopts full power current transformer to realize the speed regulation of wind power generating unit, as shown in Figure 1. 1. 5.

Figure 1. 1. 4 Structural Schematic Diagram of Direct-Driven Wind Power Generating Unit

Direct-driven wind power generating unit can be divided into electric excitation and permanent magnet according to excitation mode.

Figure 1. 1. 5 Schematic Diagram of Direct-Driven Variable Speed Constant Frequency Wind Power Generating Unit

Because of its simple structure, no exciting winding and high efficiency, permanent magnet synchronous generator is widely used in small and medium wind power generating unit. With the improvement of manufacturing technology of high performance permanent

magnet material, large capacity wind power generating unit tends to use permanent magnet synchronous generator. Permanent magnet synchronous generators are usually used in variable speed constant frequency wind power generating unit. The rotor of the generator is driven directly by the wind wheel, so the rotation speed is very low. Because the speed-increasing gear box is removed, the reliability and service life of the unit are increased; the magnetic pole composed of many high-performance permanent magnetic steels does not need complicated structure and bulky exciting winding like the electric excitation synchronous motor, thus increasing the air gap flux density and power density, and reducing the volume of the generator under the same power level.

Direct-driven wind power generating units have the following advantages:

① High power generation efficiency. There is no gear box for direct-driven wind power generating unit, which reduces transmission loss and improves power generation efficiency, especially in low wind speed environment.

② High reliability. The gear box is a component with high failure frequency in the operation of wind power generating unit. The direct-driven technology eliminates the need for the gear box and its accessories, simplifies the transmission structure and improves the reliability of the unit. At the same time, the unit operates at low speed with fewer rotating components and higher reliability.

③ Low operating and maintenance costs. The adoption of gearless direct-driven technology can reduce the number of parts and components of wind power generating unit, avoid the regular replacement of gear box oil, and reduce the operation and maintenance cost.

④ Excellent grid access performance. When the low voltage crossing of direct-driven permanent magnet wind power generating units makes the voltage of grid connection point drop, the wind power generating unit can operate continuously in the range of certain voltage drop, so as to maintain the stable operation of power grid.

2. Foundation of Wind Power Generating Unit

The foundation of the wind power generating unit is a platform that supports the connection between the tower and the bottom of the ground. The wind power generating unit is placed at an altitude of 60—100m or higher so as to obtain sufficient and stable wind power to generate electricity. The functions of the foundation of wind power generating unit are mainly reflected in the aspects: It is the main bearing component of the wind power generating unit; it is used to install and support the wind power generating unit; balance all kinds of loads generated during the operation of the wind power generating unit; ensure the safe and stable operation of the unit.

The foundation design is closely related to the geological conditions where the foundation is located. Good geological conditions can provide reliable safety guarantee for foundation construction. According to the requirements of foundation construction, the importance and complexity of tower foundation is self-evident. How to determine a safe and reasonable basic scheme under complex geological conditions is even more essential. The foundation of wind power generating unit includes onshore wind power foundation (Figure 1. 1. 6) and offshore wind power foundation (Figure 1. 1. 7).

(1) Foundation Design of Wind Power Generating Unit

① Foundation Structure and Type. According to the model and capacity of wind power generating unit, the bearing load of foundation is also different. According to the unit capacity, hub height and foundation complexity, the foundation is divided into three design lev-

Figure 1. 1. 6 Onshore Wind Power Foundation

Figure 1. 1. 7 Offshore Wind Power Foundation

els. See Table 1. 1. 1.

The foundation design of the unit shall conform to the following provisions: The foundation of all units shall meet the requirements of bearing capacity, deformation and stability; the foundation deformation calculation shall be carried out for the foundation of Level 1 and Level 2 units; the deformation checking calculation for the foundation of Level 3 unit is generally unnecessary, and deformation checking calculation shall be conducted in case of any of the following circumstances: firstly, the characteristic value of bearing capacity of the foundation is less than 130kPa or the compression modulus is less than 8MPa; secondly, the special rock and soil such as soft soil.

Table 1. 1. 1 Three Design Levels of Foundation

Design Level	Single Unit Capacity, Hub Height and Foundation Type
1	Single unit capacity greater than 1. 5MW Hub height greater than 80m Complex geological conditions or soft soil foundation
2	Foundation between Level 1 and Level 3
3	Single unit capacity less than 0. 75MW Hub height less than 60m Rock and soil foundation with simple geological conditions

Note: 1. When the foundation design level belongs to different levels according to the indexes in the table, it shall be determined according to the highest level;

2. For Level 1 foundation, if the foundation conditions are good, the foundation design level can be lowered by one through demonstration.

The foundation of wind power generating unit is cast-in-place reinforced concrete independent foundation. According to the engineering geological conditions, foundation bearing capacity, foundation load and size of wind farm site, the common foundation can be divided into block foundation and frame foundation from the view of structure form. Block foundation, i. e. solid gravity foundation, is widely used. In dynamic analysis of foundation, the deformation of foundation can be ignored, and the foundation is treated as rigid body, only considering the deformation of foundation. According to its structure profile, it can be divided into "concave" shape: and "convex" shape: as the former is shown in Figure 1. 1. 8, the

whole foundation is square solid reinforced concrete, while compared with the former, the latter shown in Figure 1. 1. 9 also belongs to solid foundation, but the difference is that the backfilled soil on the expanded base plate also becomes a part of the foundation gravity, so as to save materials and reduce costs.

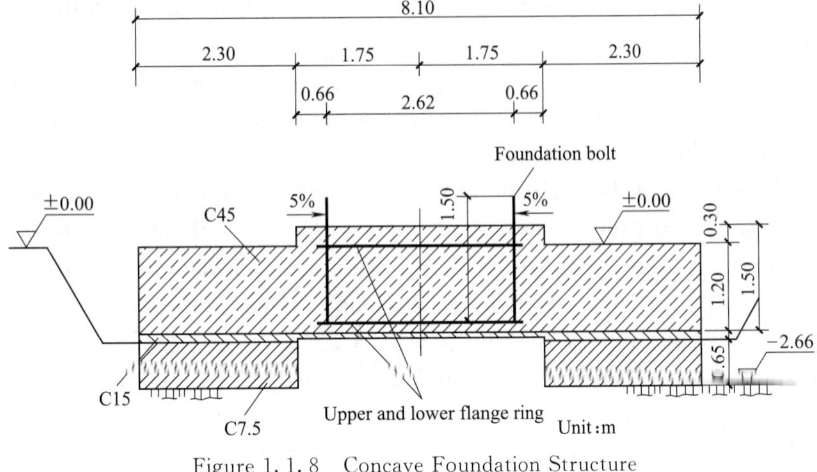

Figure 1. 1. 8 Concave Foundation Structure

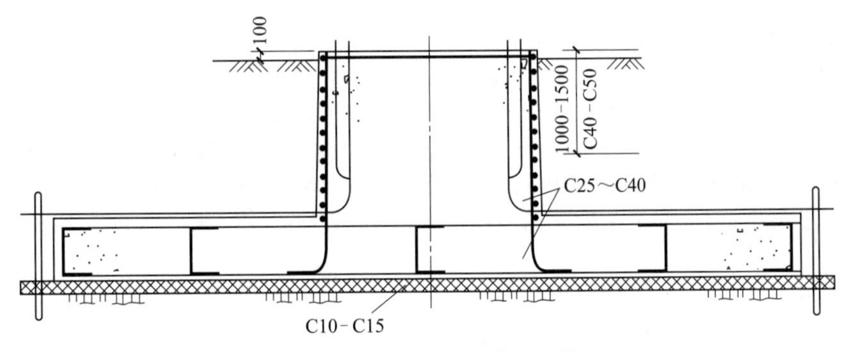

Figure 1. 1. 9 Convex Foundation Structure

The frame foundation is actually the combination of pile foundation group and flat plate girder. From the bearing capacity characteristics of single pile foundation, it can be divided into friction pile foundation and end bearing pile foundation. Friction pile foundation bears the load on pile by side friction force and pile end resistance; end bearing pile foundation bears load mainly by pile end resistance.

According to the connection mode between foundation and tower (fuselage), foundation can be divided into foundation bolt type and flange type. The former tower is fixed on the foundation bolt with nuts and nylon spring washer, flat washer, and the latter tower flange is connected with the foundation section flange by bolts. The foundation bolt type is divided into single-row bolt, double-row bolt, single-row bolt with upper and lower flange ring, etc.

② Notes in Foundation Design of Wind Power Generating Unit. The foundation of the wind power generating unit is used to install and support the wind power generating unit, balance various loads generated during the operation of the wind power generating unit to ensure the safe and stable operation of the unit. Therefore, before designing the foundation of wind power generating unit, it is necessary to carry out engineering geological investigation on the installation site of wind power generating unit to fully understand and study the cause and structure of foundation soil layer and its physical and mechanical properties, so as to

make a correct evaluation on the engineering geological conditions on site. This is a prerequisite for foundation design of wind power generating unit. At the same time, it must be noted that the installation of wind power generating unit will change the original stress state of foundation, it is necessary to apply mechanics method to study the deformation and strength of foundation soil under load, so that the foundation design meets the following two basic conditions:

a. It is required that the load acting on the foundation shall not exceed the allowable bearing capacity of the foundation, so as to ensure that the foundation has sufficient safety stock for preventing the overall damage.

b. The sedimentation of the foundation shall be controlled so that it does not exceed the allowable deformation value of the foundation, so as to ensure that the wind power generating unit will not be damaged due to foundation deformation or affect its normal operation. Therefore, the preliminary preparation of wind power generating unit foundation design is an essential link to ensure the normal operation of the unit.

③ Requirements for foundation of wind power generating unit and force analysis of foundation. When a wind power generating unit is in operation, the unit not only bears its own weight Q, but also bears the positive pressure P generated by the wind wheel, wind load q and torque M_n generated by direction adjustment of the unit. These loads are mainly balanced by foundation to ensure safe and stable operation of the unit.

Figure 1. 1. 10 shows the action state of these loads on the foundation, in which Q and G are the unit weight of the unit and foundation respectively. The overturning moment M is the resultant moment caused by factors such as eccentricity of unit weight, positive pressure P generated by wind wheel and wind load q. M_n is the torque generated during unit direction adjustment. The shearing force F is caused by the positive pressure P generated by the wind wheel and the wind load g.

In general, the shearing force F and the torque M_n produced by the wind power generating unit during the direction adjustment process are generally not very large, and are much smaller than other loads. Therefore, on the premise that the calculation effect is not affected, and also the engineering requirements can be met, the editor believes that these two items can be omitted in actual calculation. Therefore, in the foundation design of the wind power generating unit, the load generated by the unit on the foundation shall mainly consider the unit weight Q and overturning moment M of the unit. After the above simplification, the mechanical model of wind power generating unit foundation is shown in Figure 1. 1. 11.

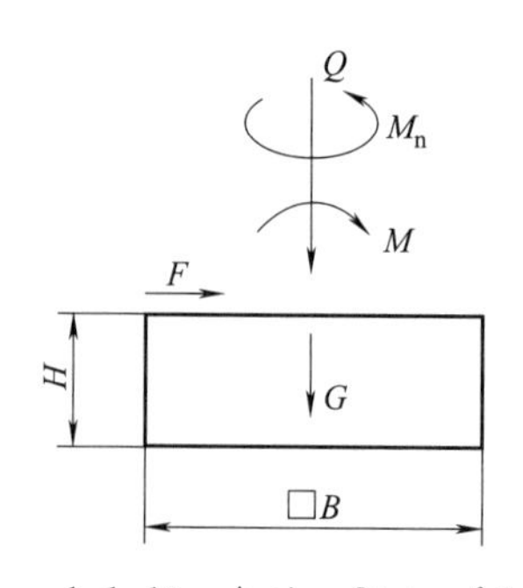

Figure 1. 1. 10 Action State of Load on Foundation

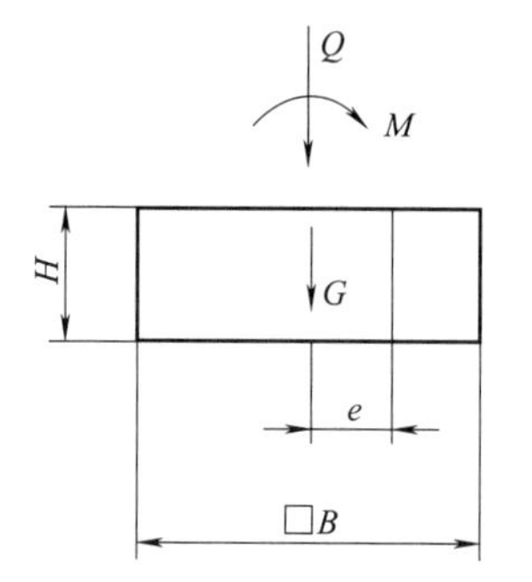

Figure 1. 1. 11 Mechanical Model of Wind Power Generating Unit Foundation

(2) Foundation of Offshore Wind Power Generating Unit

According to different geographical location and geological conditions, different design modes of offshore wind power generating unit foundation are closely related to site conditions, and their cost accounts for about 20%-30% of investment. The foundation design mode of offshore wind power generating unit mainly includes the following types: single pile foundation, gravity foundation, tripod foundation, jacket foundation, floating foundation.

① Single Pile Foundation. Single pile foundation (Figure 1. 1. 12) is the simplest in structure, so it is widely used. It is composed of welded steel pipe. The diameter of the pile is generally about 3-5m, and the wall thickness is about 1% of the pile diameter. During installation, it shall be driven into the seabed to a depth below 10-20m for fixing. Therefore, single pile foundations are suitable for water area with a hard seabed, but not if there are rocks on the seabed.

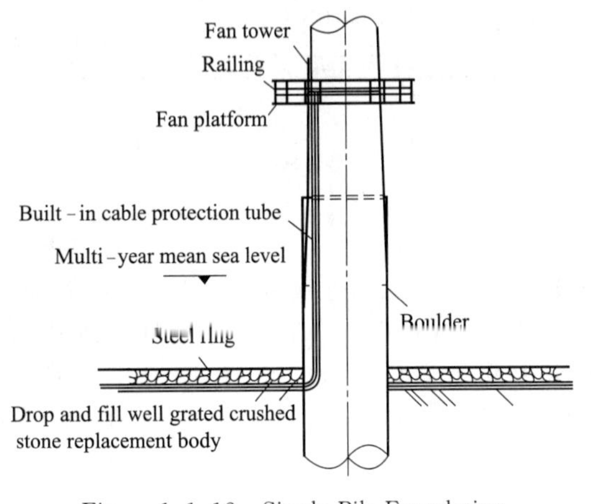

Figure 1. 1. 12 Single Pile Foundation

Single pile foundation has the advantage of simple fabrication and no seabed preparation; the disadvantage is that it is greatly restricted by the seabed geological conditions and water depth, and the construction and installation costs are relatively high, and anti-scour protection is required.

② Gravity Foundation. The gravity foundation (Figure 1. 1. 13) uses the gravity of the foundation to fix the whole system, which is not suitable for the quicksand- type seabed situation. The gravity of the foundation can be obtained by filling the inside of the foundation with reinforcement, sand, cement and rock, etc. The gravity foundation is generally reinforced concrete structure.

Advantages of gravity foundation: simple structure, low cost, high stability and reliability. Disadvantages: The seabed preparation is required in advance, the volume and weight are large, the installation is inconvenient, and the applicable water depth range is too shallow.

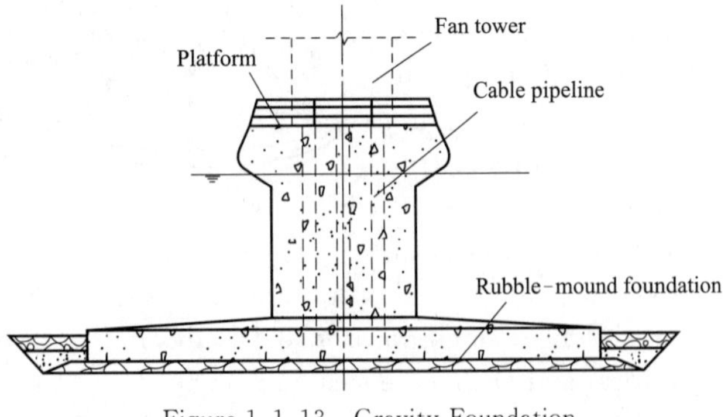

Figure 1. 1. 13 Gravity Foundation

③ Tripod Foundation. Tripod foundation (multi-pile foundation, Figure 1. 1. 14) adopts standard three-leg supporting structure, which is composed of central column, three cylindrical steel pipes inserted into seabed to a certain depth and diagonal bracing structure,

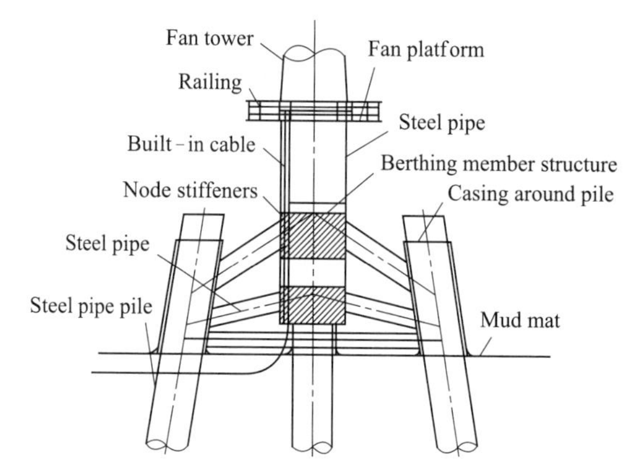

Figure 1. 1. 14 Tripod Foundation

which can resist wave and water drag forces well. The central column provides basic support for the fan tower, enhancing the rigidity and strength of the surrounding structure.

Advantages of tripod foundation: easy to manufacture, without any seabed preparation, can be used in deep sea areas without scour protection. Disadvantages: restricted by geological conditions, not suitable for shallow sea area, and higher construction and installation costs.

④ Jacket Foundation. The jacket foundation (Figure 1. 1. 15) looks like a cone-shaped space frame from the outside, and it is suitable for a wide range of water depths. It has the following advantages: The construction is convenient, the load under the action of wave and current is relatively small, and the requirements for geological conditions are not high. The disadvantage is that the cost increases rapidly with the increase of water depth.

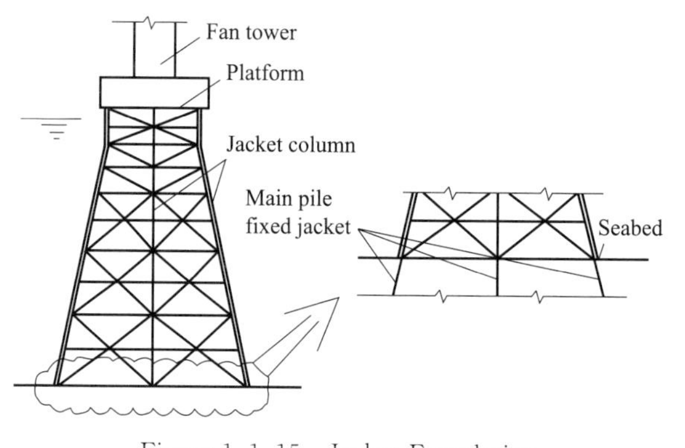

Figure 1. 1. 15 Jacket Foundation

⑤ Floating Foundation. The floating foundation (Figure 1. 1. 16) must be buoyant to support the weight of the wind power generating unit and be able to resist tilting, shaking and heave motion within acceptable limits. Floating foundations may be the best option in certain water depths.

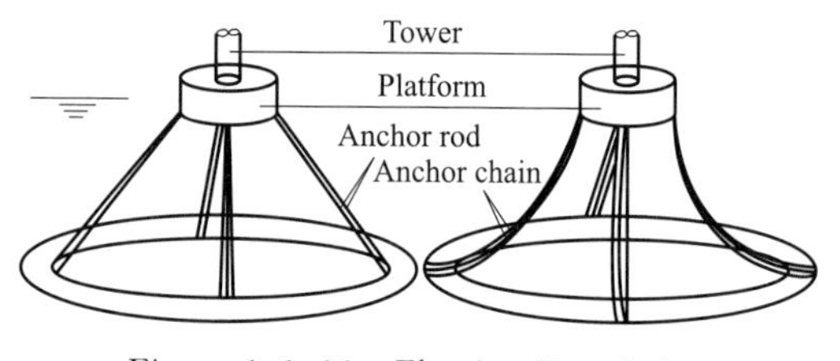

Figure 1. 1. 16 Floating Foundation

Advantages: low installation and maintenance costs, low removal costs at the end of its life; insensitive to water depth, installation depth can be more than 50m; small wave load. Disadvantages: The stability is poor, and the design of the platform and anchorage system is difficult.

3. Wind Farm Current Collection System

The wind farm power collection system, also called "wind farm current collection system", is a power connection system that collects the electric energy of the wind power generating unit and transmits it to the step-up substation or power load of the wind farm. They

can be classified according to voltage class, such as 10kV current collection system, 35kV current collection system, etc.

At present, the maturity of wind power generation technology, large-scale development and application and the trend of market development are quite outstanding. In 2020, the newly built wind power capacity reached 97GW. After 2022, onshore wind power construction enter the platform period, and most of the growth come from offshore wind power projects. In 2021, the capacity of newly built offshore wind power exceeds 10GW for the first time and will reach 30GW by 2030.

(1) Current Collection System of Onshore Wind Farm

Onshore wind farm is generally located in the area with bad environment, and the average annual load rate of wind power generating unit is relatively low. Therefore, the wind power current collection system shall have strong weather resistance, high reliability and low no-load loss.

① Electrical Primary System. According to the level of power utilization, it can be divided into primary system and secondary system. The primary system includes wind turbine, generator and frequency converter, unit step-up transformer (generator output 690V, step-up to 10kV or 35kV, and then connected to step-up substation of wind farm). The current collection system of wind farm is shown in Figure 1. 1. 17.

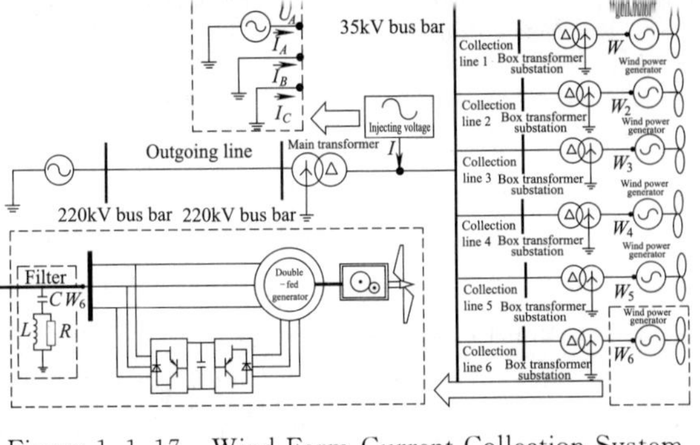

Figure 1. 1. 17 Wind Farm Current Collection System

The main wiring principle of wind farm current collection system is as follows:

a. Reliability. The maintenance of any circuit breaker shall not affect the power supply of the circuit where it is located; in case of failure of circuit breaker or bus bar, the number and time of shutdown circuits shall be minimized, and the power supply for first order load and all or most of second order load shall be guaranteed; the possibility of power cut of power station and substation shall be minimized.

b. Flexibility. During dispatching, the unit, transformer and circuit shall be flexibly cut in and out, and the power supply and load shall be flexibly allocated; during maintenance, it is convenient to stop the circuit breaker, bus bar and its relay protection device, so as not to affect the operation and power supply of the power system; during expansion, it is easy to transition from initial wiring to final wiring. Under the condition that continuous power supply is not affected and the shortest blackout, new units or transformers and circuits shall be put into operation without mutual interference, and the reconstruction workload of the system shall be minimized.

c. Economy. It requires economical investment, small occupied area and less power loss.

In the design process of the primary system of the wind farm current collection system, the design shall focus on the directions around the main wiring design, primary selection of the main wiring of the wind farm, grouping and connection of units, current collection circuit scheme, self-use of the wind farm and the main electrical equipment shall be designed according to the selected access mode to the power system and the electrical main wiring scheme.

② Electrical Secondary System. The electrical secondary system of the wind farm includes relays, contactors, control switches, automatic switches, wiring terminals, complete sets of protection and measurement and control devices. The secondary system is an indispensable part of the wind farm current collection system. It is the low-voltage electrical equipment required for monitoring, controlling, adjusting and protecting the primary equipment and providing operation and maintenance personnel with operating conditions or production command signals, so that the primary system can operate safely and economically. Among the wind farm protection devices, the complete set of protection devices and measurement and control devices are commonly used in China, as shown in Figure 1. 1. 18.

Figure 1. 1. 18 Complete Set of Protection Device and Measurement and Control Device

At present, because of the large-scale centralized development mode and the randomness of wind power output, reactive power dispatching and voltage control inside the wind farm have always been the research hotspot in relevant fields. When the wind farm is connected to the power grid in a decentralized manner, it has little impact on the overall voltage stability of the system. When the wind farm is connected to the power grid in a large-scale centralized manner, the voltage level is high and the transmission distance is long, so it has a great impact on the regional power grid system and operation. In order to reduce the negative effect of wind power grid connection on regional power grid, guideline of wind power grid connection stipulates that each wind farm shall have voltage and power factor regulation and control capacity of grid connection point to maintain stable parameters of grid connection point when wind conditions fluctuate.

(2) Current Collection System of Offshore Wind Farm

As a large-scale renewable energy with great potential, offshore wind power has become an inevitable trend of wind energy utilization in the future, supported by the requirements of energy conservation and emission reduction of all countries and the strong financial policies of governments all over the world. According to the development and planning of offshore wind power all over the world, the development of offshore wind power shows the characteristics of gradually increasing wind farm capacity and expanding offshore distance. Since Denmark built the world's first offshore wind farm in 1991, the scale of offshore wind farms has been increasing year by year. Horns Rev is the world's first large offshore wind farm with an installed capacity of 160MW and an offshore distance of 15km. In 2021, China installed a total of 16. 9 million kilowatts of new offshore wind power annually, which is 1. 8 times the total cumulative scale built previously. By the end of 2021, China's cumulative installed capacity reached 26. 39 million kilowatts, leaping to the top of the world.

Large-scale far offshore wind farms may mean more wind power generating units and longer-distance power transmission requirements. As we all know, the sea environment is harsh, so electrical equipment needs special protection measures, and the price is much higher than that of onshore. Due to special offshore conditions, construction requires special tools and equipment, so the construction, operation and maintenance costs are much higher

than that of onshore. In order to realize the economical and reliable grid-connected operation of offshore wind farm, some special requirements should be put forward for the electrical system of offshore wind farm.

In order to collect the electric energy generated by the wind power generating unit scattered around the wind farm, the offshore wind farm electrical system connects the units through submarine cables in a certain manner and transmits the electric energy to the power grid. Structurally, it can be divided into three parts: current collection system, offshore booster platform and transmission system. The current collection system connects the wind power generating units to each other through medium voltage submarine cables and connects to the corresponding booster station. The offshore booster platform connects each "string" of fan in a certain form of main wiring, and raises the voltage level as required. The power transmission system connects the wind farm to the system grid connection point through high voltage submarine cables. Figure 1. 1. 19 shows electrical system structure diagram of a typical large-scale offshore wind farm. At present, there are three types of structures commonly used: AC system, AC/DC hybrid system, and DC system.

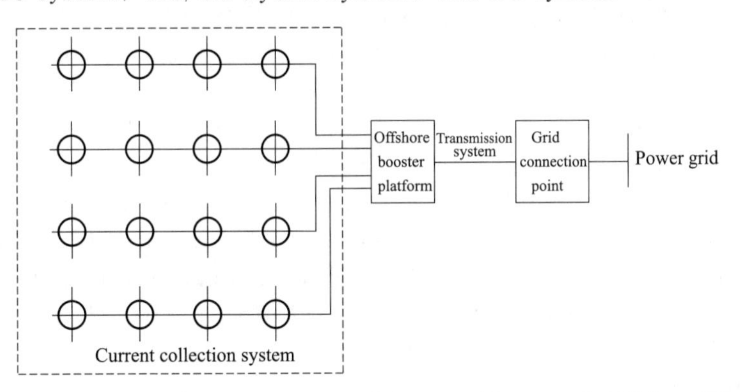

Figure 1. 1. 19 Electrical System Structure Diagram of Large-scale Offshore Wind Farm

① AC System. At present, most offshore wind power generating units adopt 690V terminal voltage. In order to reduce the loss of electric energy transmission inside the wind farm, it is common to install box transformer at the outlet of wind power generating unit to raise the voltage level. Considering the factors of equipment cost and transmission loss, it is generally believed that 30-36kV is the best voltage level for the connection between fans in AC electrical system. When the capacity of offshore wind farm is less than 100MW and the offshore distance is less than 15km, there is usually no need to install offshore substation, but it is directly connected to onshore substation through medium voltage line and then connected to the power grid. When the wind farm is large and the offshore distance is relatively long, the offshore substation can be used to raise the voltage level and connect to grid connection point through the high voltage transmission line.

② AC/DC Hybrid System. Large-scale offshore wind farm grid connection not only needs to consider the economic problem of submarine cable power transmission of dozens of kilometers, but also needs to consider its influence on the stability of power grid operation. High voltage DC transmission connection is a wind farm grid connection mode which not only meets the guideline of wind farm grid connection, but also has high economy. AC/DC hybrid system connects the offshore wind power generating units into string and connects them to the offshore converter station by using AC current collection system, and then con-

nects the wind farm to the power grid by means of high voltage DC transmission. See Figure 1. 1. 20 for the wiring method. Based on the current development of DC power transmission technology, when the capacity of offshore wind farm is more than 100MW and the offshore distance is more than 90km, it is more economical to adopt voltage source converter based high voltage direct current transmission (VSC-HVDC). When the capacity of the wind farm is more than 350MW and the offshore distance is more than 100km, the traditional HVDC transmission mode can be considered.

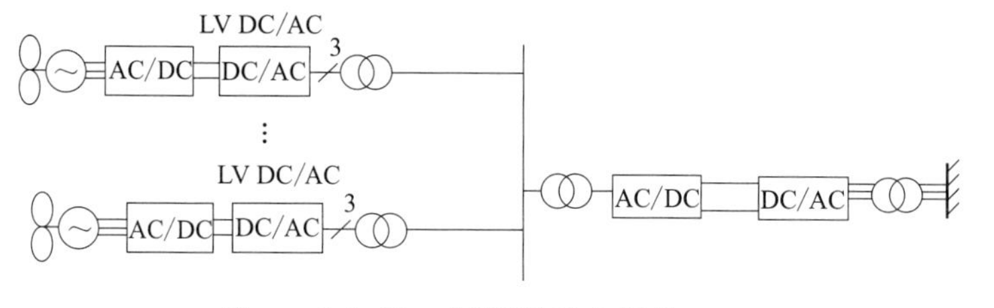

Figure 1. 1. 20 AC/DC Hybrid System

③ DC System. With the rapid development of VSC-HVDC technology, DC transmission also has certain competitiveness at lower voltage level and shorter transmission distance. Especially after the DC transmission has been engineered in offshore wind farm, DC transmission has gained more and more attention in the current collection system. In DC current collection system, the wind power generating unit raises the voltage to the medium voltage level through a set of AC/DC/DC converters. In order to connect with the high voltage DC transmission line of offshore wind farm, there are two main design ideas for the DC current collection system: parallel connection and series connection. Parallel connection uses DC/DC converter station to raise medium voltage DC to high voltage level, such as 150kV, and then connects to power grid through DC transmission line through onshore DC/AC converter station. In series connection, submarine cable is used to connect wind power generating units in series to obtain N times of DC voltage to achieve the purpose of boosting voltage. Then it is also connected to the power grid through the high voltage DC transmission line and onshore DC/AC converter station. Specific wiring is shown in Figure 1. 1. 21 and Figure 1. 1. 22. At present, the DC system is still in the stage of design and research.

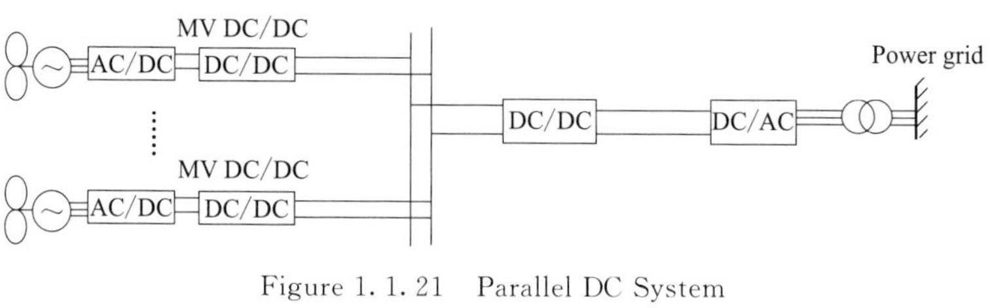

Figure 1. 1. 21 Parallel DC System

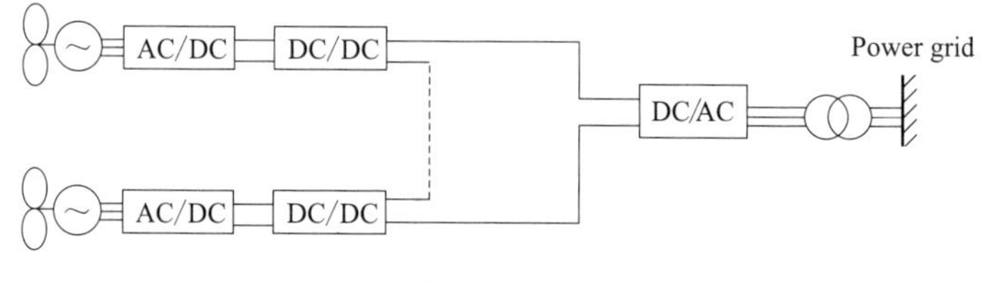

Figure 1. 1. 22 Series DC System

There are two main differences between parallel DC current collection system and AC

current collection system:

a. Number of Line. The current collection system of offshore wind farm adopts cable line, i. e. the AC current collection system usually adopts three-core submarine cable, while the DC current collection system adopts single-core submarine cable. Therefore, AC current collection system requires only 1 cable while parallel DC current collection system requires 2 cables. This affects not only the cost of submarine cables, but also the cost of laying submarine cables.

In terms of cable cost, compared with DC cable, AC cable is usually three-phase four-wire system or three-phase five-wire system, with higher insulation safety requirements and complex structure, so the cost is much higher. However, for AC/DC cables of the same cross section, the load that can be carried by 30kV DC submarine cable and 35kV AC submarine cable is not much different. That is to say, for the same wind power generating unit, the output current of 30kV DC system is almost twice of that of 35kV AC system, which means, the cross section of DC cable required for the string structure composed of the same number of wind power generating units may be much larger than the required AC cable.

In terms of installation costs, the laying of submarine cables is expensive. However, the outer diameter of AC submarine cables with the same conductor cross section is much larger than that of DC cables with the same conductor cross section, which makes it more difficult and complicated to coil and transport AC cables on ships. At the same time, the weight of AC submarine cable with the same length is much larger than that of DC cable, so the laying of AC submarine cable is generally higher than that of DC submarine cable. However, considering that the submarine cable laying requires an interval of 25-50m, the submarine cable laying workload of DC current collection system may be larger than that of AC current collection system, and the lease cost of wind farm sea area required is relatively high.

b. Medium Voltage Converter. DC/DC transformers are required in parallel DC current collection systems to raise the lower voltage level at the fan outlet to the medium voltage level. Although there are many design methods for DC/DC converters, under the current power electronic technology conditions, it is necessary to adopt DC/DC converter structure with electrical isolation, which is similar to DC/AC-transformer-AC/DC form when the DC step-up ratio is larger than 10.

4. Wind Farm Grid Connection Control

(1) Working Principle of Grid-Connected Wind Power Generation System

There are two different types of wind power generation: Off-grid type for independent operation and grid-connected type for connected power system operation. Off-grid wind power generation is small in scale and can solve power supply problems in remote areas through energy storage devices such as batteries or combining with other energy generation technologies. Larger wind farms are made up of dozens or even hundreds of wind power generators, often connected to the power grid.

Grid-connected wind power generation system (Figure 1. 1. 23) usually has large unit capacity, and wind farm is formed by multiple wind power generating units to transmit electric energy to power grid in a centralized manner. The frequency of grid-connected wind power generation shall always be equal to the grid frequency. The system can be divided into constant speed constant frequency wind power generation system and variable speed constant frequency wind power generation system according to the operation mode of generator.

① Constant Speed Constant Frequency Wind Power Generation System. Three-phase synchro-

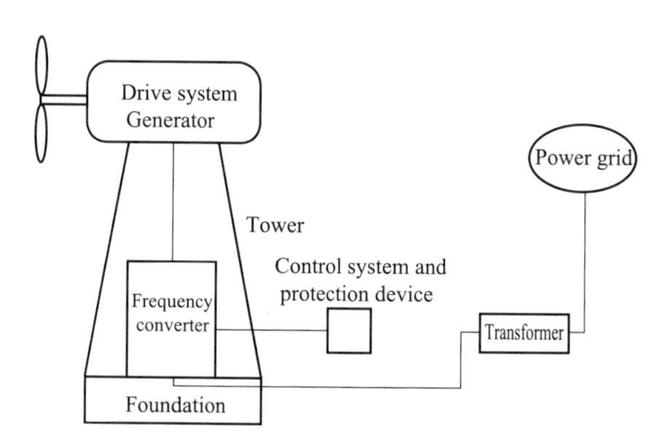

Figure 1. 1. 23 Structural Schematic Diagram of Grid-Connected Wind Power Generation System

nous generator and squirrel cage induction generator (SCIG) are mainly used in constant speed constant frequency wind power generation system. In the grid-connected wind power generating unit with fixed pitch, SCIG is generally adopted, and the wind wheel controlled by stall at fixed pitch makes it run stably at a speed slightly higher than the synchronous speed n [generally between (1-1.05)n]. Figure 1. 1. 24 shows the structure diagram of constant speed constant frequency wind power generation system using SCIG. Since SCIG outputs active power to power grid, it needs to absorb lagging reactive power from grid to establish rotating magnetic field with rotating speed of v_1 as well, which increases the burden of reactive power of grid, resulting in power factor reduction of grid. For this purpose, shunt capacitor banks of appropriate capacity are provided between the SCIG unit and the power grid to compensate reactive power. Within the whole operating wind speed range (3-25m/s), the velocity of air flow is constantly changing. In order to improve the efficiency at middle and low wind speed, three-phase (cage type) asynchronous double-speed generator is widely used for fixed pitch wind power generating unit.

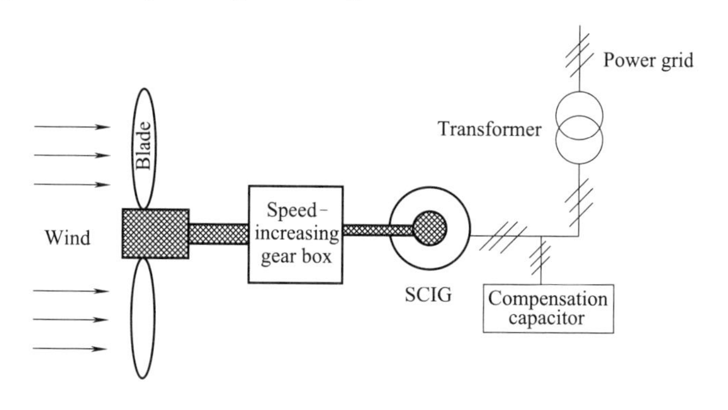

Figure 1. 1. 24 Constant Speed Constant Frequency Wind Power Generation System Using SCIG

Constant speed constant frequency wind power generation system has the advantages of simple motor structure, low cost and high reliability. Its main disadvantages are: the operating range is narrow; the wind energy cannot be fully utilized (its wind energy utilization factor cannot be kept at the maximum value); when the wind speed jumps, the components such as the spindle, gear box and generator are subjected to large mechanical stresses.

② Variable Speed Constant Frequency Wind Power Generation System. In order to overcome the disadvantages of constant speed constant frequency wind power generation systems, various variable speed constant frequency wind power generation systems based on variable pitch technology began to enter the market in the mid-1990s. Their main characteristics are as follows: When the wind speed is lower than that of the rated, adjust the torque of the generator to make the rotation speed change with the wind speed, keep the tip speed

ratio of the wind wheel at the optimum value, and maintain the operation of the wind power generating unit under the maximum wind energy utilization rate; when the wind speed is higher than that of the rated, adjust the pitch angle to limit the power absorbed by the wind turbine not exceeding the maximum value; the constant frequency electric energy is obtained through the combination of generator and power electronic conversion device. At present, the variable speed constant frequency wind power generating unit mainly adopts wound rotor double-fed induction generator, and low speed synchronous generator direct-driven wind power generation system has received widespread paid attention.

a. Variable Speed Constant Frequency Wind Power Generation System Based on Wound Rotor Double-Fed Induction generator. The rotor side of the wound rotor double-fed induction generator (DFIG) is excited by AC excitation through electric ring and electric brush, which can input or output electric energy. Figure 1. 1. 25 shows the structure diagram of variable speed constant frequency wind power generation system based on wound rotor double-fed induction generator, in which the rotor winding of DFIG is connected to the power grid through reversible transducer, and wide range variable speed constant frequency power generation operation is realized by controlling the frequency of rotor excitation current. Its working principle is as follows:

Assuming that the magnetic field speed of stators n_0, i. e., the synchronous speed is n_0, the rotor is fed with three-phase low frequency excitation current to form a low-speed rotating magnetic field. The mechanical rotation speed n_1 of the generator rotor is superimposed with the rotation speed n_2 of the magnetic field, equal to the synchronous speed of stator n_0, i. e.

$$n_1 \pm n_2 = n_0$$

Thereby, the power frequency voltage corresponding to the synchronous speed n_0 is induced in the DFIG stator winding. When generator speed n_1 changes with wind speed, adjust the frequency of rotor excitation current, change n_2 to compensate the change of n_1 (general variation range is 30% of n_0, which can be adjusted bi-directionally), so as to keep the frequency of output electric energy constant.

The variable speed constant frequency scheme shown in Figure 1. 1. 25 is realized in the rotor circuit, and the power flowing through the rotor circuit is determined by the operation range of DFIG rotating speed, and the slip power is generally only 1/4-1/3 of the rated power, thus significantly reducing the capacity and cost of the transducer. In addition, by adjusting the active and reactive components of the rotor excitation current, the active and reactive power of the generator can be adjusted independently to adjust the power factor of the power grid and compensate the reactive power demand of the power grid. In fact, the generator and the power system form a "flexible connection" due to the adjustable frequency, amplitude and phase AC excitation of the DFIG rotor.

b. Direct-Driven Wind Power Generation System Based on Low-Speed Synchronous Generator. In direct-driven wind power generation system, the wind wheel is directly connected with permanent magnet (or electric excitation) synchronous generator, thus eliminating the common speed-up gear box. Figure 1. 1. 26 is the structural schematic diagram of permanent magnet direct-driven variable speed constant frequency wind power generation system. Wind energy is converted into alternating current with variable frequency and amplitude in permanent magnet synchronous generator (PMSG) stator winding through wind turbine and PMSG, and then input into full power transducer (which usually adopts con-

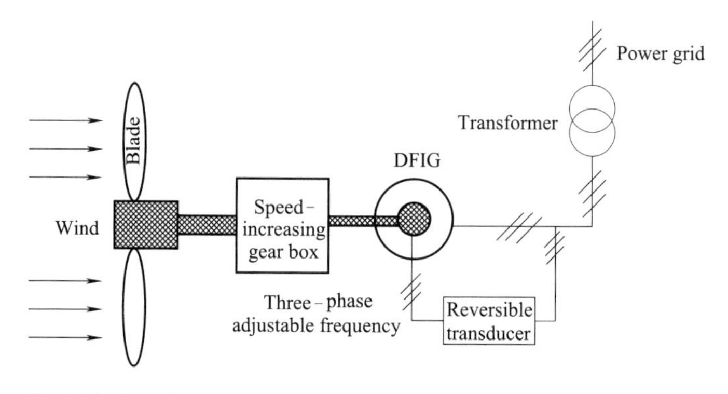

Figure 1. 1. 25　Variable Speed Constant Frequency Wind Power Generation System Based on DFIG

trollable PWM rectification or uncontrollable rectification followed by DC/AC conversion), first rectifies into DC, and then converts into three-phase power frequency AC for output by three-phase inverter. The active power and reactive power of the system are controlled by the full power transducer on the stator side, and the electromagnetic torque of the generator is controlled to adjust the rotor speed to realize the maximum power tracking. Compared with the wind power generation system based on DFIG, this system can be connected to the power grid in a wider speed range, but the capacity requirement of its full power transducer is large. Compared with the wind power generation system with gear box, this system improves the efficiency and reliability, reduces the operation noise, but the generator speed is low. In order to obtain a certain power, the generator should have larger electromagnetic torque, so its volume is large and the cost is high.

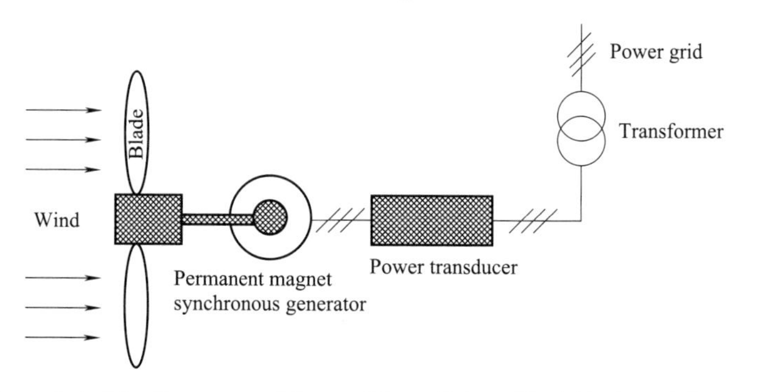

Figure 1. 1. 26　Permanent Magnet Direct-Driven Variable Speed Constant Frequency Wind Power Generation System

(2) Overview of Wind Power Grid Connection Mode

At present, the grid connection mode of wind power generation can be roughly divided into three modes: Grid connection of asynchronous power generating unit, grid connection of synchronous power generating unit and grid connection of double-fed power generating unit.

① Grid Connection of Asynchronous Power Generating Unit. Because the wind turbine is power machinery running at low speed, the speed is increased between the wind turbine and the rotor of the asynchronous generator through the speed-increasing gear transmission, so as to reach the speed suitable for the operation of the asynchronous generator. Generally, 4-pole or 6-pole generator is usually selected for asynchronous generator running in parallel with power grid, so asynchronous generator speed must exceed 1500r/min or 1000r/min be-

fore it can operate in power generation state to supply power to power grid. According to the theory of the generator, the load of asynchronous generator is adjusted by slip ratio when it is connected to power grid, and its output power is almost linear with rotating speed. Therefore, the speed regulation requirements for the unit are not as strict and precise as that of the synchronous generator, and synchronous equipment and full-step operation are not required. As long as the rotating speed is close to the synchronous speed, it can be connected to the power grid. However, at the moment of grid connection, the asynchronous generator will have relatively large impulse current (about 4~7 times of rated current of asynchronous generator), and make the grid voltage drop instantaneously. With the continuous increase of the single unit capacity of wind power generating units, the impact of impulse current on the safety of generator components and power grid becomes more and more serious. Excessive impulse current may disconnect the automatic switch in the main circuit where the generator is connected to the power grid; while a large drop in the grid voltage may cause the low voltage protection to act, resulting in that the asynchronous generator cannot be connected to the power grid at all.

At present, asynchronous generator grid connection methods used in wind power generation systems are as follows:

a. Direct Grid Connection. This method requires that the phase sequence of the generator is the same as that of the grid when it is connected to the grid. When the speed of the wind driven asynchronous generator approaches the synchronous speed, it can be automatically connected to the power grid. However, as mentioned above, direct grid connection will result in large impulse current and grid voltage drop, so this grid connection method is only suitable for situation asynchronous generator capacity is below 100 kilowatts and grid capacity is large.

b. Step-Down Grid Connection. Step-down grid connection is achieved by connecting a resistor or reactor in series between the asynchronous generator and the power grid, or by connecting an autotransformer to reduce the amplitude of the impulse current at the instant of grid connection and the voltage drop of the power grid. Because components such as resistors and reactors consume power, they must be switched off quickly when the generator enters stable operation state after it is connected to the power grid. This method of grid connection is suitable for units with large capacity and above 100 kilowatts.

c. Thyristor Soft Grid Connection. The thyristor soft grid connection method is to connect the stator of asynchronous generator and the power grid through a bidirectional thyristor connected in series for each phase. All three phases are controlled by thyristors. Both ends of the bidirectional thyristors are connected in parallel with the moving contact of the grid-connected automatic switch. The purpose of bidirectional thyristor connection is to control the instantaneous impulse current of generator grid connection within the allowable limit. The process of grid connection is as follows: After receiving the start command sent by the microprocessor in the control system, firstly check whether the phase sequence of the generator is consistent with that of the power grid. If the phase sequence is correct, send the release command and start the wind power generating unit. When the generator speed approaches synchronous speed (99%-100% synchronous speed), the control pin of bidirectional thyristor is opened synchronously from 180° to 0° simultaneously; meanwhile, the conduction angle of bidirectional thyristor increases gradually from 0° to 180° at the same time. At this time, the grid-connected automatic switch does not act and the moving contact

is not closed, and the asynchronous generator is smoothly connected to the power grid through the thyristor; as the generator speed continues to rise, the slip ratio of the generator tends to zero gradually. When the slip ratio is zero, the grid-connected automatic switch acts, the moving contact is closed, the bidirectional thyristor is short-circuited, and the output current of the asynchronous generator will flow into the power grid through the closed automatic switch contact instead of the bidirectional thyristor. After the generator is connected to the power grid, the compensation capacitor shall be connected at the generator immediately to increase the power factor of the generator to above 0.95.

② Grid Connection of Synchronous Power Generating Unit. Synchronous generator has been widely used in power system because it can not only output active power but also provide reactive power, stable cycle wave and high power quality.

Grid Connection of Direct-Driven AC Permanent Magnet Synchronous Power Generating Unit. The low-speed alternator is directly driven by the wind turbine, which is connected to the power grid through IGBT inverter with fast working speed, low driving power and reduced conduction voltage. The characteristics of grid connection operation of this system are as follows:

a. Because the gear box is not used, the horizontal axial length of the unit is greatly reduced, the mechanical transmission path for electric energy production is shortened, and the loss and noise caused by the rotation of the gear box are avoided.

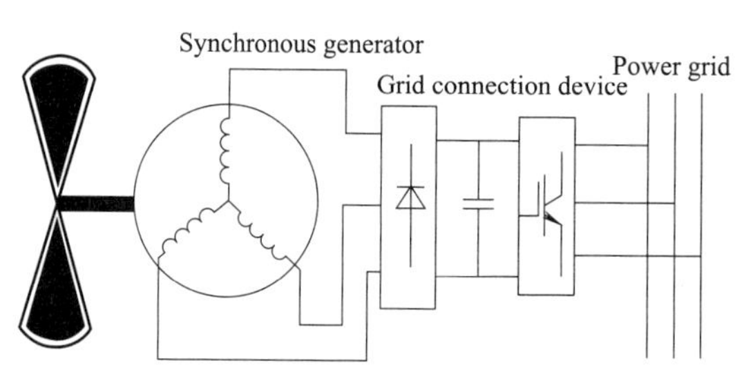

Figure 1. 1. 27 Grid Connection Circuit of AC Synchronous Generator

b. Because the generator has a large surface, the heat dissipation condition is more favorable, so that the temperature rise of the generator during operation is reduced and the fluctuation of the temperature rise of the generator is reduced.

Figure 1. 1. 27 shows a typical electric energy device conversion circuit using AC synchronous generator. The whole grid-connected power generation system is mainly composed of synchronous generator and grid-connected device, etc.

The output AC current of the three-phase synchronous generator is rectified into DC by the uncontrollable rectifier, and then sent to the input terminal of DC/AC inverter through DC filter, which is inverted into electric energy whose voltage, frequency, phase angle, power factor and harmonic wave meet the requirements of the power grid, and then is connected to the power grid after passing through the AC filter.

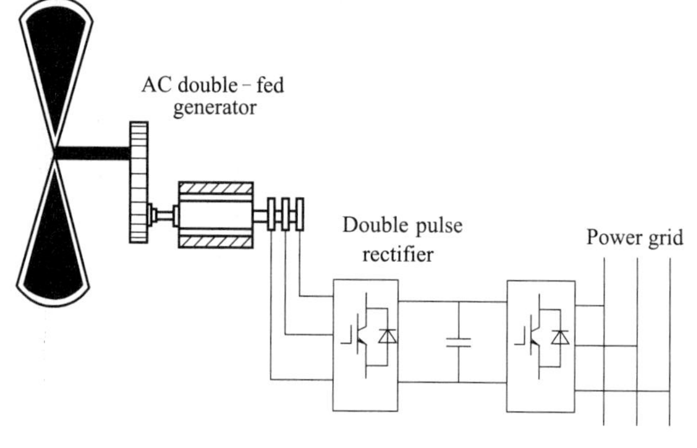

Figure 1. 1. 28 Grid Connection Circuit of AC Double-Fed Generator

③ Grid Connection of Double-Fed Power Generating Unit. Figure 1. 1. 28 shows a typical electric energy conversion circuit for AC double-fed generator. The whole grid-connected power generation system is mainly

composed of double-fed generator and double pulse rectifier, etc.

The characteristic of this grid connection mode is to add pulse rectifier at generator side and power grid side respectively. In case of low wind speed, the AC voltage output by generator is boosted by generator side pulse rectifier, which can meet the normal operation of pulse rectifier at grid side.

(3) Wind Farm Substation

A wind farm substation is a place where some equipment is assembled to cut off or switch on, change or adjust the voltage. The substation of the wind farm is mainly equipped with transformer, power distribution device, reactive power compensation equipment and control equipment, etc., which are used to raise the electric energy voltage collected by the wind farm current collection line and control the collection and transmission of the power generated by the wind power generating unit, as shown in Figure 1. 1. 29.

Figure 1. 1. 29　Step-up Substation of Wind Farm

The main structures of the substation include: main control room, indoor power distribution device, outdoor power distribution device, reactive power compensation, lightning protection earthing system, main transformer, relay protection of wind farm substation, integrated automation system of wind farm substation, etc., as shown in Figure 1. 1. 30.

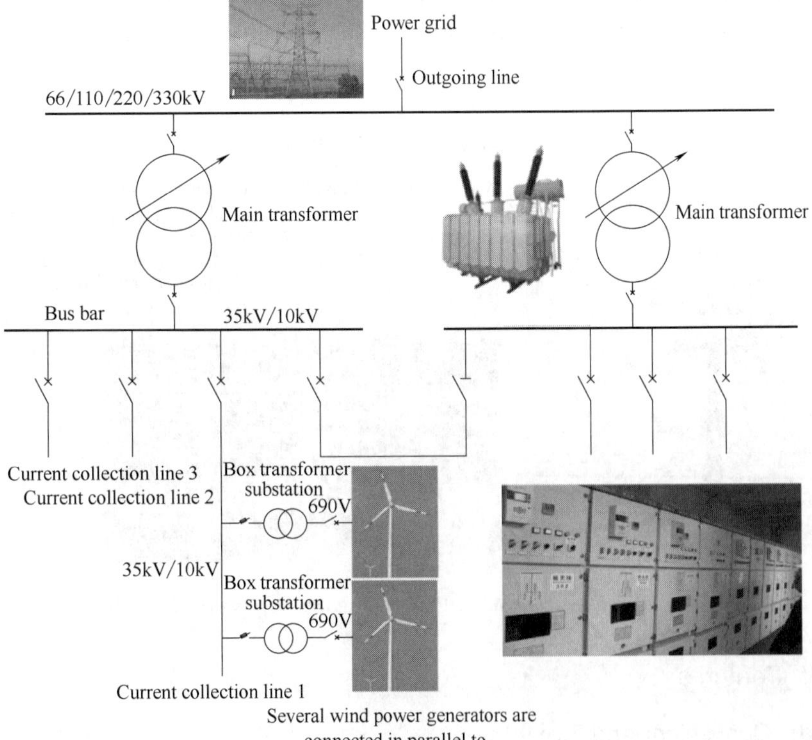

Figure 1. 1. 30　Substation Structure

Part Ⅱ

Large-scale Wind Power Generation System Operational Training Platform

Ⅰ. Product Overview

THWPWG-3B Large-scale Wind Power Generation System Operational Training Platform is shown in Figure 2. 1. 1. This training platform is designed with reference to MW-level fan control system. It has visual physical simulation objects and creates a good training environment, including a set of wind power generating unit simulation objects and four control cabinets, respectively: Fan object model (MW fan structure, active yaw and independent pitch function can be realized), control cabinet of energy control • monitoring management • weather station, control cabinet of yaw pitch control system, control cabinet of energy conversion storage control system, control cabinet of grid-connected inverter control system. It can realize the training of the control process of the fan pitch system, electrical system and yaw system and the demonstration of the whole machine operation of the fan. It can be used for electrical control operation and troubleshooting of electric variable pitch yaw device of wind power generation, training of wind power grid connection control technology for students in vocational colleges, and also can be used for skill identification training of wind power maintenance workers.

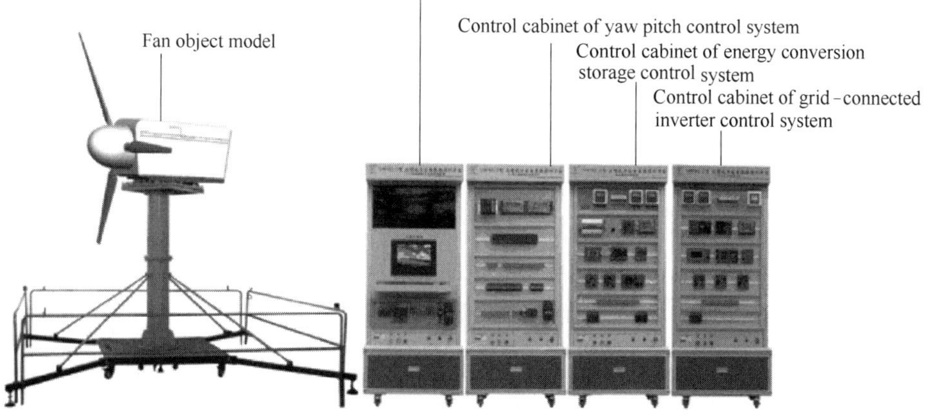

Figure 2. 1. 1 THWPWG-3B Large-scale Wind Power Generation System Operational Training Platform

Ⅱ. Product Features

1. Vivid, Convenient and Flexible

The mechanical structure of pitch, yaw and braking device of wind power generating unit adopts vivid semi-physical model. Parts of electrical control component adopt mesh plate

installation structure mode, which is convenient and flexible to install.

2. Temperature Monitoring

The object system is equipped with an environment simulation system, which can simulate the ambient temperature of −40-+80℃. The generator, frequency converter and speed-increasing gear box are equipped with multiple temperature sensors for temperature monitoring.

3. Simulate Wind Speed and Direction

It has the function of simulating wind speed and wind direction, reflecting the working characteristics of pitch and yaw mechanism according to the wind speed and wind direction change.

4. Grid-Connected Inverter

The grid-connected inverter control system reverses DC 24V to AC 36V, 50Hz (safety voltage), rises to AC 220V, 50Hz through step-up transformer, and realizes grid-connected power generation function by connecting with single-phase electric supply. The main controller adopts 32-bit fixed-point TMS320F2812 chip of TI company. The output power factor of the inverter is close to 1 and the output current is sine wave. The control strategy adopts double closed-loop control structure, the inner loop is grid-connected current loop and the outer loop is DC voltage loop. The synchronization of grid connection adopts digital phase lock technology, which has high precision and is easy to realize. At the same time, the output power factor correction control is realized.

5. Electric Quantity Measurement, Maximum Power Tracking and Battery Management

The energy conversion storage control system performs the functions of electric quantity measurement, maximum power tracking, energy storage and battery management. The maximum power tracking microprocessor adopts 51 series single chip microcomputer, which is universal and can be downloaded online. It is convenient for users to program and debug. The hardware is completely open. Users can write different MPPT algorithms to realize maximum power tracking, and send regulation parameters to PWM driver module for regulation. PIC series single chip microcomputer is used in PWM driving CPU to receive regulation parameters and output PWM signals with different duty ratio, control main circuit and realize power regulation. The intelligent charge and discharge controller can adjust the charging state and current according to the battery voltage, protect the battery from over-charging or over-discharging, and prolong the service life of the battery.

6. Safety Protection

The operational training platform has perfect safety device. It is equipped with removable safety railing on the periphery, which can ensure that users wont enter the dangerous area. Meanwhile, the infrared reflection warning switch is installed. When users enter the dangerous area accidentally, the system will automatically cut off and brake.

Ⅲ. Technical Performance

1. Rated Working Voltage

Three-phase four-wire AC 380V±10%, 50Hz.

2. Working Environment

Temperature −10-+40℃, relative humidity <85% (25℃), altitude<4000m.

3. Device Capacity

Less than 2.5kV • A.

4. Overall Dimension

3400mm×3400mm×3800mm (fan object model);

880mm×600mm×2100mm (control cabinet of energy control • monitoring management • weather station);

880mm×600mm×2100mm (control cabinet of yaw pitch control system);

880mm×600mm×2100mm (control cabinet of energy conversion storage control system);

880mm×600mm×2100mm (control cabinet of grid-connected inverter control system).

Ⅳ. Product Structure and Composition

1. Fan Object Model

It is mainly composed of variable pitch motor, yaw motor, absolute value encoder, wind power generator, prime mover, gear box, blade, rotor shaft, nacelle, tower, hub, etc.

2. Control Cabinet of Energy Control • Monitoring Management • Weather Station

It is mainly composed of LED display screen, industrial personal computer, PLC, low-voltage electrical apparatus, switch and other devices. LED screen can display wind speed, wind direction, yaw angle, pitch angle and other information. The monitoring software of wind power generation system is installed on industrial panel PC, which can realize monitoring and management of wind power generation system, simulation of wind power generating unit and control of wind power generation system.

3. Control Cabinet of Yaw Pitch Control System

It is mainly composed of slave PLC, frequency converter, absolute value encoder, AC gear motor, control button and other devices. It can complete the installation of pitch yaw control system, the programming and debugging of PLC control program for manual and automatic pitch yaw, and the communication between slave PLC and master PLC.

4. Control Cabinet of Energy Conversion Storage Control System

It is mainly composed of DC voltage and current sampling module, temperature alarm module, PWM driving module, CPU core module, human-computer interaction module, communication module, lightning protector, intelligent charge and discharge controller, battery pack, switching power supply, DC voltmeter and DC ammeter, etc.

5. Control Cabinet of Grid-Connected Inverter Control System

It is mainly composed of core module, interface module, liquid crystal display module, keyboard interface module, drive module, DC voltage booster module, DC voltage sampling module, AC voltage sampling module, AC current sampling module, temperature alarm module, communication module, switching power supply, DC motor, square indicator light, DC voltmeter, DC ammeter, multi-functional digital display meter, transformer, etc.

Ⅴ. Safe Operation Specifications

In order to successfully complete the operational training project and ensure the safe, reliable and long-term operation of the equipment during training, the operational training

personnel shall strictly abide by the following safety regulations.

1. Preparation before Training

① Read the instruction carefully before training to get familiar with relevant parts of the system.

② Read the system operation instructions and precautions carefully before training.

③ Read the user manual of frequency converter carefully before training to understand the usage of frequency converter.

④ Before training, ensure that the power supply of control cabinet of each system is disconnected.

⑤ Get familiar with the operation steps of this training according to the relevant contents in the training instruction before training.

2. Notes during Operational Training

① Check whether each power supply is normal before use.

② Before wiring, be sure to be familiar with the function and wiring position of each unit module of the device.

③ The main power supply must be disconnected before the training wiring, and live wiring is strictly prohibited.

④ Power on after wiring is completed and checked without error.

⑤ Proficient in control principle and program control method of yaw pitch motor.

⑥ There is AC 220V access point in the control cabinet. Pay attention to safety during training.

⑦ The training platform shall be kept clean and tidy all the time, and sundries shall not be placed randomly to avoid faults such as short circuit.

⑧ After the training is completed, the power switch shall be turned off in time and the training platform shall be cleaned timely.

⑨ Power on and off the system in strict accordance with correct operation steps to avoid damage to the system caused by misoperation.

⑩ In the process of operating the system, after the battery switch of the energy conversion storage control system is turned on, there is a process of waiting for the self-test initialization of the intelligent charge and discharge controller; the next operation can be carried out only after the "red light" of the intelligent charge and discharge controller goes off.

⑪ In the process of training, pay attention to prevent falling from high place and squeezing injury during equipment installation.

⑫ In the process of training, pay attention to the safety of strong current where there is "danger" sign.

3. Steps of Training

The following steps shall be achieved during training:

(1) Preview report, detailed and complete, get familiar with the equipment

Before the commencement of operational training, the instructor shall check the students' preview report and require students to understand the purpose, content and safe operation steps of this training. Only after meeting this requirement can the training be started.

The instructor shall make a detailed introduction to the training device. Students must be familiar with various equipment used for this training, and make clear the functions and using methods of these equipment.

(2) Establish teams and divide labor reasonably

Each training is carried out by a team consisting of 2-3 persons.

(3) Trial operation

Before formal training, get familiar with the operation of the device, then connect the power supply according to certain safe operation specifications, and observe whether the equipment is normal. If the equipment is abnormal, the power supply shall be cut off immediately and the fault shall be eliminated; if everything is normal, the training can be started formally.

(4) Be conscientious and responsible, and carry the training through to the end

After the training is completed, the instructor shall check the training materials. After being approved by the instructor, turn off all power supplies according to safe operation steps, arrange the articles used in training and put them back to their original position.

4. Summary of Training

This is the last and most important stage of operational training. We should analyze the training phenomenon and write the training report. Each training participant shall independently complete a training report, which shall be prepared with a serious and practical attitude.

The training report is a training summary and experiences based on the problems found in the training, which is written through self-analysis and research or analysis and discussion among the team members. It shall be concise, clear in writing and clear in conclusion.

The training report shall include the following contents:

① Training name, major, class, student number, student name, name of the same team, etc.

② Training purpose, contents and steps.

③ Model and specification of training equipment.

④ Arrangement of training materials.

⑤ Analyze and summarize the training results with theoretical knowledge, and draw correct conclusions.

⑥ Analyze and discuss the phenomena and problems encountered in the training, write out the experience, and put forward own suggestions and improvement measures.

⑦ The training report should be written on the report paper of certain specification and kept neat and tidy.

⑧ Each training person shall independently complete a training report and submit it to the instructor for review on time.

Part Ⅲ

Operational Training Project

Project Ⅰ　Cognizance of Fan Object Structure

Project Description

Consult the instruction manual of THWPWG-3B large-scale wind power generation system operational training platform, understand the structure composition of fan object model, complete the cognition of fan object model, understand the basic principles of wind power generation, and master the functions of various wind power equipment in operation and control by using the knowledge learned and relevant installation manuals.

Competency Objectives:

① Understand the structure of wind power generating unit.

② Understand common wind power equipment and record equipment model and parameters.

③ Understand the basic principles of wind power generation and master the functions of various wind power equipment in operation and control.

Project Environment

THWPWG-3B large-scale wind power generation system operational training platform, as shown in Figure 2. 1. 1.

① The mechanical structure of pitch, yaw and braking device of wind power generating unit adopts vivid semi-physical model. Parts of electrical control component adopt mesh plate installation structure mode, which is convenient and flexible to install.

② The object system is equipped with an environment simulation system, which can simulate the external ambient temperature. The generator, power motor and speed-increasing gear box are all equipped with multiple temperature sensors for temperature monitoring.

③ The device has the function of simulating wind speed and wind direction, and reflects the working characteristics of pitch and yaw mechanism according to the wind speed and wind direction change.

Project Principle and Basic Knowledge

Wind power generation uses wind energy to generate electricity. Wind power generating unit is machinery that converts wind energy into electric energy. Wind wheel is the main component of wind power generating unit, consisting of blades and hubs. Blades have good aerodynamic shape. Under the action of air flow, they can generate aerodynamic force to make the wind wheel rotate, convert wind energy into mechanical energy, and then drive the generator which the help of the speed increase of gear box to convert mechanical energy

into electric energy. In theory, the best wind wheels can only convert about 60% of the wind energy into mechanical energy. The efficiency of wind wheel of modern wind power generating unit can reach more than 50%. Before the output of unit reaches rated power, its power is proportional to the cube of wind speed, that is, the wind speed increases by 1 time and the output power increases by 7 times. It can be seen that the efficiency of wind power generation has great relationship with local wind speed.

Wind power generating units can be divided into grid-connected type and off-grid type according to different application situations. Off-grid wind power generating unit is also called independent operation wind power generating unit, which is applied in the area without power grid, and the general power is relatively small. The independent operation wind power generating unit generally needs to form an independent operation wind power generation system together with the storage battery and other control devices. Such independent operation system can be a few kilowatts or even dozens of kilowatts to solve the power supply system of a village, or several tens to several hundred watts of small wind power generating unit to solve the power supply of a household. Grid-connected wind power generating unit is mainly composed of two parts: Wind turbine section——converting wind energy into mechanical energy; generator section——converting mechanical energy into electric energy.

Ⅰ. Basic Structure of Wind Power Generating Unit

Wind power generating unit is composed of wind wheel, pitch system, transmission system, yaw system, hydraulic system, braking system, generator, control and safety system, nacelle, and tower, etc.

The functions of each main component are as follows:

① Wind wheel: Wind wheel is composed of blades and hubs. A blade is a unit that absorbs wind energy and is used to convert the kinetic energy of the air into mechanical energy for the rotation of the impeller. The function of a hub is to hold the blades together and bear various loads transmitted from blades, which are then transmitted to the generator rotating shaft.

② Pitch system: The pitch system changes the pitch angle of blades to make the blades in the best state of absorbing wind energy at different wind speeds. When the wind speed exceeds the cut-out wind speed, blades will be feathered and braked.

③ Transmission system: The transmission system transfers mechanical energy to the generator. Generally, the system includes low-speed shafts, high-speed shafts, a gear box, couplings, and brakes. However, not every type of wind power generating unit will have all these components. Hubs of some units are directly connected to the gear box, so low-speed transmission shafts are not needed. There are also units designed without a gear box, and the blades of which are directly connected to the generators.

④ Yaw system: The yaw system adopts the form of active wind gear drive, and cooperates with the control system to keep the impeller in the windward state all the time, making full use of the wind energy and improving the power generation efficiency. At the same time, the system provides necessary locking torque to ensure safe operation of the unit.

⑤ Hydraulic System: The main function of the hydraulic system of a wind power generating unit is braking (high-speed shaft, low-speed shaft, and yawing). The hydraulic system typically consists of an electric motor, an oil pump, an oil tank, filters, pipelines, and various hydraulic valves.

⑥ Braking system: The currently widely-used braking system of the horizontal-axis wind power generator generally consists of an air braking system and a mechanical braking system. The air braking system is mainly divided into two categories: control by blade tip spoilers of fixed pitch wind power generator and active pitch control of variable pitch wind power generator. The air braking system can slow down the wheel speed, but it cannot bring the wind wheel to a complete stop. The mechanical braking system is responsible for stopping the fan.

⑦ Generator: A generator is a component that converts the mechanical energy of an impeller into electric energy. The rotor is connected with frequency converter, which can supply adjustable frequency voltage to the rotor circuit, and the output speed can be adjusted within ±30% of synchronous speed.

⑧ Control and safety system: The control and safety system contains a computer that continuously monitors the status of the wind turbine and controls the yaw and pitch devices. In order to prevent any failure (i. e. overheating of gear box or generator), the controller can automatically stop the rotation of the wind turbine and give an alarm.

⑨ Nacelle: The nacelle contains the key equipment of wind turbine, including gear box and generator. Maintenance personnel can enter the nacelle through the wind turbine tower. At the left end of the nacelle is the rotor of the wind turbine, i. e. rotor blade and shaft.

⑩ Tower: The wind turbine tower contains the nacelle and rotor. Usually, tall towers have an advantage because the higher the distance from the ground, the greater the wind speed.

Ⅱ. Main Parameters of Wind Power Generating Unit

① Diameter of wind wheel: In general, the larger the power of the wind power generating unit is, the larger the diameter of wind wheel is.

② Number of blades: There are 2-4 pieces for high speed power generation unit and more than 4 pieces for low speed unit;

③ Blade material: High-strength and low-density composite materials are often used in modern times;

④ Wind energy utilization coefficient: Generally between 0. 15-0. 5;

⑤ Starting wind speed: Generally 3-5m/s;

⑥ Shutdown wind speed: Usually 15-35m/s;

⑦ Output power: Modern wind power generating unit is usually several hundred kilowatts to several megawatts;

⑧ Generator: Divided into DC generator and AC generator;

⑨ Other: tower height, etc.

Project Implementation

Ⅰ. List of Instruments, Equipment and Tools

① THWPWG-3B large-scale wind power generation system operational training platform;

② Ladder;

③ Mobile phone;

④ Word Document.

Ⅱ. Safe Operation Specification

① When knowing the equipment of wind power generating unit, it is necessary to use ladder to carry out aerial work. It is required that the ladder shall be stably placed on the flat ground;

② When getting up and down the ladder and standing on the ladder for operation, step on firmly, concentrate on the operation, and don't play;

③ Working in a group of 4 during operation. 2 peoples shall stand on 2 ladders to cooperate in assembly and disassembly, and the other 2 peoples shall be responsible for safety protection and transferring tools and parts on the ground;

④ Beware of falling and missing your step. Beware of falling tools or parts.

Ⅲ. Operational Training Steps

① List the basic structure of the wind power generating unit that you know, and describe the main functions of each component.

② The fan object model is mainly composed of pitch motor, yaw motor, absolute value encoder, wind power generator, prime mover, gear box, blade, rotor shaft, nacelle, tower, hub, etc.

Take THWPWG-3B Large-scale Wind Power Generation System Operational Training Platform as an example, find out various equipment in the list according to the technical parameters of the equipment and its function in the operation and control of the wind power generating unit, take pictures with mobile phone, and insert the pictures into the equipment positions in Table 3. 1. 1.

Table 3. 1. 1 Technical Parameters of Main Equipment of Fan Object Model

Number	Name	Specification	Quantity	Equipment Position and Photo	Remark
1	Fan object model	3400mm×3400mm×3800mm	1 set		
2	AC reducer	Pitch motor 80YB25GV22-GM10-GK75RC Output power: Single-phase 25W Rated voltage: 220V Rated current: 0. 25A Rated speed: 1300r/min Reduction ratio of intermediate reduction gearbox: 1 : 10 Reduction ratio of right angle center control reduction gearbox: 1 : 75	3 sets		
3		Yaw motor 90YB90GY38-90GM10-90GF60HE Output power: Three-phase 90W Rated voltage: 380V Rated current: 0. 41A Rated speed: 1300r/min Reduction ratio: 1 : 600	3 sets		

Continued

Number	Name	Specification	Quantity	Equipment Position and Photo	Remark
4	Horizontal reducer	GH28-0. 75-70S/380 Output power: Three-phase 750W Rated voltage: 380V Rated current: 1. 97A Rated speed: 1400r/min Reduction ratio of reduction gearbox: 1 : 70	1 set		
5	Slewing bearing	011. 20. 280	1 piece		
6	Planetary gear speed-increasing box	NB300L2-20-PC-V01A Speed increasing ratio: 19. 75 : 1 Output type: V01A Continuous input torque: 850N · m Maximum input torque: 1200N · m	1 set		
7	Conductive slip ring	SRH50120-6P/36S 380V AC/24V DC Rated current: 6 rings 10A/36 rings 5A Insulation resistance ≥ 500MΩ/500V DC Working speed: 0-500r/min Working temperature: −20-+80℃ Working humidity: 60% RH	1 piece		
8	Three-phase permanent magnet generator	24V-300W-400r Rated output voltage: 24V Rated power: 300W Rated speed: 400r/min	1 set		
9	Absolute value encoder	Blade angle absolute value encoder TRD-MA-512N Output signal: Gray code (maximum 9bit, this equipment adopts 8bit 256 lines) Maximum corresponding frequency: 30kHz Maximum permissible speed: 3000r/min Supply voltage: DC 10. 8-26. 4V Output form: Open collector of NPN or PNP	3 pieces		
10	Incremental encoder	J38S-6G-600BZ-C5-24	1 piece		

Continued

Number	Name	Specification	Quantity	Equipment Position and Photo	Remark
11	Hall sensor	SM12-31010NA Switch category: Inductive(S) Appearance: Cylindrical(M) Working voltage: 24V Detection distance: 10mm Output mode: Three-wire DC NPN negative logic normally open output Output status: Normally open(A) Standard detection object: Permanent magnet Working ambient temperature: −25-+70℃	5 pieces		
12	U-type photoelectric switch	Measuring simulated wind speed SU003-3K Switch category: Inductive(S) Working voltage: 24V Output mode: Three-wire DC NPN negative logic normally open output Output status: Normally open(A) Working ambient temperature: −25-+70℃	1 piece		
13	Signal lamp	Working voltage DC/24V(green)	36 pieces		
14	Reflection sensor	ZR-350N Switch category: Inductive(S) Working voltage: 24V Detection distance: 1200mm Output mode: Three-wire DC NPN negative logic normally open output Output status: Normally open(A) Working ambient temperature: −25-+70℃	4 pieces		
15	SMD PT100	Analog quantity output Measuring temperature: −50-+420℃	6 pieces		
16	Shock-resistant pressure gauge	YN-100ZQ/10MPa Measuring pressure range 0-10MPa	1 set		
17	Two-position three-way solenoid valve	3WE6E61B/CG24N9Z6L 24VDC two-position three-way valve	1 set		

Continued

Number	Name	Specification	Quantity	Equipment Position and Photo	Remark
18	Direct-acting relief valve	DBDH6P10B/100 7 pressure level adjustable relief valves	1 set		
19	Two-position four-way solenoid valve	24V DC two-position four-way valve	1 set		
20	Micro hydraulic power unit	Integrated hydraulic station motor Two oil return ports, one oil inlet port Rated voltage:380V 50Hz Rated power:750W Pressure:10MPa Capacity:6L	1 set		Use as a whole, [illegible] perfect sealing
21	Pressure sensor	TP212-10M423 Analog quantity output:0-5V Measuring pressure:10MPa	1 set		
22	Fluid storage tank	330 121 407 Storage of coolant 2L	1 set		
23	Water pump	Power supply:Single-phase 220V Rated power:95W Rated current:0.46A Rated flow: 16L/min, inlet water pressure above 1m	1 set		
24	Solenoid valve	24 VDC water flow stop valve	1 set		

Project Work

According to the application and control of PLC/single chip microcomputer to the equipment, try to divide some devices into two groups: Input equipment and output equipment, complete Table 3.1.2 and Table 3.1.3.

Table 3. 1. 2 Name and Function of Input Equipment

Input Equipment Name	Function	Input Equipment Name	Function

Table 3. 1. 3 Name and Function of Output Equipment

Output Equipment Name	Function	Output Equipment Name	Function

Project Ⅱ Assembly and Disassembly of Blade and Pitch Bearing

Project Description

Look up the installation manual of wind power generating unit impeller and pitch system to understand the structural composition of blade and pitch bearing and assembly process of relevant parts and components, and then complete the assembly and disassembly of blade and pitch bearing of 1. 5MW wind power generating unit simulator with knowledge learned and relevant installation manuals.

Competency Objectives:

① Grasp the application of assembly tools and detection tools for wind power generating unit blades and pitch bearings;

② Grasp the diagonal tightening method of technological disassembly and assembly;

③ Have the ability to read process requirements and component assembly capabilities.

Project Environment

This operational training task mainly involves the impeller part of the fan object model of this training platform. To complete this task, you need to refer to the Equipment Instruction Manual of THWPWG-3B Large-scale Wind Power Generation System Operational Training Platform to know and understand the components and structure composition of blade and pitch system, as well as the usage specifications and precautions of relevant tools.

Ⅰ. Structure of Wind Wheel

Wind wheel, also called impeller, is mainly composed of blades, hub and pitch system, with three blades. The blade is also called paddle, and the limit switch is the limit sensor. The main components and structure of wind wheel are shown in Figure 3. 2. 1-Figure 3. 2. 3.

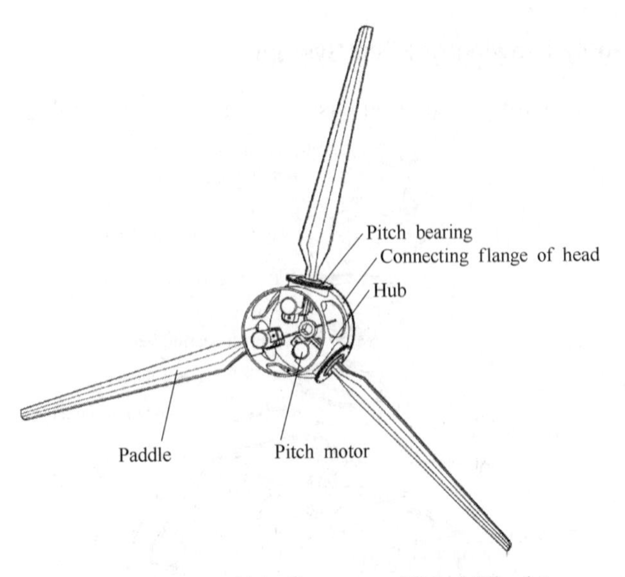

Figure 3. 2. 1 Main Structure of Wind Wheel 1

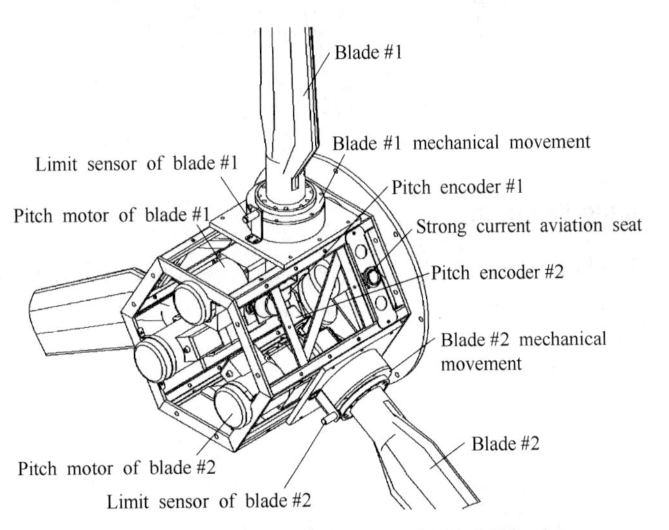

Figure 3. 2. 2 Main Structure of Wind Wheel 2

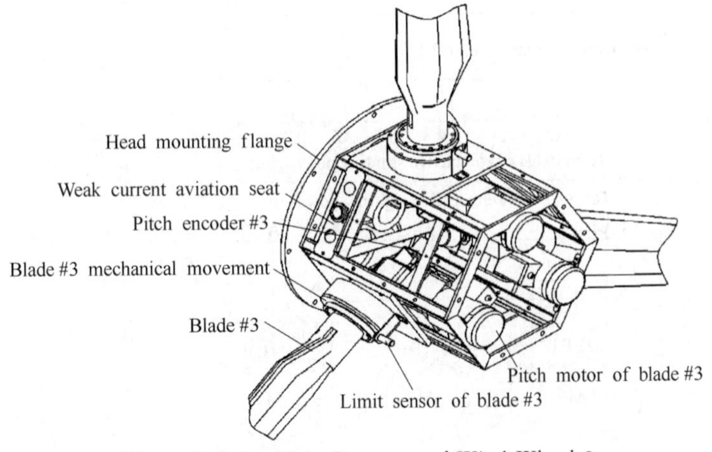

Figure 3. 2. 3 Main Structure of Wind Wheel 3

Ⅱ. General Assembly Drawing of Pitch System

Refer to Figure 3. 2. 4 for the general assembly drawing of pitch system.

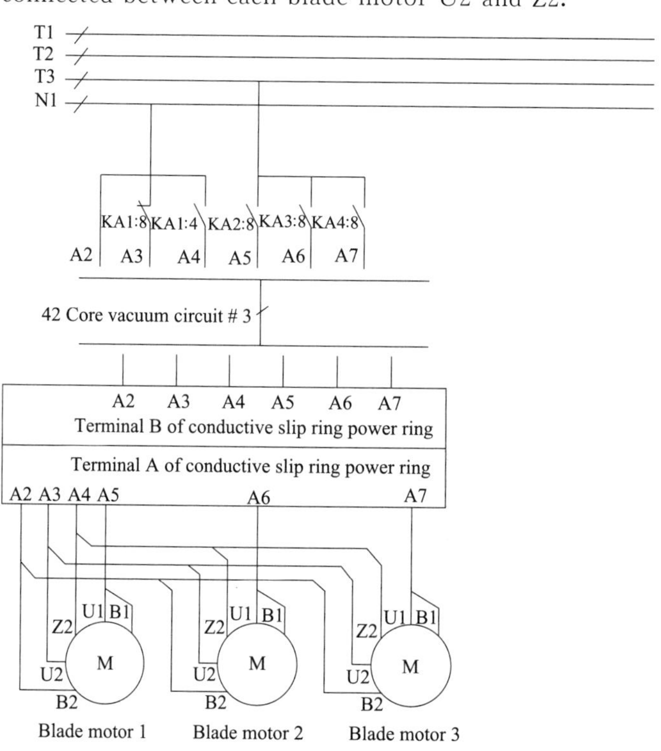

Figure 3. 2. 4 General Assembly Drawing of Pitch System

Ⅲ. Control Schematic Diagram

The principle of cybernetics for the pitch system is shown in Figure 3. 2. 5, the operating capacitance is connected between each blade motor U2 and Z2.

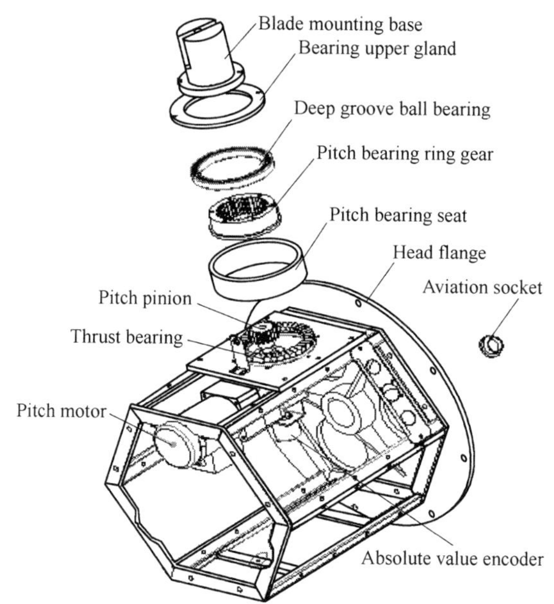

Figure 3. 2. 5 Control Schematic Diagram of Pitch System

Project Principle and Basic Knowledge

Ⅰ. Blade

1. Function of Blade

Wind power generating unit is a kind of unit and system which converts wind energy into mechanical energy and then into electric energy. The former transformation is realized by wind wheel, and the latter is realized by generator. The wind wheel mainly consists of two parts: blades (generally 3 pieces) and hub (Figure 3. 2. 6). Blades are the only critical components that convert wind energy into mechanical energy. The blade has an aerodynamic shape which determines the aerodynamic performance of the whole unit. A blade with a good aerodynamic profile can make the energy conversion efficiency of the unit higher and obtain more wind energy. The blade bears a large load (wind power and mass force) while converting energy. In nature, the wind conditions are complex and changeable, therefore, the load on the blade is also complicated. The main source of load of the whole wind power generating unit is the blade, so the blade must have sufficient strength and rigidity.

It can be seen that the material, structure and technology of blades are very critical. The material and structure guarantee the strength and rigidity of blades, and the weight shall be light, and appropriate process and method shall be provided to guarantee the production of large-sized components with complex profile and aerodynamic principle.

Figure 3. 2. 6 Blade and Hub

2. Performance Index of Blade

Main economic and technical index of blades include: Application scope of blades (type of wind power generating unit, applicable wind farm class), diameter of wind wheel, length of blade, maximum speed of rotation, weight of blade, maximum wind energy utilization coefficient of wind wheel, safe wind speed (extreme wind speed), designed service life, etc.

Example: Main technical parameters of blade of a 1. 5MW 40. 25m wind power generating unit (shown in Table 3. 2. 1).

Table 3. 2. 1 Technical Parameters of Blade of a 1. 5MW 40. 25m Wind Power Generating Unit

Environmental conditions	
Operating temperature	−30-+50℃
Ambient temperature	−40-+50℃
Lightning protection requirements	Class Ⅰ (IEC 61400-24)

Continued

System parameters	
Design grade	IEC Class Ⅲ
Number of blades	3
Working life	20 years
Diameter of wind wheel	82-83m
Direction of rotation	Clockwise(looking upwind)
Power control	Variable speed variable pitch
Operating speed range of wind wheel	≤20r/min
Rated speed of rotation	17. 3r/min
Cut-in wind speed	3m/s
Cut-out wind speed	25m/s
Rated wind speed	10-11m/s
Safe wind speed	52. 5m/s
Technical parameters of blade	
Blade length	40. 25m
Rated power	1500kW
Blade weight	6400kg
Maximum wind energy utilization coefficient C_p	0. 4935(D=82. 5m)
First order natural frequency(flapping direction)	0. 7-0. 9Hz
First order natural frequency(shimmy direction)	1. 4-1. 65Hz
Surface protection	The outer surface of the blade is painted
Position of blade center of gravity	-12. 0m(from blade root)
Blade material	GFRP(epoxy)
Maximum chord length	3. 24m
Maximum torsional angle	12. 8°
Connection mode	
Bolt hole position tolerance	ϕ0. 5mm
Diameter of bolt hole circle	ϕ1800mm
Number of bolt holes	54/64
Thread specification	M30
Outer diameter of blade cylindrical section	ϕ1895mm

3. Production Process of Blade

The quality of blades is mainly determined by the aerodynamic performance of blades, namely the maximum wind energy utilization coefficient $C_{p_{max}}$, thrust coefficient C_T, and then the structural strength and weight of blades. These key properties are not only closely related to aerodynamic shape design, performance and load calculation, material selection, structural design and strength and rigidity calculation, mold design and manufacture, but also depend on the process control in the production process. The blade structure is shown in Figure 3. 2. 7.

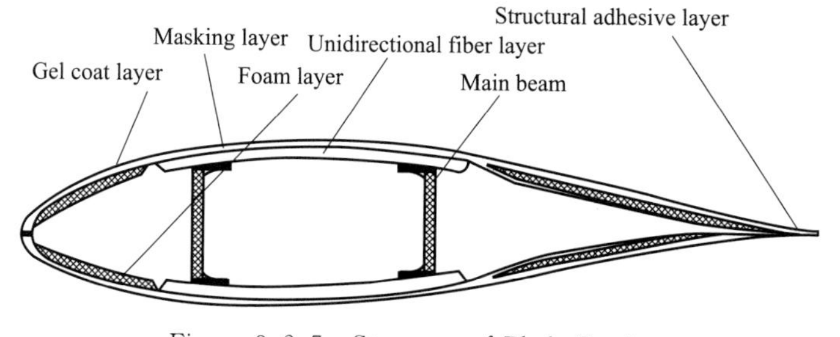

Figure 3. 2. 7 Structure of Blade Section

At present, the production process of composite material blade is to form the blade masking layer, main beam, web and other parts on each special mold and tool respectively, then these parts are glued together on the main mold, and then the integral blade is made after being mold closing pressed and solidified. The overall manufacturing process route is shown in Figure 3. 2. 8.

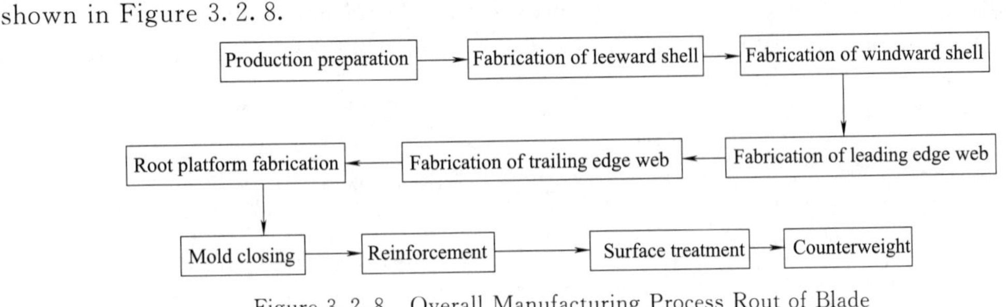

Figure 3. 2. 8 Overall Manufacturing Process Rout of Blade

The key technology in blade production is molding process, which can be roughly divided into hand lay-up molding, compression molding, pultrusion molding, fiber winding resin transfer molding, prepreg molding, vacuum suction molding, in which compression molding, resin transfer molding and vacuum suction molding is a closed mold molding process. At present, there are two types of typical molding processes that are more reliable: One is prepreg (Figure 3. 2. 9) and the other is vacuum suction (also called vacuum injection). The controllability and quality stability of prepreg molding process (Figure 3. 2. 10)

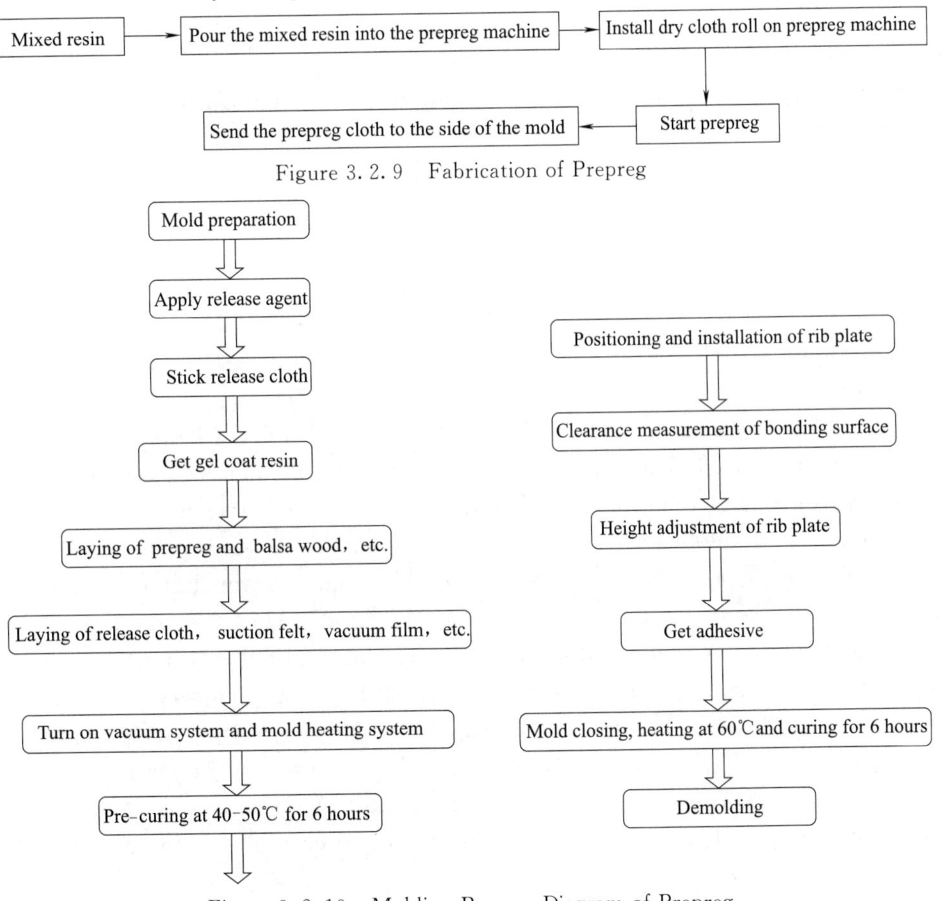

Figure 3. 2. 9 Fabrication of Prepreg

Figure 3. 2. 10 Molding Process Diagram of Prepreg

are better, but under the condition of relatively mature training of technicians and workers, the quality stability control of vacuum suction process (Figure 3.2.11) can meet the requirements, and the economy of vacuum suction process is much better than that of prepreg process. Taking the vacuum suction production process as an example, the blade production process roughly includes blanking, manufacture of main beam and wing beam, layering, vacuum suction molding, mold closing and lifting, model modification, inspection, leveling and delivery, etc.

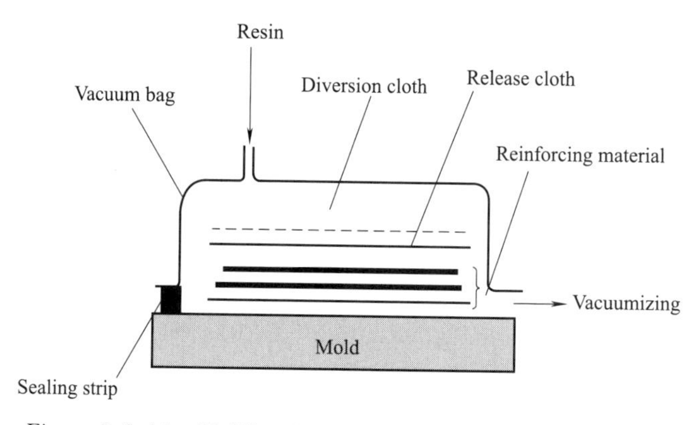

Figure 3.2.11 Molding Process Diagram of Vacuum Suction

Ⅱ. Pitch System

1. Summary

All components of the pitch system are mounted on the hub. During normal operation of the wind power generating unit, all components rotate with the hub at a certain speed.

The pitch system controls the rotation speed of the wind wheel by controlling the angle of blades, and then controls the output power of the wind power generating unit, and can safely shut down the unit by means of aerodynamic braking.

The blade (root) of the wind power generating unit is connected with the hub through the pitch bearing, and each blade shall have its own relatively independent electrically controlled synchronous pitch drive system. The pitch drive system meshes with the internal gear of the pitch bearing through a pinion.

During normal operation of the wind power generating unit, when the wind speed exceeds the rated wind speed of the unit (when the wind speed is between 12m/s and 25m/s), in order to control the power output, and limit the pitch angle between 0° and 30° (the pitch angle is automatically adjusted according to the change of wind speed), the rotation speed of the wind wheel is kept constant by controlling the blade angle. Any shutdown caused by any condition will cause the blade to feather to the 90° position (when the emergency feathering command is executed, the blade will be feathered to the 90° limit position).

The pitch system sometimes needs to be powered by backup battery for pitch operation (for example, after failure of main power supply of pitch system), so the pitch system must be equipped with backup battery to ensure safe shutdown (blade feathering to 90° limit) in case of serious failure or major accident of the unit. A redundant limit switch (for the 95° limit) is also required to ensure safe braking of the pitch motor if the main limit switch (for the 90° limit) fails.

When the standby power supply has not been used for a long time due to unit failure or

other reasons, the main controller of the wind power generating unit needs to check the status of the backup battery and the normality of the pitch operation function supplied by the backup battery.

Each pitch drive system is equipped with an absolute value encoder installed at the non-driving end (motor tail) of the motor, and a redundant absolute value encoder is installed beside the internal gear of the pitch bearing at the blade root, which records the pitch angle through a pinion meshing with the internal gear of the pitch bearing.

The main controller of the wind power generating unit receives the signals from all encoders, while the pitch system effciency only applies the signals from the encoder at the tail of the motor. Only when the encoder at the tail of the motor fails can the main controller of the unit control the pitch system to apply the signals of redundant encoders.

2. Function of Pitch System

Automatically adjust the angle between blades and wind direction according to the wind speed to realize the maximum effciency utilization of wind energy (below rated wind speed) or a constant speed (above rated wind speed), by using aerodynamic principle, the blades can be feathered 90° parallel to the wind direction, so as to shut down the wind power generating unit.

3. Composition of Main Components

Main components of pitch system is shown in Table 3. 2. 2.

Table 3. 2. 2 Main Components of Pitch System

Component Name	Quantity
Electric cabinet(central control box, shaft control box)	1 set(4 pieces)
Pitch motor(with pitch system main encoder: A encoder)	3 sets
Backup battery box	3 sets
Mechanical limit switch	3 sets(6 pieces)
Limit switch bracket related connections	3 sets
Redundant encoder: B encoder	3 sets
Redundant encoder bracket, measuring pinion and related connections	3 sets
Connecting cables and cable connectors between components	1 set

Note: With the development of wind power generation technology and power electronic technology, at present, the pitch control of some units has replaced the backup battery box with super capacitor, which is placed in the shaft control box, so there are only four control boxes in the hub. Some manufacturers not only replace the backup battery box with super capacitor, but also do not use central control boxes. The slip ring cable is directly connected to the three shaft control boxes respectively.

4. Connection of Various Components of Pitch System

Connection block diagram of various components of pictch system is shown in Figure 3. 2. 12, and mechanical connection of pictch system is shown in Figure 3. 2. 13.

(1) Central control box

The pitch central control box performs the connection between the shaft control box in the hub and the nacelle control cabinet in the nacelle, as shown in Figure 3. 2. 14.

The nacelle control cabinet provides electric energy and control signals to the pitch central control cabinet through a slipring. In addition, the connection of Profibus-DP for data exchange between the control system of the wind power generating unit and the pitch controller is also realized through this slip ring.

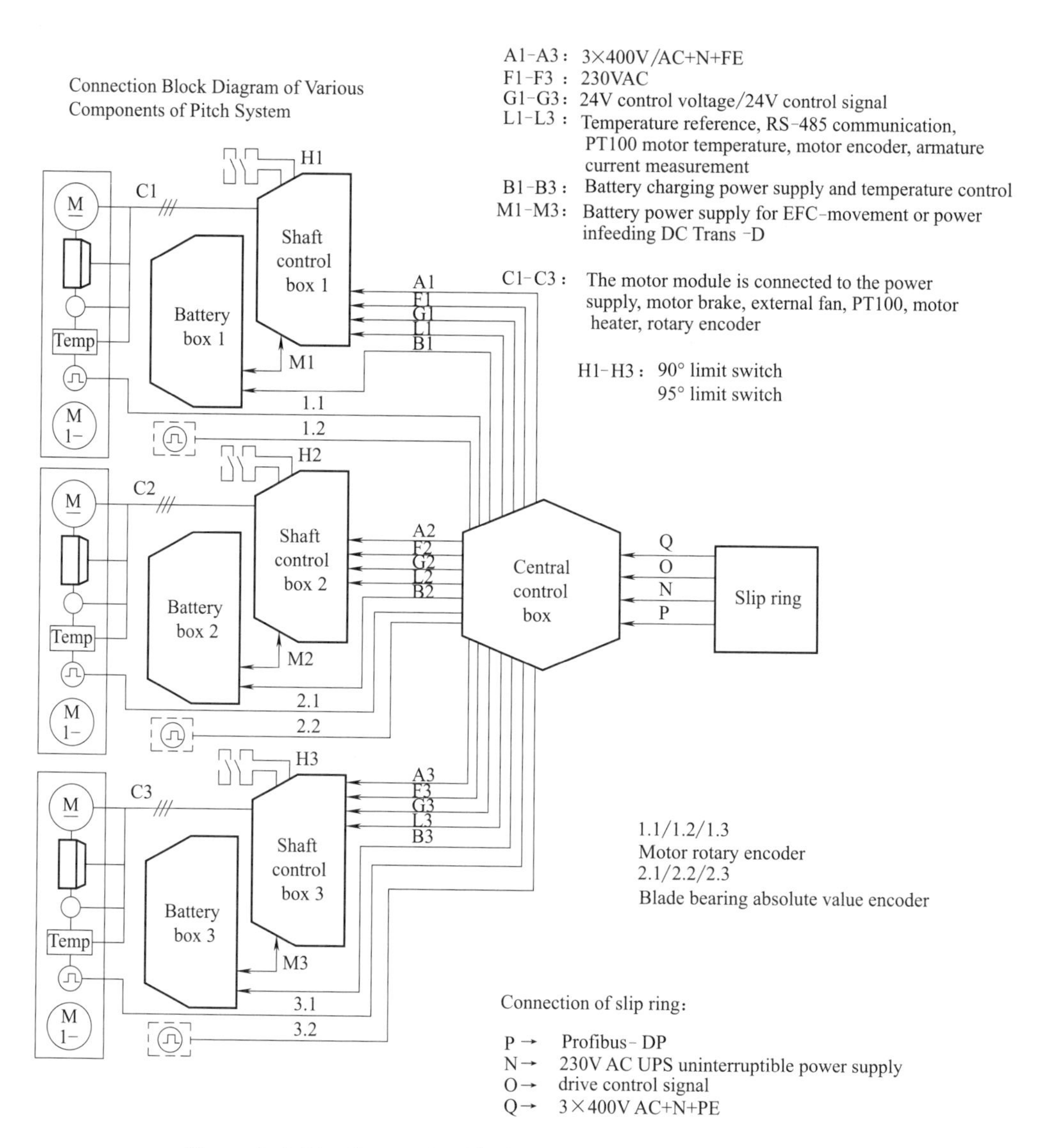

Figure 3. 2. 12 Connection Block Diagram of Various Components

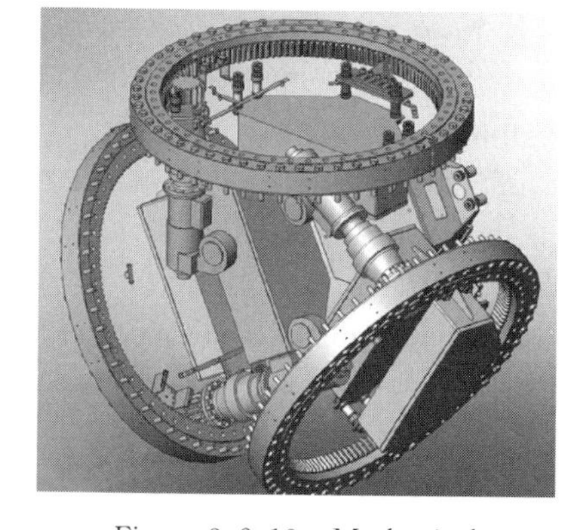

Figure 3. 2. 13 Mechanical Connection of Pitch Mechanism

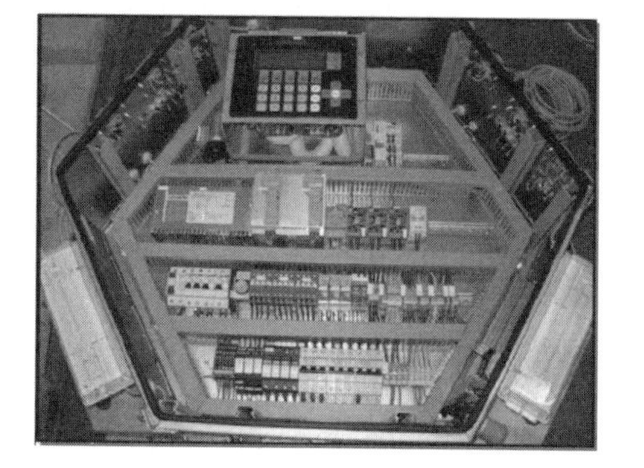

Figure 3. 2. 14 Central Control Box

The pitch controller is in the pitch central control box and is used to control the position of the blades. In addition, the charging process of the battery pack in the three battery boxes is controlled by the central charging unit installed in the pitch central control box.

There are three shaft control boxes in the pitch system, one for each blade, as shown in Figure 3. 2. 15. The converter in the box controls the speed and direction of the pitch motor.

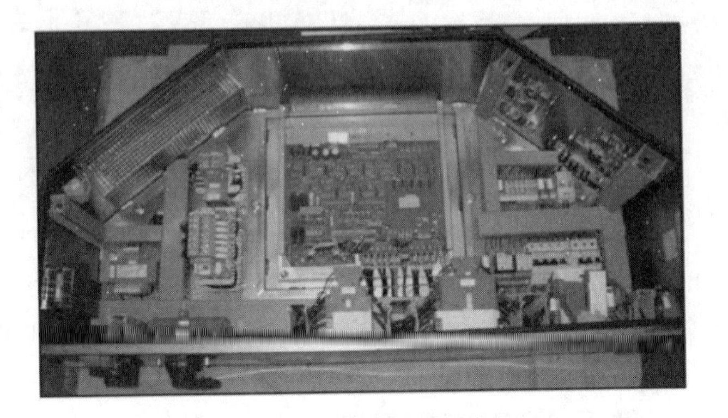

Figure 3. 2. 15 Shaft control box

(2) Battery box

As with the shaft control box, each blade is assigned a battery box, as shown in Figure 3. 2. 16. In the event of a power failure or reset of the EFC signal (emergency feathering control signal), the battery power controls the rotation of each blade to the feathered position.

(3) Pitch motor

The pitch motor (Figure 3. 2. 17) is a DC motor. Under normal conditions, the motor is controlled by the converter of the shaft control box. In case of emergency feathering, the battery-powered motor acts.

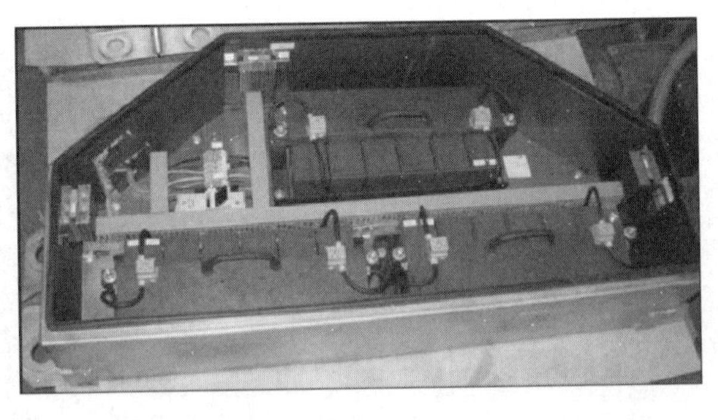

Figure 3. 2. 16 Battery Box

Figure 3. 2. 17 Pitch Motor

(4) Absolute value encoder and redundant encoder

Most large wind power generating units of MW-level and above adopt the electric pitch system. An absolute value encoder (A encoder) is installed at the rear of each pitch motor and connected to the pitch control box as the blade position feedback signal. A redundant encoder (B encoder) is installed at the blade root, which is connected to the pitch control box as a redundant feedback signal for blade position, as shown in Figure 3. 2. 18.

Figure 3. 2. 18　Absolute Value Encoder and Redundant Encoder

(5) Limit switch

Each blade corresponds to two limit switches: 90° limit switch and 95° limit switch. The 95° limit switch is used as the redundant switch, as shown in Figure 3. 2. 19.

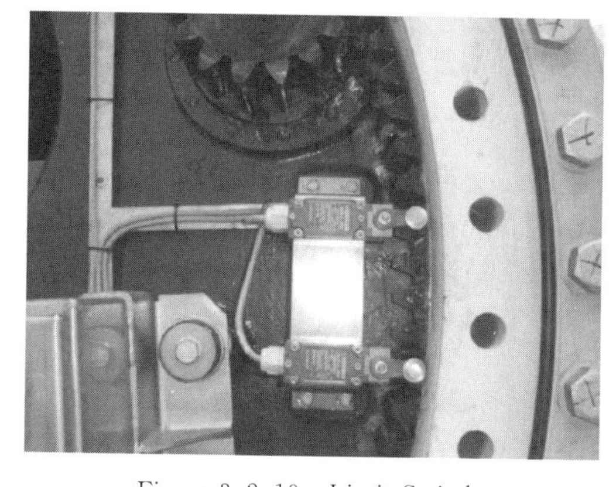

Figure 3. 2. 19　Limit Switch

(6) Connecting cable between components

The pitch central control box, shaft control box, battery box, pitch motor, redundant encoder and limit switches are connected by cables. To prevent confusion when connecting cables, cables have their own number.

Project Implementation

Ⅰ. List of Instruments, Equipment and Tools

① Fan object model of THWPWG-3B large-scale wind power generation system operational training platform;

② One set of hexagonal wrench;

③ One needle nose pliers;

④ One cross screwdriver and one flat screwdriver respectively.

Ⅱ. Safe Operation Specification

① Ensure that all power supplies are turned off;

② When assembling and disassembling blade and pitch bearing, use ladder to carry out

aerial work. Ladder must be stably placed on flat ground;

③ When getting up and down the ladder and standing on the ladder for operation, step on firmly, concentrate on the operation, and do not play;

④ Work in a group of 4 during operation. 2 peoples shall stand on 2 ladders to cooperate in assembly and disassembly, and the other 2 peoples shall be responsible for safety protection and transferring tools and parts on the ground;

⑤ Beware of falling and missing your step. Beware of falling tools or parts. Tools and parts shall not be placed on the equipment;

⑥ The personnel responsible for protection shall not stand under the parts to be removed;

⑦ Pay attention to the placement of parts and tools during disassembly and assembly. The parts and tools shall be classified and placed neatly without interfering with operation. Special attention shall be paid to the placement of connecting bolts, nuts, gaskets and other small parts to avoid loss and damage.

Ⅲ. Operational Training Steps

① The two peoples in charge of assembly and disassembly shall correctly use tools and remove the blade, blade mounting base, pitch limit switch, bearing upper gland, deep groove ball bearing, pitch bearing ring gear, pitch bearing base, pitch pinion and other parts in turn according to the diagonal loosening method. Pay attention to the angle position of the blade when removing the blade;

② Assemble and restore the removed parts according to the diagonal tightening method. Note that the angle position of the blade shall also be restored;

③ During the assembly and disassembly of the above two peoples, the other 2 peoples shall be responsible for holding the ladder, paying attention to the safety of the personnel working on the ladder and themselves at any time, and delivering the parts of the tool box for the personnel in charge of assembly and disassembly, as shown in Figure 3. 2. 20;

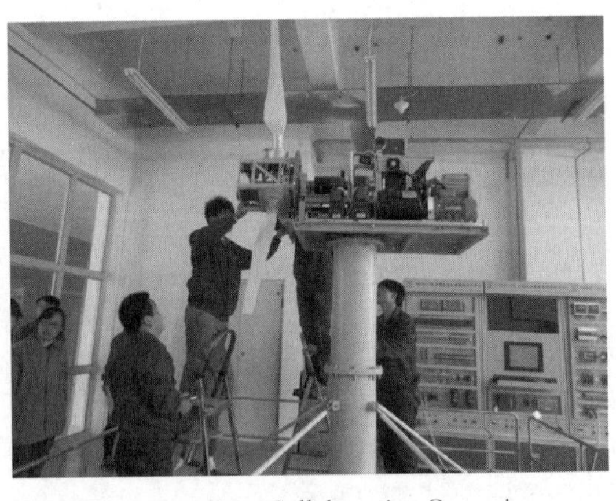

Figure 3. 2. 20 Collaborative Operation

④ Two teams exchange tasks and reassemble again;

⑤ Write training report.

Project Work

(1) Describe what is diagonal tightening method according to personal experience dur-

ing training operation.

(2) What are the advantages of super capacitor compared with backup battery box (storage battery)?

Project Ⅲ Installation and Commissioning of Energy Storage System

Project Description

Energy conversion storage control system is mainly composed of DC voltage and current sampling module, temperature alarm module, PWM driving module, CPU core module, human-computer interaction module, communication module, lightning protector, intelligent charge and discharge controller, battery pack, switching power supply, DC voltmeter and DC ammeter, etc.

Competency Objectives:

① Have the ability to master the principle of each functional module;

② Have the ability to understand circuit schematic diagram and wiring diagram;

③ Have the ability to realize debugging after the completion of system installation.

Project Environment

The energy storage control unit is mainly composed of DC voltmeter, ammeter, circuit breaker, switching power supply, MPPT controller, battery pack and intelligent charge and discharge controller.

Ⅰ. Function Description of Each Module

① DC voltmeter and ammeter: Monitoring electric quantity parameters, such as fan output voltage, fan output current, etc.

② Circuit breaker: It mainly completes the disconnection and connection of each circuit branch.

③ Connection module: It plays a connection role between the control object and the controller, facilitating the plugging and unplugging of the connecting wire.

④ Switching power supply: Provide stable DC power supply for circuits and equipment.

⑤ MPPT controller: It mainly completes the maximum power tracking (MPPT) algorithm.

⑥ Battery pack: When there is no wind, it can supply electric power, or act as a buffer when the generated power changes suddenly, store electric energy, and electric peak regulation etc. to expand the application scope of the system, so as to improve the additional economic value of the grid-connected system.

⑦ Intelligent charge and discharge controller: Adjust the charging state and current according to the battery voltage, so as to prevent the battery from over-charging or over-discharging and prolong the service life of the battery.

Its outline is shown in Figure 3.3.1.

See Table 3.3.1 for technical parameters of main equipment of energy conversion storage control system.

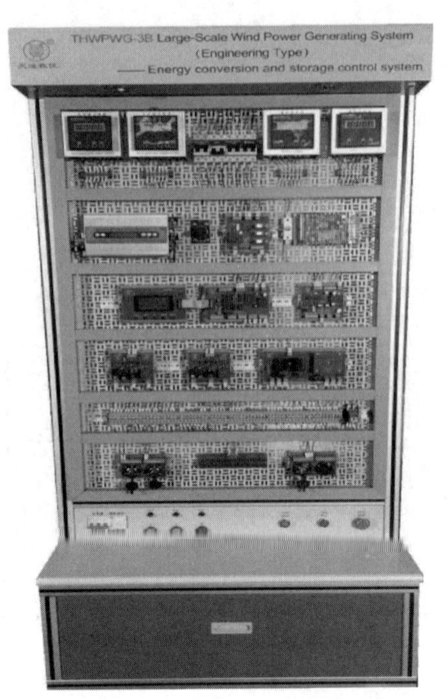

Figure 3. 3. 1 Control Cabinet of Energy Conversion Storage Control System

Table 3. 3. 1 Technical Parameters of Main Equipment of Control Cabinet of Energy Conversion Storage Control System

Serial Number	Name	Specification	Quantity
1	Control Cabinet of Energy conversion storage control system	880mm×620mm×2120mm	1 set
2	DC voltage and current sampling module	Input:0-60V,0-5A output:0-5V	1 piece
3	PWM driving module	PWM wave frequency:19. 2kHz Duty ratio adjustment range:0-90%	1 piece
4	Human-computer interaction module	Resolution:128×64	1 piece
5	Intelligent charge and discharge controller	Power:600W Rated voltage of battery:12/24V automatic switching	1 set
6	Battery pack	Voltage:12V capacity:24A • h	4 pieces
7	Switching power supply	Rated input voltage:AC 220V Rated output voltage:DC 24V Rated power:35W	1 piece
8	DC voltmeter	Input voltage range:0-500V Accuracy:0. 5%±5 words Communication:RS 485 communication interface	2 pieces
9	DC ammeter	Input current range:0～5A Accuracy:0. 5%±5 words Communication:RS 485 communication interface	2 pieces

Ⅱ. Circuit Block Diagram

Circuit block diagram of energy storage control unit is shown in Figure 3. 3. 2.

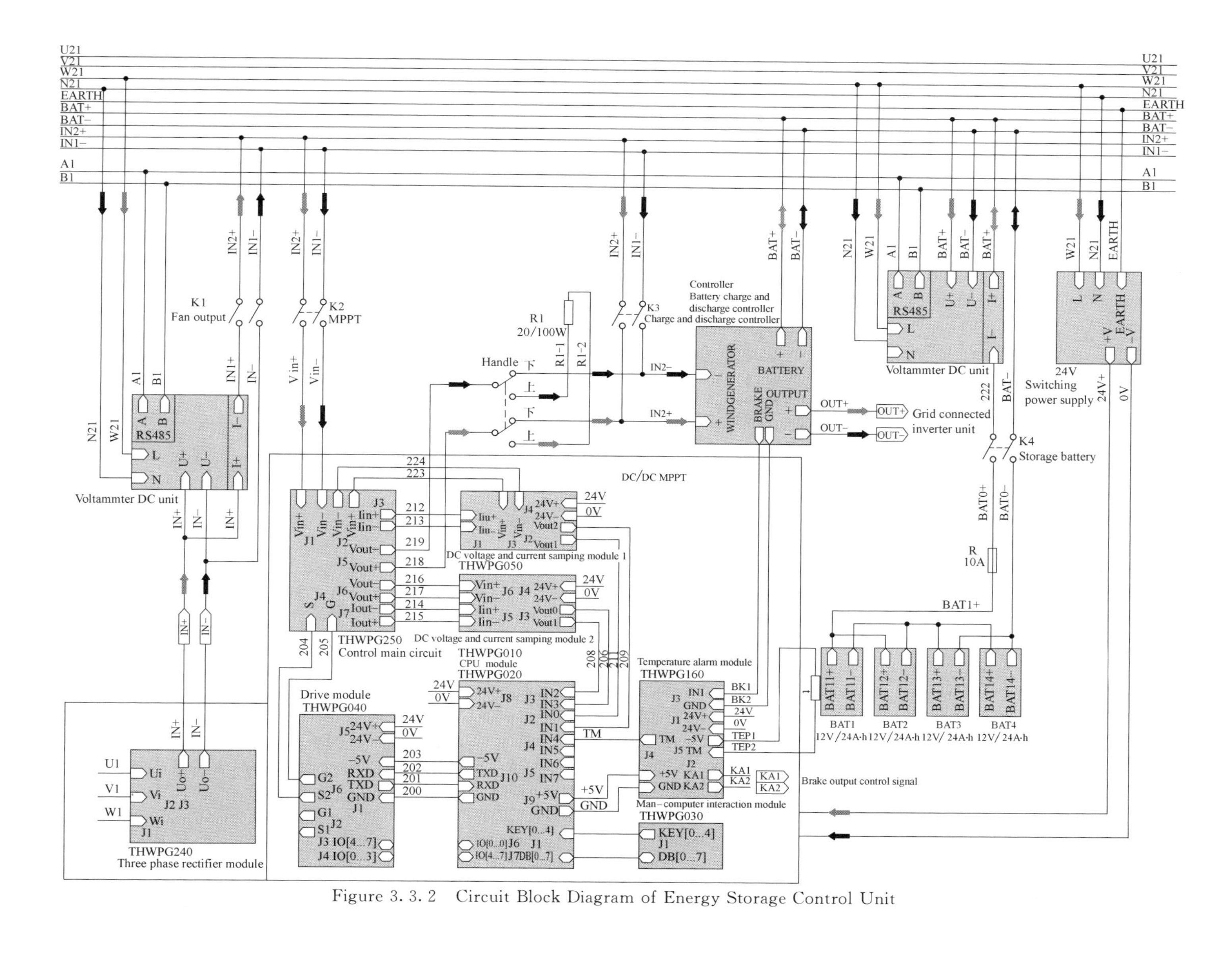

Figure 3. 3. 2 Circuit Block Diagram of Energy Storage Control Unit

Project Principle and Basic Knowledge

Ⅰ. DC Unit Module

DC voltmeter and ammeter: Monitoring electric quantity parameters, such as fan output voltage, fan output current, etc.

(1) Main Functions

① Measurement and display of DC voltage (current).

② RS485, serial output, analog quantity output and upper and lower limit alarm and control (contact output).

③ The address, baud rate, display value, upper and lower limit alarm threshold and return difference can be set manually or by the upper computer.

(2) Technical Index

① Features.

Modular design, single display, display and other functions can be selected according to user needs.

It adopts the latest PIC chip and has strong anti-interference capability.

② General instructions.

Enclosure material: Flame retardant plastics.

Isolation: Power/input/output isolated from each other.

③ Performance index.

Maximum display: ±19999.

Display resolution: 0.001.

Maximum input range: Voltage: DC 0-600V current: DC 0-5A.

Input mode: Single-ended input.

Accuracy class: Class 0.5.

Absorbed power: <1.2V • A.

Measurement speed: About 5 times/sec.

Output load capacity: ≤300Ω.

Power supply: AC 220V±20%, 50/60Hz.

Power frequency withstand voltage: Between power/input/output: AC 2kV/min • 1mA.

Burst interference (EFT): 2kV 5kHz.

Surge impulse voltage: 2kV 1.2/50μs.

Insulation resistance: ≥100MΩ.

(3) Key Operation Instruction

See Table 3.3.2 for key operation instruction.

Table 3.3.2 Key Operation Instruction Table

Display Character	Corresponding Definition	Data Range
PASS	Accuracy adjustment	0-9999
BAUd	Baud rate	1200,2400,4800,9600,19200
dISP	Display value	0-19999
ADDr	Local address	1-32

Continued

Display Character	Corresponding Definition	Data Range
rHNu	Relay output upper threshold	When upper limit threshold > lower limit threshold, alarm outside the interval When upper limit threshold < lower limit threshold, alarm in the interval When upper limit threshold = lower limit threshold, turn off the alarm function
rLHu	Relay output lower threshold	
SAVE	save and exit	
E	Exit without saving	
r_rE	Return difference	
DECi	Decimal point position	

Depending on the selected function, the menu will increase or decrease accordingly. set, ← and → keys are available. Press "set" key to enter the main menu of debugging program, LED displays PASS, press "←" key to display dISP, DECi, SAVE and E, press "→" key to display reversely and circularly, press "←" or "→" key to the item to be set, and press "set" key to enter modification menu.

When the following items are adjusted to the required values, press the "set" key to return to the main menu.

① dISP: LED displays 00000 (if the decimal point display bit has been adjusted, the decimal point will be displayed in the corresponding position). At the same time, the modified bit flashes, press "→" to add 1 to the modified bit, press "←" key to move the modified bit one digit to the left. In this way, adjust it to the required value (the displayed value corresponds to the value displayed when full input, for example: Enter DC 100V to set the display to 100. 00).

② DECi: LED displays the current decimal point position, press "→" key to move the decimal point one digit to the right, move to the last digit, the unit indicates to flip (ignore if there is no unit display), press "←" key. Move the decimal point one digit to the left to the highest, the unit indicates to flip (ignore if there is no unit).

③ Local address (ADDr): The LED flashes to display the local address. Press the "←" key to increase the value by 1, press the "→" key to decrease the value by 1, and adjust it to the corresponding value (1-32).

④ Baud rate (Baud): The LED flashes to display the current set baud rate (1200-19200), press "←" key or "→" key to change the value.

⑤ rHNu/rLNu: The LED displays the current operation threshold, and the modified bit flashes at the same time. Press "→" key to plus one by modified bit, and press "←" key to move the modified bit one digit to the left. In this way, adjust it to the actual operation threshold.

⑥ Return difference setting of relay operation (r _ rE): The LED displays the return difference of current operation (the return difference value is the last two digits of actual measurement). At the same time, the modified bit flashes. Press "→" key to plus one by modified bit, and press "←" key to move the modified bit one digit to the left. In this way, adjust it to the actual operation return difference.

⑦ Accuracy adjustment (PASS): Display accuracy adjustment. Corresponding password is required to modify accuracy without user adjustment.

Press "set" while S is displayed to save the changes and exit. Press "set" while E is dis-

played to ignore modification and exit directly. Note: Under any menu level, after pressing the key, the system will automatically exit the setting menu after no operation time is longer than 60 seconds.

(4) Physical Picture

DC unit module is shown in Figure 3.3.3.

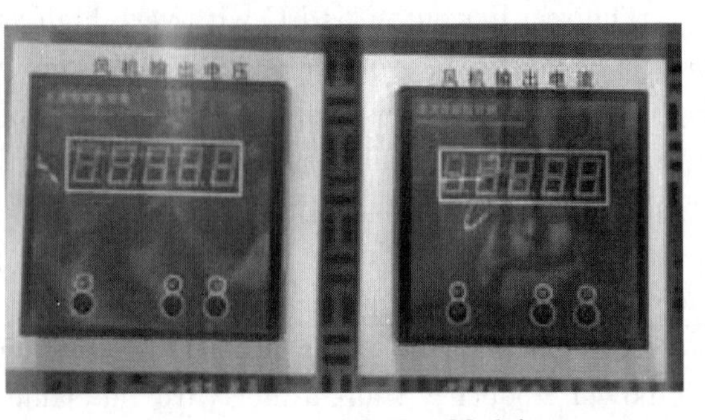

Figure 3.3.3 DC Unit Module

(5) Port Definition

See Table 3.3.3 and Table 3.3.4 for the port definition of DC voltmeter and DC ammeter.

Table 3.3.3 DC Voltmeter

Serial Number	Definition	Description
1	U+	Measured voltage input
2	U−	
3	A	RS485
4	B	
5	L	AC 220V power input
6	N	

Table 3.3.4 DC Ammeter

Serial Number	Definition	Description
1	I+	Measured current input
2	I−	
3	A	RS485
4	B	
5	L	AC 220V power input
6	N	

Ⅱ. Switching Power Supply

Modern power technology is used to control the on/off time ratio of switching transistor to maintain a stable output voltage, typically consisting of pulse width modulation (PWM) control IC and MOSFET. The switching power supply (Figure 3.3.4) used in this system provides +24V DC power supply for MPPT controller function modules.

Figure 3.3.4 Switching Power Supply

(1) Features

Switch: Power electronic devices operate in switch state.

High frequency: Power electronic devices operate at high frequency.

DC: The output of switching power supply is DC.

(2) Structural Composition

Most commonly used switching power supplies DC power for electronic equipment. The DC voltage required by electronic equipment is usually several volts to more than ten volts, while the voltage supplied by AC electric supply is 220V (110V) and the frequency is 50Hz (60Hz). The function of switching power supply is to convert an AC power frequency of a high voltage level into a DC power of a low voltage level. The AC power frequency is rectified directly after entering the switching power supply, thus eliminating the large volume and weight of the power frequency rectifier transformer.

The rectifier output is DC with very high voltage. The rectified voltage is filtered by capacitor, the average voltage is 300-310V. The DC power of high voltage level is sent to the input end of the inverter, and transformed into high voltage and high frequency AC by the inverter. At present, the conversion working frequency of the inverter of switching power supply is several tens kHz to several hundred kHz. The AC power output by the inverter can be connected to the primary side of the high frequency step-down transformer. Since the frequency of the high frequency AC power generated by the inverter is much higher than the power frequency, the volume of the high-frequency transformer is much smaller than that of the power frequency transformer with the same capacity, which fundamentally reduces the volume and weight of the whole power supply.

After the high frequency AC generated by the inverter is stepped down by the high-frequency transformer, and the low voltage DC energy conforming to the load requirements is transformed through rectification and voltage stabilization, which is supplied to the load. It can be seen from the structure of switching power supply that there is no compensating pipe in the circuit, so no extra energy will be consumed. All electronic devices work in switch state. If the conduction voltage drop of switching device and stray resistance of circuit are ignored, the efficiency of circuit shall be 1. The power frequency rectifier circuit is generally an uncontrolled rectifier circuit. According to the power capacity, it can be single-phase rectifier. Generally, single-phase bridge structure is adopted. Three-phase AC power supply can be used for large capacity switching power supply.

(3) Classification

In the technical field of switching power supply, the development of related power electronic devices and the development of switching frequency conversion technology are at the same time, both two promote the development of switching power supply toward light, small, thin, low noise, high reliability and anti-interference with the growth rate of more than two digits each year.

Switching power supply can be divided into AC/DC and DC/DC. DC/DC converter has realized modularization now, and its design technology and production process have been mature and standardized at home and abroad, and have been recognized by users. However, the modularization of AC/DC has encountered more complicated technical and process manufacturing problems due to its own characteristics.

(4) Selection

In terms of input anti-interference performance of switching power supply, it is difficult to pass general input interference such as surge voltage due to multi-stage series connection. Compared with linear power supply, the technical index of output voltage stability has a greater advantage, which can reach (0.5-1)%. As a power electronic integrated device, switching power supply shall be selected with attention to:

① Selection of output current. Since the working efficiency of switching power supply is high, generally above 80%, the maximum absorbed current of the electric equipment shall be accurately measured or calculated for the selection of its output current, so that the selected switching power supply has a high cost performance ratio. The output calculation formula is usually as follows: $I_s = KI_f$. Where: I_s—rated output current of switching power supply; I_f—maximum absorbed current of electric equipment; K—margin coefficient, generally 1.5-1.8.

② Earthing. Switching power supply will generate more interference than linear power

supply. For electric equipment sensitive to common mode interference, earthing and shielding measures shall be taken. Electromagnetic compatibility (EMC) measures shall be taken for switching power supply according to EMC limits such as ICE1000, EN61000, FCC, etc. Therefore, switching power supply shall generally be equipped with EMC filter. For example, the HA series switching power supply of Harvest can meet the above electromagnetic compatibility requirements only by connecting its FG terminal to the earth or to the user's enclosure.

③ Protection circuit. Switching power supply shall be designed with overcurrent, overheating and short circuit protection functions. Therefore, switching power supply with complete protection function shall be preferred in design, and technical parameters of protection circuit shall match with working characteristics of electric equipment to avoid damage to electric equipment or switching power supply.

(5) Technical Index

Input: 115V AC 0.8A; 230V AC 0.45A; 50/60Hz.

Output: +24V 1.5A.

Size: 130mm×98mm×36mm.

(6) Port Definition

See Table 3.3.5 for the port definition of switching power supply.

Table 3.3.5 Port Definition Table

Serial Number	Definition	Description	Serial Number	Definition	Description
1	L	AC power input	4	V−	0V
2	N		5	V+	+24V output
3	PE	Ground wire			

Ⅲ. Sealed Lead-Acid Battery

1. Structural Features

Although this battery is also lead-acid battery, it has many advantages compared with the original lead-acid battery, especially favored by users who need to install battery supporting equipment together.

This is because VRLA battery is completely sealed, will not leak acid, and will not release acid mist when charging and discharging like old lead-acid battery, which will corrode equipment and pollute environment, so from the structural characteristics people also call VRLA battery closed (sealed) lead-acid battery. For the sake of distinction, the old lead-acid battery is called open lead-acid battery. Because VRLA battery is not only completely sealed from the structure, but also has a valve that can control the gas pressure inside the battery, the full name of VRLA lead-acid battery is "valve-regulated closed lead-acid battery".

2. Technical Parameters

The electrical properties of lead-acid battery are measured by the following parameters: battery electromotive force, open-circuit voltage, cut-off voltage, working voltage, discharge current, capacity, internal resistance, service life (floating charge life, charge-discharge cycle life), energy, storage performance, etc.

(1) Battery electromotive force, open-circuit voltage, working voltage

When the battery conductor is externally connected, the electrochemical reaction of the

positive electrode and the negative electrode proceeds spontaneously. If the conversion of electric energy and chemical energy in the battery is balanced, the difference between the equilibrium electrode potential of the positive electrode and that of the negative electrode is the battery electromotive force, which is numerically equal to the open-circuit voltage at which the stable value is reached.

The product of battery electromotive force and unit electric quantity, indicating the maximum electric work that can be done by unit electric quantity. However, battery electromotive force and open-circuit voltage have different meanings. The battery electromotive force can be calculated by thermodynamics or through measurement according to the reaction in the battery, with clear physical significance; the terminal voltage of the battery in open-circuit state is called open-circuit voltage. The open-circuit voltage of the battery is equal to the difference between the positive electrode potential and the negative electrode potential of the battery. The battery working voltage is the terminal voltage through which the battery has current flowing (closed circuit). The working voltage at the initial discharge of the battery is called initial voltage. After the battery is connected to the load, the working voltage of the battery is lower than the open-circuit voltage due to ohmic resistance and polarization overpotential.

(2) Capacity

Battery capacity refers to the amount of energy stored in the battery and is indicated by symbol C. The commonly used unit is ampere-hour (A • h) or milliampere-hour (mA • h). Battery capacity can be divided into rated capacity (nominal capacity) and actual capacity.

Discharge rate: The discharge rate is divided into time rate and current rate according to the discharge current of battery. The discharge time rate refers to the length of the voltage time from the discharge to the end under certain discharge conditions. According to IEC standard, discharge time rates have 20, 10, 5, 3, 1, 0.5 hour rate, etc, respectively expressed as: 20Hr, 10Hr, 5Hr, 3Hr, 2Hr, 1Hr, 0.5Hr, etc.

Discharge cut-off voltage: Lead-acid battery discharged at a certain discharge rate, the lowest voltage that can be recharged at 25℃ ambient temperature is called the discharge cut-off voltage. Most fixed batteries specify an cut-off voltage of 1.8V/battery at 10Hr discharge (25℃). The value of the cut-off voltage depends on the discharge rate and demand. In general, in order to make the battery operate safely, the value of the cut-off voltage is slightly higher when discharging at small current less than 10Hr, and the value of the cut-off voltage is slightly lower when discharging at large current larger than 10Hr. In the communication power supply system, the cut-off voltage of battery discharge is determined by the basic voltage requirements of the communication equipment.

Discharge current rate: It is designed to compare the discharge current of battery with different nominal capacity. Generally, 10Hr rate current is taken as the standard, represented by I10, while 3Hr rate and 1Hr rate discharge current are respectively represented by I3 and I1.

Rated capacity: Fixed lead-acid batteries are specified as rated capacity that can be reached by discharging to the cut-off voltage at a rate of 10Hr at 25℃. The rated capacity at 10Hr rate is represented by C10. The current value at 10Hr rate is C10/10. Capacity at other hour rates is expressed as follows: The 3Hr rate capacity (A • h) is represented by C3. The measured capacity (A • h) is the product of discharge current and discharge time (h) at the ambient temperature of 25℃. C3 and I3 of valve-regulated lead-acid fixed battery shall be C3=0.75 C10 (A • h), I3=2.5 I10 (A), one hour constant capacity (A • h) shall be re-

presented by C1, and the measured value of C1 and I1 shall be C1=0.55 C10 (A • h), I1=5.5 I10 (A).

Actual capacity: The battery output power under certain conditions. It is the product of discharge current and time in A • h.

(3) Internal resistance

The battery internal resistance includes ohmic internal resistance and polarization internal resistance, and polarization internal resistance includes electrochemical polarization and concentration polarization. The internal resistance makes the terminal voltage lower than the battery electromotive force and open-circuit voltage when discharging, and the terminal voltage is higher than the battery electromotive force and open-circuit voltage when charging. The internal resistance of the battery is not constant and varies with time during charging and discharging because the composition of the active substance, electrolyte concentration and temperature are constantly changing. Ohmic resistance obeys Ohm's law; polarization resistance increases with increasing current density, but not linearly, usually linearly with increasing logarithm of current density.

(4) Cycle life

The battery undergoes a charge and discharge, which is called a cycle. Under certain discharge conditions, the number of cycles the battery can withstand before operating to a specified capacity value is called cycle life. The cycle times of various batteries are different. The traditional fixed lead-acid battery is 500-600 times, and the starting lead-acid battery is 300-500 times. The cycle life of valve-regulated sealed lead-acid battery is 1000-1200 times. Factors affecting cycle life: The first is the performance of the manufacturer's products and the second is the quality of maintenance work. The life of fixed lead battery can also be measured by floating charge life (year). For starting lead-acid battery, according to the standards issued by China, the unit number of over-charging endurance capacity and cycle endurance capacity is used to express the life, but not the cycle number.

(5) Energy

The energy of a battery refers to the electric energy that the battery can give under a certain discharge regime, usually expressed in watt-hour (W • h). Battery energy is divided into theoretical energy and actual energy. The theoretical energy Wtheory can be expressed by the product of theoretical capacity and battery electromotive force (E), i. e., $W_{theory}=C_{theory}E$. The actual energy of the battery is the product of the actual capacity "C_{actual}" and the average working voltage "$U_{average}$" under certain discharge conditions, i. e., $W_{actual}=C_{actual}U_{average}$.

Specific energy is often used to compare different battery systems. Specific energy refers to the electric energy output per unit mass or volume of a battery in W • h/kg or W • h/L, respectively. Specific energy can be divided into theoretical specific energy and actual specific energy. The former refers to the theoretical energy that can be output when 1kg of battery reaction substance is completely discharged. The actual specific energy is the actual energy that can be output by 1kg of battery reaction substance. Because of various factors, the actual specific energy of the battery is far less than the theoretical specific energy.

The relationship between the actual specific energy and the theoretical specific energy can be expressed as follows:

$$W_{actual}=W_{theory} \cdot K_V \cdot K_R \cdot K_m$$

Where K_V is Voltage efficiency; K_R is Reaction efficiency; K_m is Mass efficiency. Voltage

efficiency refers to the ratio of the working voltage of the battery to the battery electromotive force. When the battery is discharged, the working voltage is less than the battery electromotive force due to electrochemical polarization, concentration polarization and ohmic voltage drop. The reaction efficiency indicates the utilization rate of the active substance. The specific energy of battery is a comprehensive index, which reflects the quality level of battery, and also indicates the technology and management level of the manufacturer.

(6) Storage performance

During storage of the battery, there are impurities in the battery, which can be combined with the negative active substance to form a microbattery, resulting in dissolution of negative electrode metal and release of hydrogen. Another example, if the standard electrode potential of impurity dissolved from the positive electrode grid in solution is between the positive and the negative electrode potential, it will be oxidized by the positive electrode and reduced by the negative electrode. Therefore, the existence of harmful impurities causes the active substance of positive electrode and negative electrode to be gradually consumed, resulting in battery capacity loss, which is called self-discharge. Battery self-discharge rate is expressed as a percentage of decrease in content per unit time: That is, the percentage of the capacity difference before storage (C10′)(C10″) and the storage time T (day, month).

3. Technical Index

Single battery capacity: 12V 24A • h. single battery size: 165mm×125mm×175mm.

4. Physical Picture

Battery pack is shown in Figure 3. 3. 5.

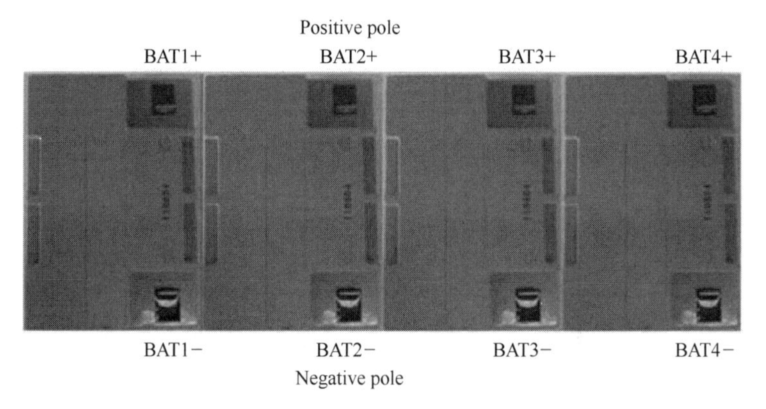

Figure 3. 3. 5 Battery Pack

5. Port Definition

See Table 3. 3. 6 for the port defintion of battery pack.

Table 3. 3. 6 Port Definition Table

Number	Definition	Description	Serial Number	Definition	Description
1	BATX+	Battery positive output	2	BATX−	Battery negative output

Ⅳ. MPPT Controller

MPPT controller tracks the maximum power of the fan, which improves the working efficiency of the fan effectively and improves the working performance of the system at the same time, as shown in Figure 3. 3. 6.

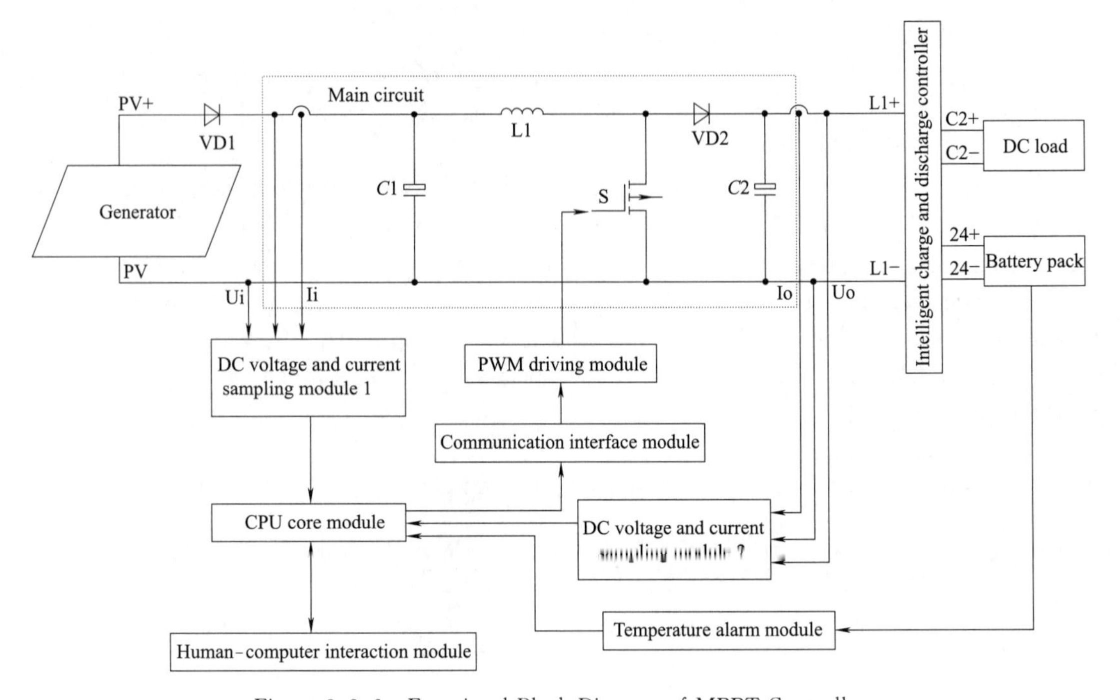

Figure 3. 3. 6 Functional Block Diagram of MPPT Controller

1. Description of Each Function Module of MPPT Controller

(1) DC voltage and current sampling module 1

Through voltage Hall sensor and current sensor, the output voltage and current of fan array battery are converted into voltage and current signal satisfying the requirements of the input terminal of single chip microcomputer.

(2) Human-computer interaction module

Input and output terminals of CPU core module.

(3) Over temperature alarm module

Check the battery temperature. If the temperature exceed the reference temperature, the LCD will display that the battery temperature is too high.

(4) PWM driving module

Receive the power regulation parameter (duty ratio parameter) output by the maximum power tracking microprocessor and output PWM signals with different duty ratios, isolate the PWM signals output by the PWM microprocessor from the DC/DC circuit, convert the PWM signals into driving signals satisfying the switching tube requirements, and improve the driving capability.

(5) DC voltage and current sampling module 2

Through voltage Hall sensor and current sensor, the output voltage and current are converted into voltage and current signal satisfying the requirements of the input terminal of the single chip microcomputer.

(6) Main circuit

The duty ratio of the main circuit can be adjusted by collecting the power of the fan array battery in real time, and the impedance of the load can be adjusted equivalently, so that the power taken by the load can be changed, which can always follow the maximum power output point of the fan array battery, equivalent to an impedance converter.

(7) Intelligent charge and discharge controller

Adjust the charging state and current according to the battery voltage, so as to prevent the battery from over-charging or over-discharging and prolong the service life of the battery.

2. CPU Core Module

Conduct program design and debugging of maximum power tracking according to voltage and current sampling signals of fan array. This part of hardware is completely open to users. Users can write different MPPT algorithms to realize maximum power tracking, and send regulation parameters (duty ratio parameters) to PWM driving module through serial port for adjustment. The microprocessor adopts 51 series single chip microcomputer with on-line download function, which is convenient for users to program and debug and realize MPPT control algorithm.

(1) Schematic Diagram

The principle of CPU core module is shown in Figure 3. 3. 7.

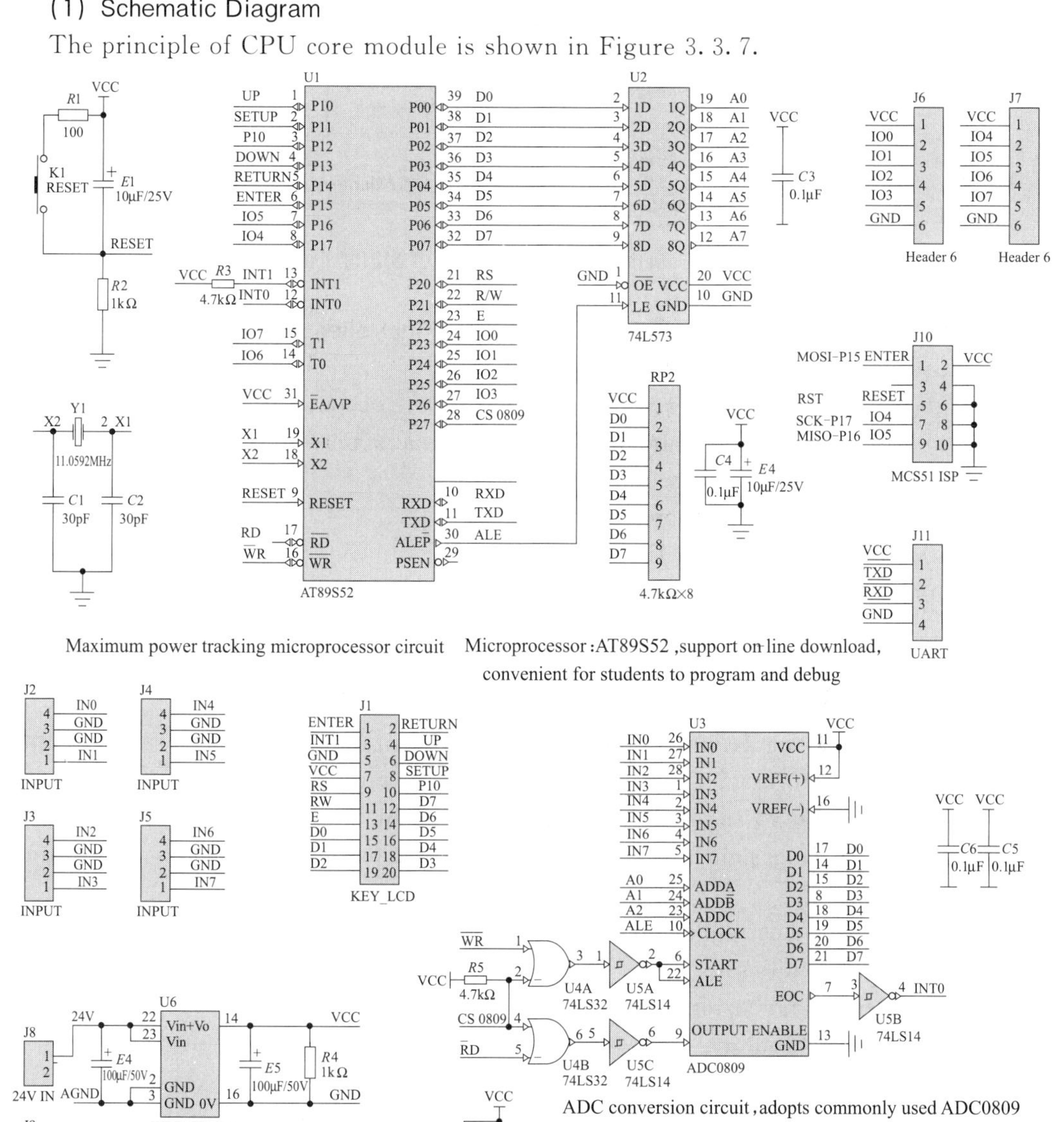

Figure 3. 3. 7 Schematic Diagram of CPU Core Module

(2) Physical Picture

CPU core module is shown in Figure 3. 3. 8.

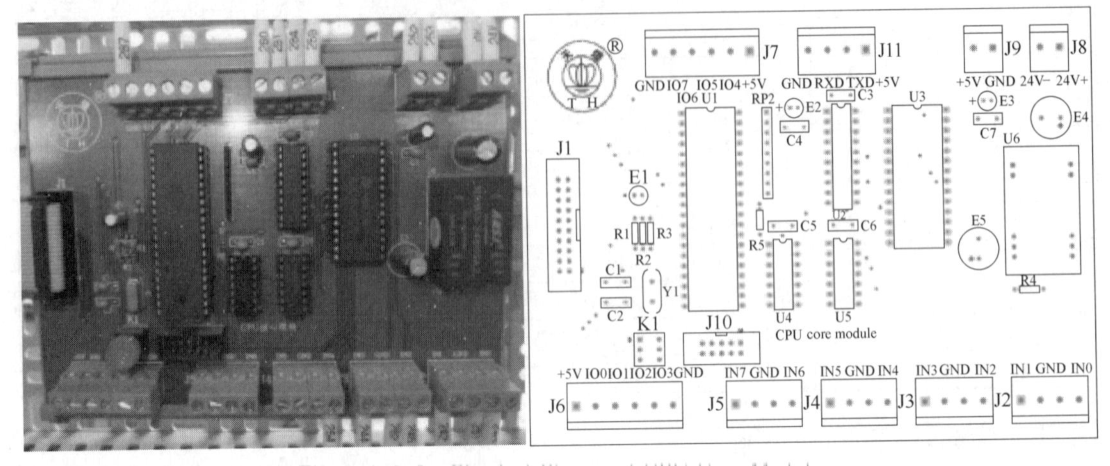

Figure 3. 3. 8 Physical Picture of CPU Core Module

(3) Port Definition

The port definition of CPU core module is shown in Table 3. 3. 7.

Table 3. 3. 7 Port Definition Table

Number	Name	Description	Extension Interface	Remark
1	J2:IN0	AD sampling input 0, voltage signal within 5V		
2	J2:GND			
3	J2:GND	AD sampling input 1, voltage signal within 5V		
4	J2:IN1			
5	J3:IN2	AD sampling input 2, voltage signal within 5V		
6	J3:GND			
7	J3:GND	AD sampling input 3, voltage signal within 5V		
8	J3:IN3			
9	J4:IN4	AD sampling input 4, voltage signal within 5V		
10	J4:GND			
11	J4:GND	AD sampling input 5, voltage signal within 5V		
12	J4:IN5			
13	J5:IN6	AD sampling input 6, voltage signal within 5V	√	
14	J5:GND		√	
15	J5:GND	AD sampling input 7, voltage signal within 5V	√	
16	J5:IN7		√	
17	J6:+5V	+5V supply output	√	
18	J6:IO0	Digital input/output interface 0, voltage signal within 5V	√	
19	J6:IO1	Digital input/output interface 1, voltage signal within 5V	√	
20	J6:IO2	Digital input/output interface 2, voltage signal within 5V	√	
21	J6:IO3	Digital input/output interface 3, voltage signal within 5V	√	
22	J6:GND	Earth	√	
23	J7:+5V	+5V supply output	√	
24	J7:IO4	Digital input/output interface 4, voltage signal within 5V	√	
25	J7:IO5	Digital input/output interface 5, voltage signal within 5V	√	
26	J7:IO6	Digital input/output interface 6, voltage signal within 5V	√	
27	J7:IO7	Digital input/output interface 7, voltage signal within 5V	√	

Continued

Number	Name	Description	Extension Interface	Remark
28	J7:ND	Earth	√	
29	J11:+5V	+5V supply output		
30	J11:TXD	Serial port sending end		
31	J11:RXD	Serial port receiving end		
32	J11:GND	Earth		
33	J9:+5V	Isolated power supply DC/DC output 5V positive electrode	√	
34	J9:GND	Isolated power supply DC/DC output 5V negative electrode	√	
35	J8:24V+	Isolated power supply DC/DC input 24V positive electrode		
36	J8:24V−	Isolated power supply DC/DC input 24V negative electrode		
37	J1	To human-computer interaction module		20 cable

(4) Notes of Programming

MPPT controller contains intelligent charge and discharge controller. Since the intelligent charge and discharge controller needs at least 1 minute to complete startup after power-on, the maximum power tracking program shall be designed with a time delay of 1 minute to ensure normal operation of intelligent charge and discharge controller.

Serial port communication is adopted for communication between CPU core module microprocessor and PWM driving module microprocessor, baud rate: 9600bit/s; 8 bit data bit; 1 bit stop bit; no parity bit; one 8 bit indicates duty ratio, data range: 0x00-0xFF (0x00: Duty ratio is 0%; 0xFF: Duty ratio is 99%).

Four analog signals to be converted are connected to the IN0, IN1, IN2, and IN3 input channels of the ADC0809: Assembly uses accumulator A and external data storage to transfer instructions for selecting channels and reading AD values: MOVX@DPTR, A; MOVX A, @DPTR. The access addresses for each channel are as follows:

① Fan output voltage (IN0) address: 0x7FF8; fan output battery (IN1) address: 0x7FF9;

② Boost converter output current (IN2) address: 0x7FFA; Boost converter output voltage (IN3) address: 0x7FFB;

③ Assembly instructions for selecting channels: MOVX@DPTR, A where DPTR is the address and A can be any value [0x00-0xFF];

④ Assembly instruction for reading AD values: MOVX A, @DPTR where DPTR is the address and A is the AD conversion value.

Four analog signals to be converted are connected to the IN0, IN1, IN2, and IN3 input channels of the ADC0809: The C language requires each channel to be predefined as an external data area, defined as follows:

① #define II_CURRENT XBYTE [0x7FF8]/* Fan output current */

② #define UI_VOL XBYTE [0x7FF9] /* Fan output voltage */

③ #define IO_CURRENT XBYTE [0x7FFA]/* Boost converter output current */

④ #define UO_VOL XBYTE [0x7FFB] /* Boost converter output voltage */C statement for selecting channels: XXXX=X where XXXX is the predefined name and X can be any value [0x00-0xFF].

⑤ C statement for reading AD values: $V_{ariable}$=XXXX, where XXXX is the predefined name and the content of the variable is the AD conversion value.

Ⅴ. Human-Computer Interaction Module

Human-computer interaction module is the input and output terminals of CPU core module.

1. Schematic Diagram

The principle of human-computer interaction module is shown in Figure 3. 3. 9.

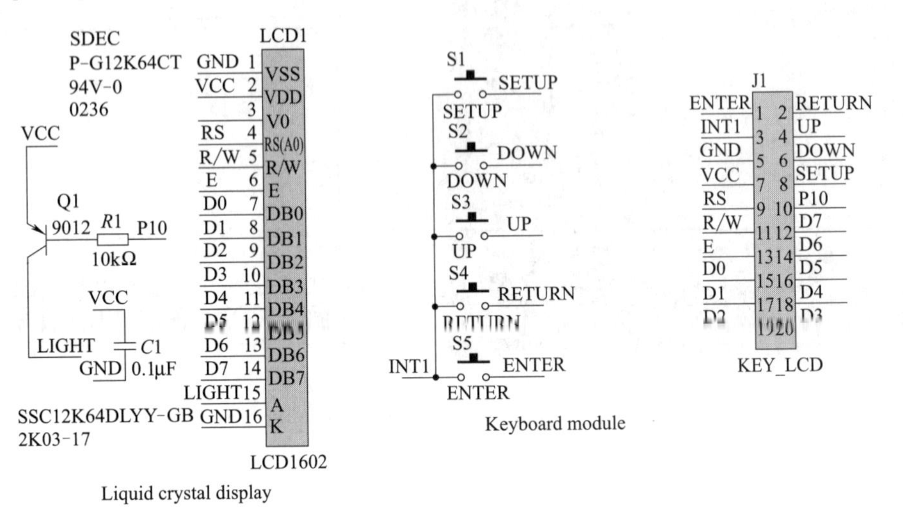

Figure 3. 3. 9 Schematic Diagram of Human-Computer Interaction Module

2. 12864 LCD Screen

Liquid crystal is a kind of polymer material. Because of its special physical, chemical and optical properties, liquid crystal was widely used in light and thin display in the middle of 20th century.

The main principle of liquid crystal display is to stimulate liquid crystal molecules to generate dots, lines and surfaces by current and form a picture with back lamp tube. For simplicity of description, various liquid crystal displays are commonly referred to directly as LCD.

Various types of LCD are usually named according to the number of rows of display characters or the number of rows and columns of LCD dot matrix. For example: 1602 means that each line displays 16 characters and can display two lines in total; similar names include 0801, 0802, 1601, etc. This kind of LCD is usually character mode LCD, that is, only ASCII code characters can be displayed, such as numbers, capital and lowercase letters, symbols, etc. 12232 LCD belongs to graphic mode LCD, which means that the liquid crystal consists of 122 columns and 32 rows, that is, there are 122×32 dots to display various graphics. We can control through program to display or not display any of these 122×32 dots. Similar names include 12864, 19264, 192128, 320240, etc. According to customer requirements, manufacturers can design any number of combinations of dot matrix LCD.

LCD has small volume, low power consumption, simple display operation, but it has a fatal weakness, its use temperature range is very narrow, universal LCD normal working temperature range is 0-+55℃, storage temperature range is −20-+60℃, even for a wide temperature level LCD, its normal working temperature range is only −20-+70℃, storage temperature range is −30-+80℃, therefore in the design of corresponding products, be

sure to consider carefully, select appropriate LCD.

3. Physical Picture

Human-computer interaction module is shown in Figure 3. 3. 10.

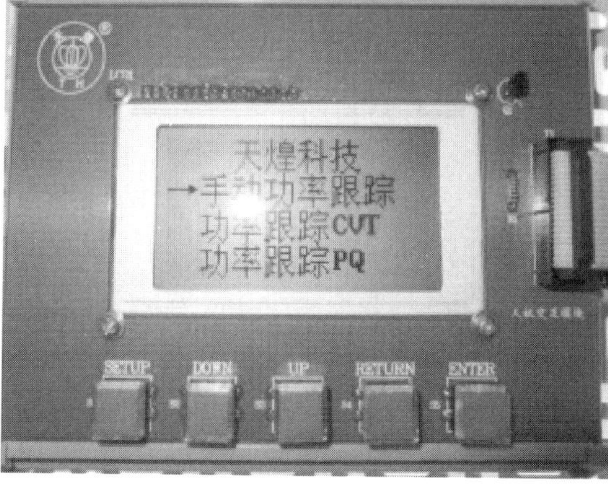

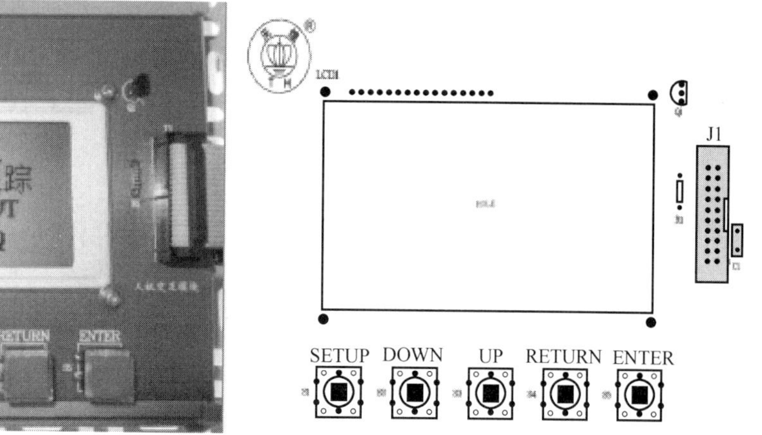

Figure 3. 3. 10 Human-Computer Interaction Module

4. Port Definition

The port definition of human-computer interaction module is shown in Table 3. 3. 8.

Table 3. 3. 8 Port Definition Table

Number	Name	Description	Serial Number	Name	Description
1	ENTER	OK key	2	RETURN	Back key
3	INT1	Key common terminal	4	UP	Up key
5	GND	Earth	6	DOWN	Down key
7	VCC	Power supply	8	SETUP	Setting key
9	RS	Register select	10	P10	Backlight control
11	R/W	Read/write signal	12	D7	Bit 7
13	E	Enable signal	14	D6	Bit 6
15	D0	Bit 0	16	D5	Bit 5
17	D1	Bit 1	18	D4	Bit 4
19	D2	Bit 2	20	D3	Bit 3

Ⅵ. PWM Driving Module

Receive power regulation parameters (duty ratio parameters) output by CPU module, CPU processes and outputs PWM signal, which is isolated to main circuit module through optocoupler.

1. Schematic Diagram

The principle of PWM driving module is shown in Figure 3. 3. 11.

2. Physical Picture

PWM driving module is shown in Figure 3. 3. 12.

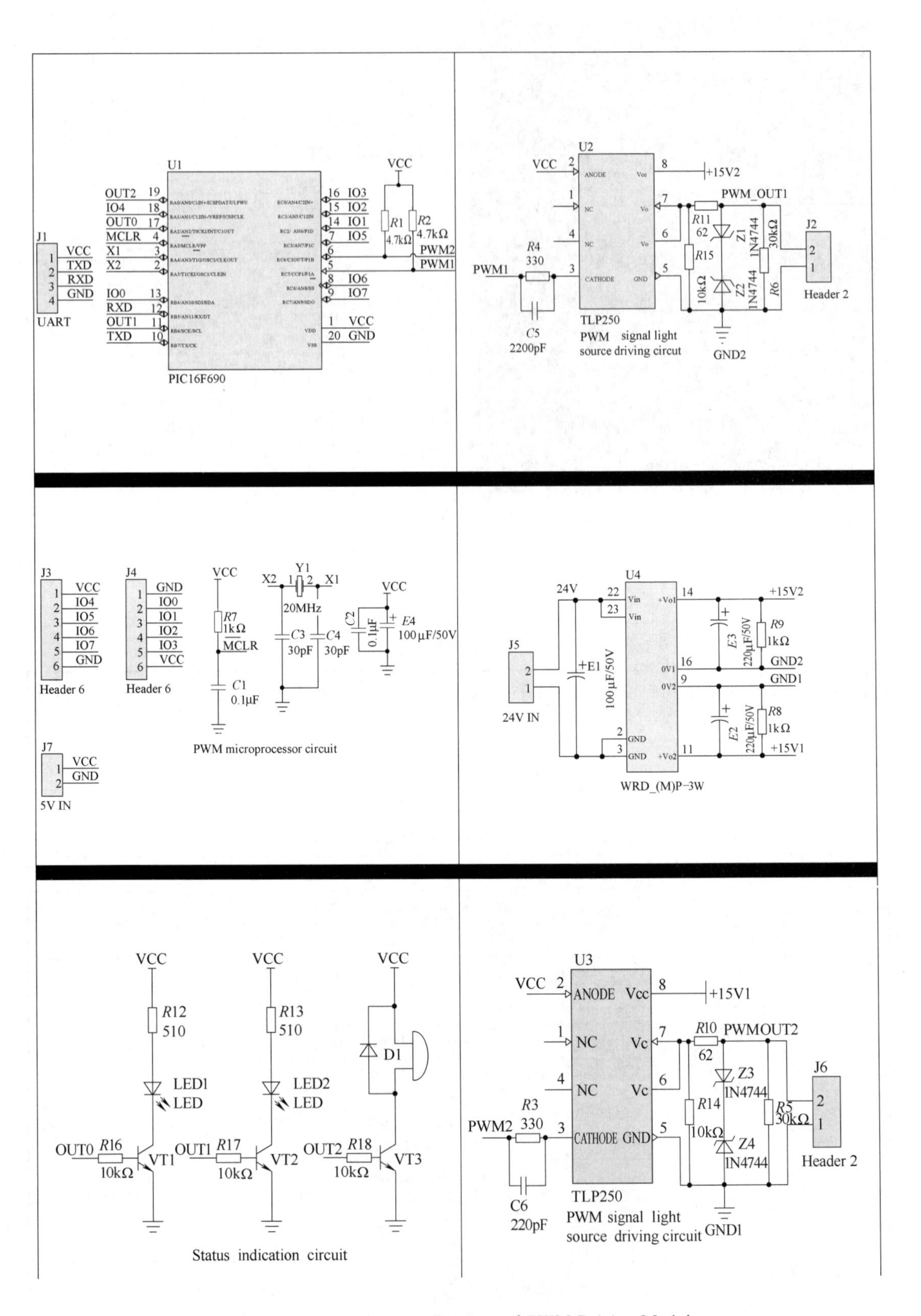

Figure 3. 3. 11 Schematic Diagram of PWM Driving Module

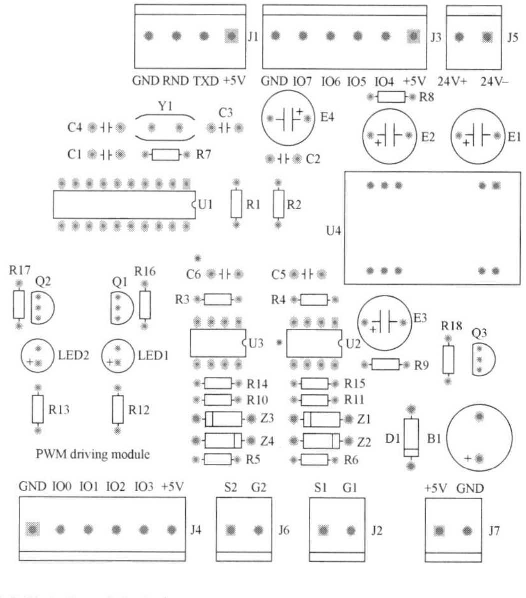

Figure 3. 3. 12 PWM Driving Module

3. Port Definition

The port definition of PWM driving module is shown in Table 3. 3. 9.

Table 3. 3. 9 Port Definition Table

Number	Name	Description	Extension Interface	Remark
1	J4:GND	Earth	√	
2	J4:IO0	Digital input/output interface 0, voltage signal within 5V	√	
3	J4:IO1	Digital input/output interface 1, voltage signal within 5V	√	
4	J4:IO2	Digital input/output interface 2, voltage signal within 5 V	√	
5	J4:IO3	Digital input/output interface 3, voltage signal within 5V	√	
6	J4:+5V	+5V supply output	√	
7	J3:+5V	+5V supply output	√	
8	J3:IO4	Digital input/output interface 4, voltage signal within 5V	√	
9	J3:IO5	Digital input/output interface 5, voltage signal within 5V	√	
10	J3:IO6	Digital input/output interface 6, voltage signal within 5V	√	
11	J3:IO7	Digital input/output interface 7, voltage signal within 5V	√	
12	J3:GND	Earth	√	
13	J1:+5V	+5V supply input		
14	J1:TXD	Serial port sending end		
15	J1:RXD	Serial port receiving end		
16	J1:GND	Earth		
17	J7:+5V	+5V supply output	√	
18	J7:ND	Earth	√	

Continued

Number	Name	Description	Extension Interface	Remark
19	J5:24V+	Isolated power supply DC/DC input 24V positive electrode		
20	J5:24V-	Isolated power supply DC/DC input 24V negative electrode		
21	J6:S1	Isolated PWM driving signal 1 negative terminal		
22	J6:G1	Isolated PWM driving signal 1 positive terminal		
23	J2:S1	Isolated PWM driving signal 2 negative terminal	√	
24	J2:G1	Isolated PWM driving signal 2 positive terminal	√	

Ⅶ. temperature Alarm Module

1. Schematic Diagram

The principle of temperature alarm module is shown in Figure 3. 3. 13.

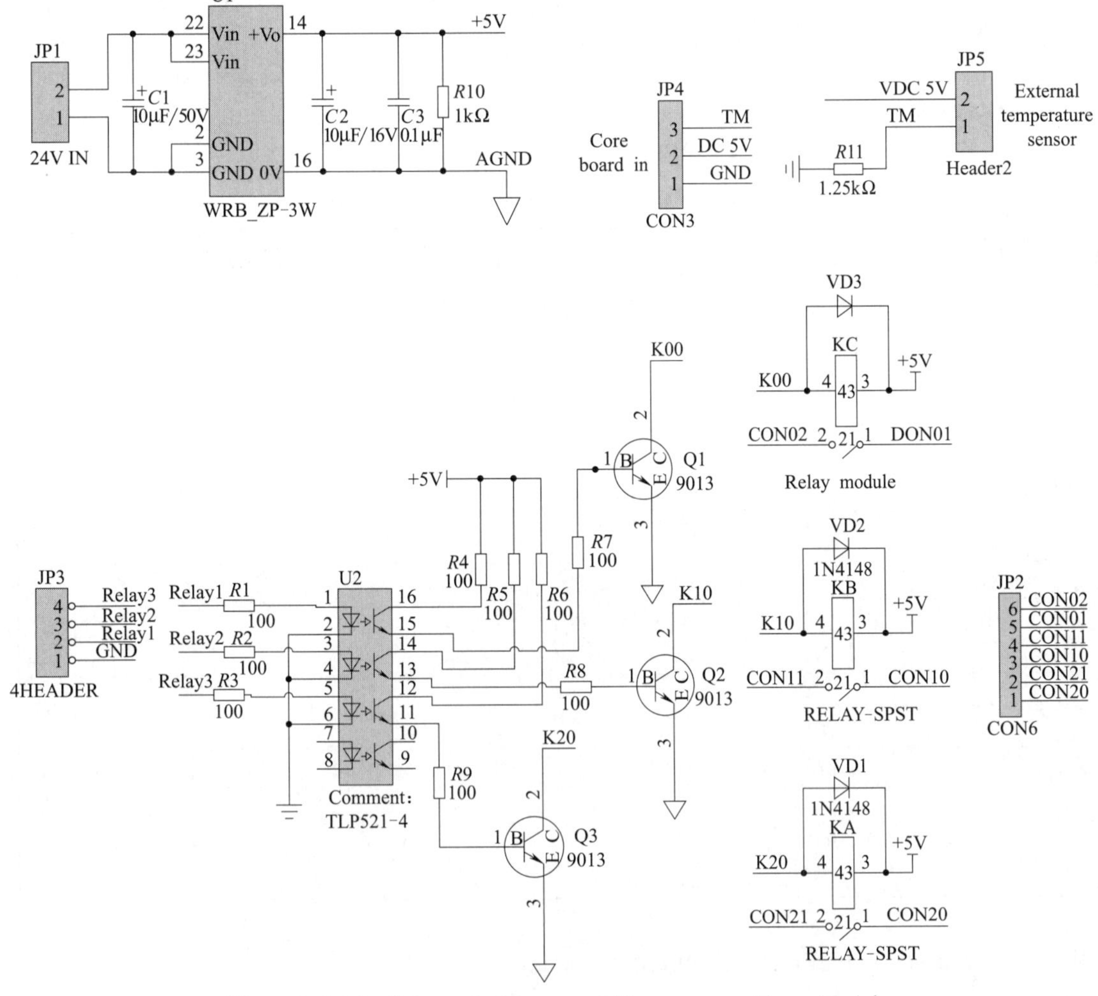

Figure 3. 3. 13 Schematic Diagram of Temperature Alarm Module

2. Physical Picture

temperature alarm module is shown in Figure 3. 3. 14.

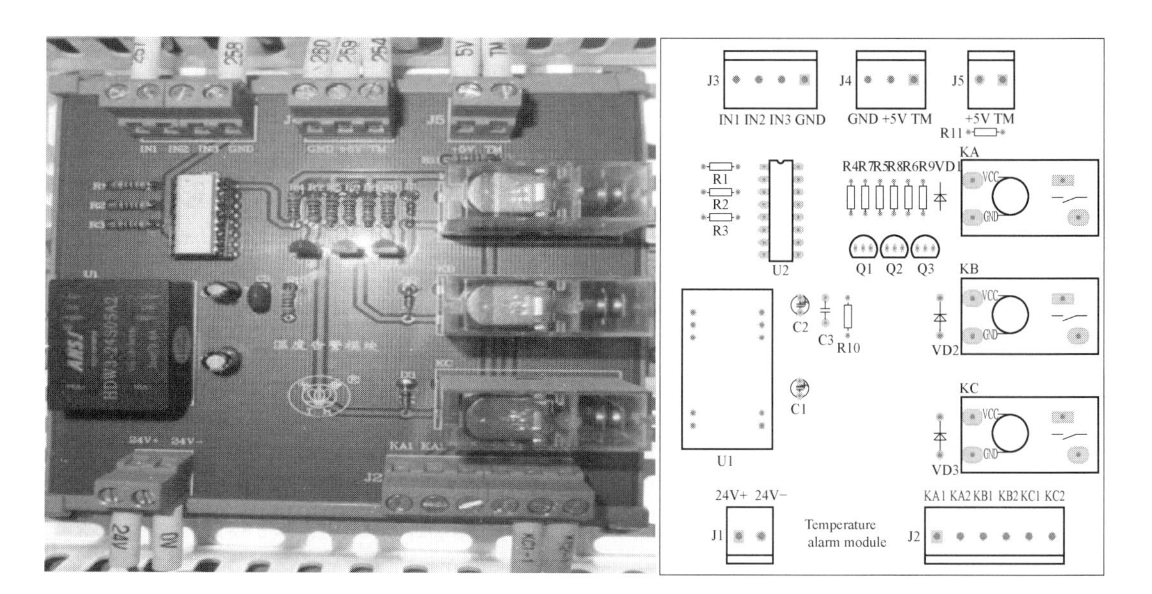

Figure 3. 3. 14 Temperature Alarm Module

3. Port Definition

The port definition of temperature alarm module is shown in Table 3. 3. 10.

Table 3. 3. 10 Port Definition Table

Number	Name	Description	Extension Interface	Remark
1	J1:24V+	Isolated power supply DC/DC 24V input		
2	J1:24V−			
3	J2:KA1	KA relay	√	
4	J2:KA2		√	
5	J2:KB1	KB relay	√	
6	J2:KB2		√	
7	J2:KC1	KC relay		
8	J2:KC2			
9	J3:IN1	KA relay control signal input	√	
10	J3:IN2	KB relay control signal input	√	
11	J3:IN3	KC relay control signal input		
12	J3:GND	Earth		
13	J4:GND	+5V supply input		
14	J4:+5V			
15	J4:TM	Temperature analog quantity output		
16	J5:+5V	Connected to temperature sensor		
17	J5:TM			

Ⅷ. DC Voltage and Current Sampling Module 1

Through voltage Hall sensor and current sensor, the output voltage and current of fan array are converted into voltage signal satisfying the requirement of the input terminal of single chip microcomputer.

1. Schematic Diagram

The principle of DC voltage and current sampling module 1 is shown in Figure 3. 3. 15.

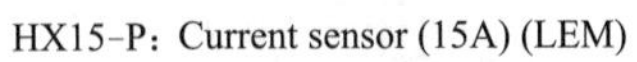

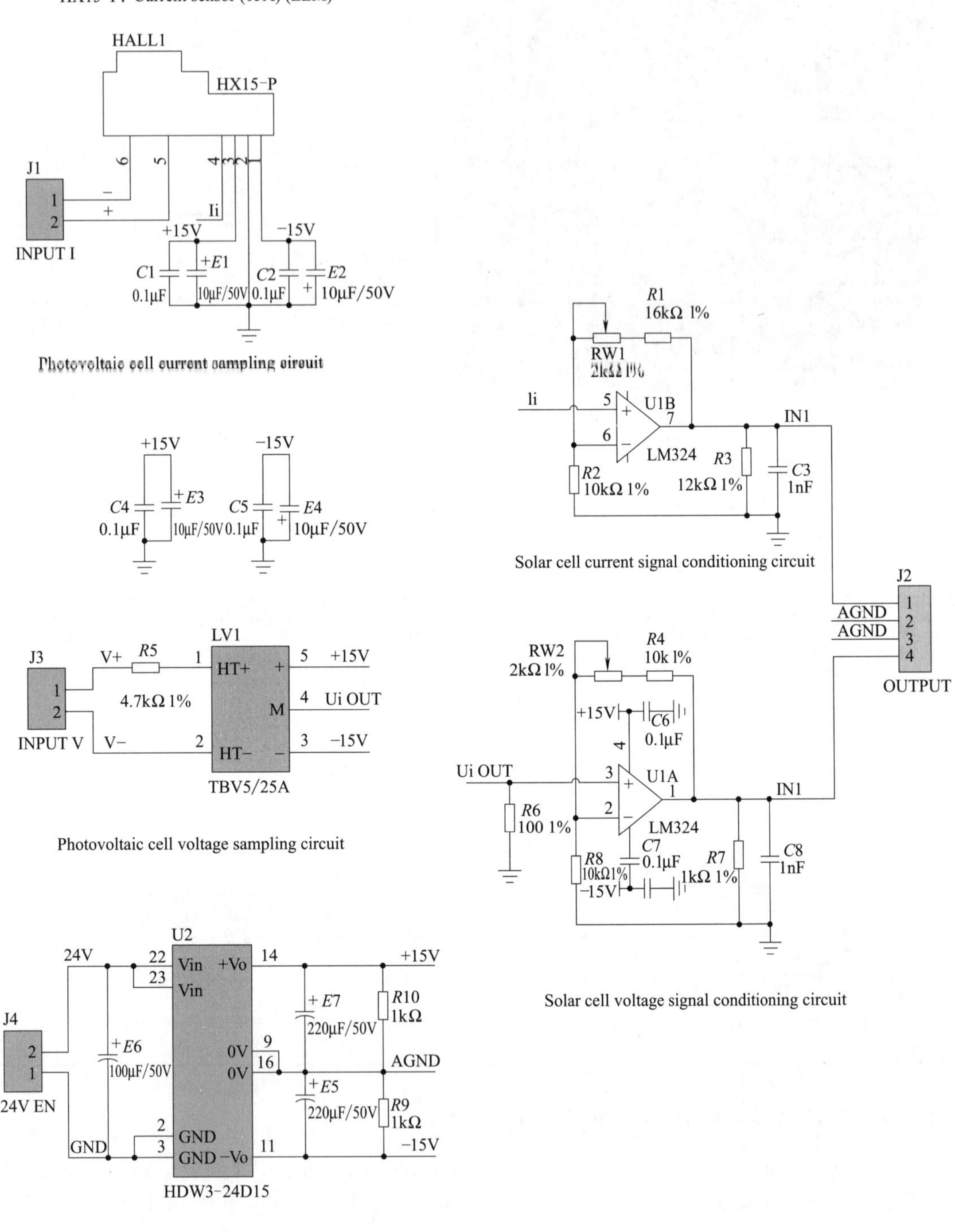

Figure 3. 3. 15 Schematic Diagram of DC Voltage and Current sampling Module 1

2. Physical Picture

DC voltage and current sampling module 1 is shown in Figure 3. 3. 16.

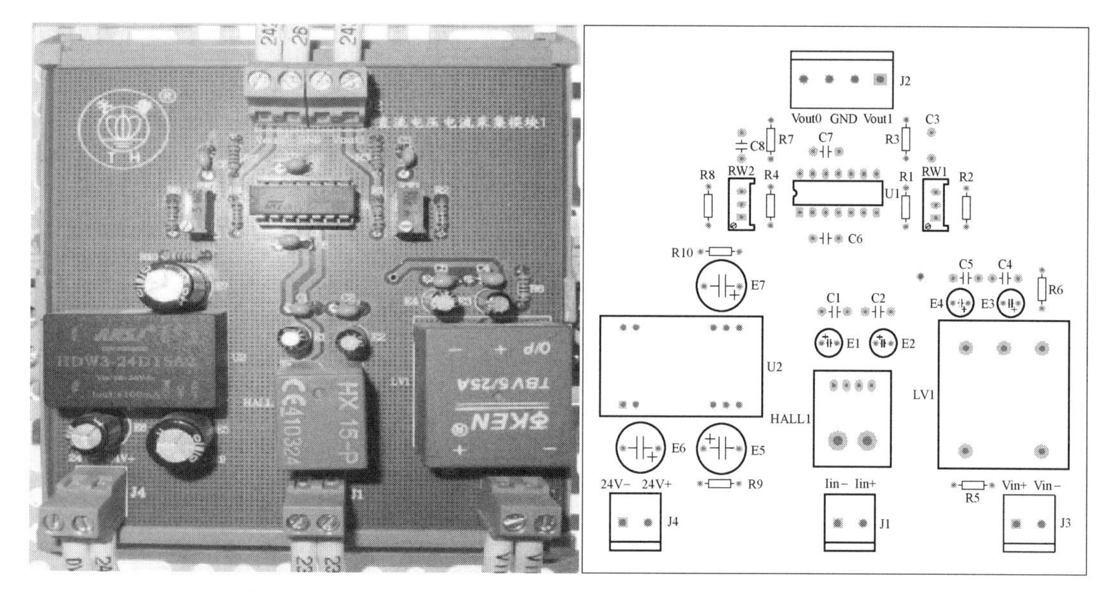

Figure 3. 3. 16 DC Voltage and Current Sampling Module 1

3. Port Definition

The port definition of DC voltage and current Sampling module 1 is shown in Table 3. 3. 11.

Table 3. 3. 11 Port Definition Table

Number	Name	Description
1	J4:24V+	Isolated power supply DC/DC 24V input
2	J4:24V−	
3	J1:Iin−	Current sampling input
4	J1:Iin+	
5	J3:Vin−	Voltage sampling input
6	J3:Vin+	
7	J2:Vout0	Current sampling conditioning output
8	J2:GND	
9	J2:GND	Voltage sampling conditioning output
10	J2:Vout1	

Ⅸ. DC Voltage and Current Sampling Module 2

Through voltage Hall sensor and current sensor, the output voltage and current are converted into voltage signal satisfying the requirement of the input terminal of single chip microcomputer.

1. Schematic Diagram

The principle of DC voltage and current sampling module 2 is shown in Figure 3. 3. 17.

2. Physical Picture

DC voltage and current sampling module 2 is shown in Figure 3. 3. 18.

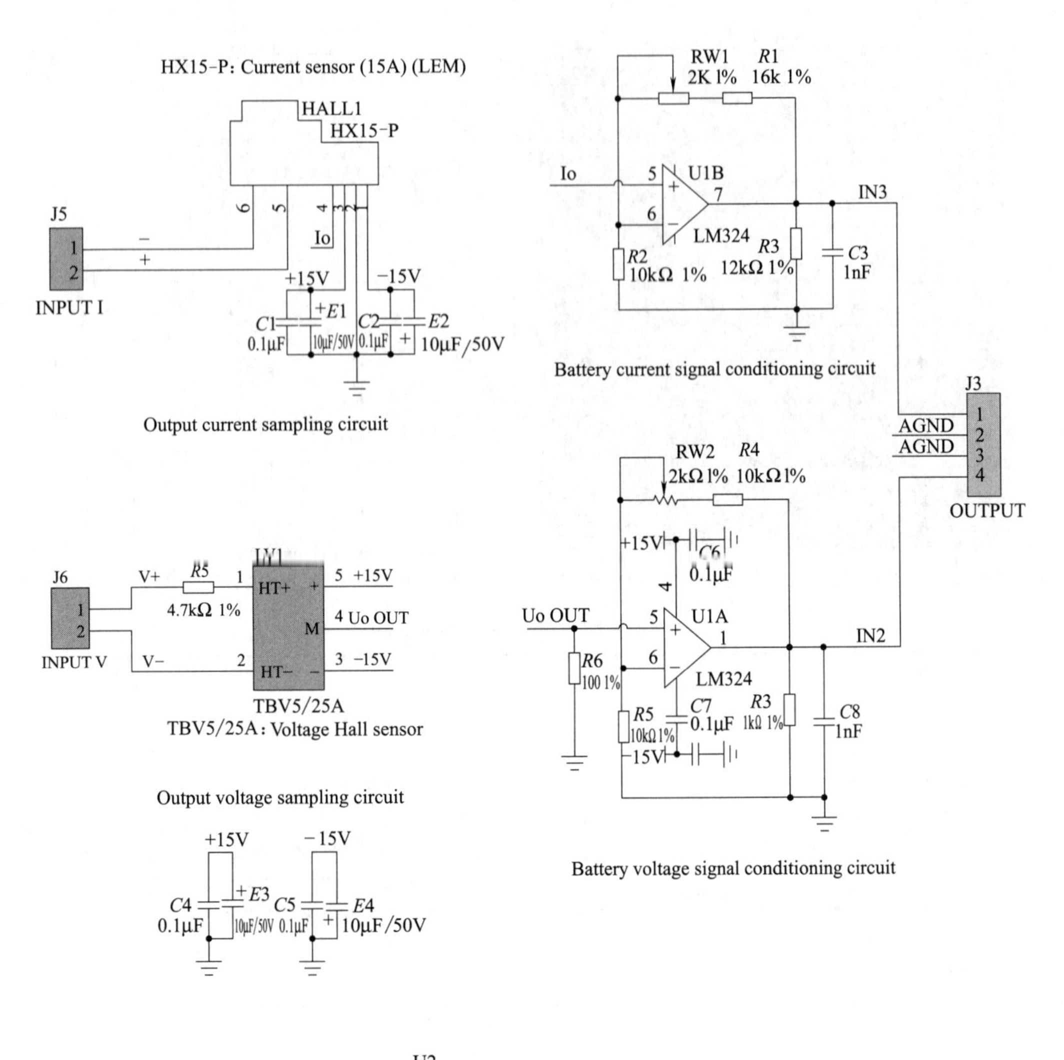

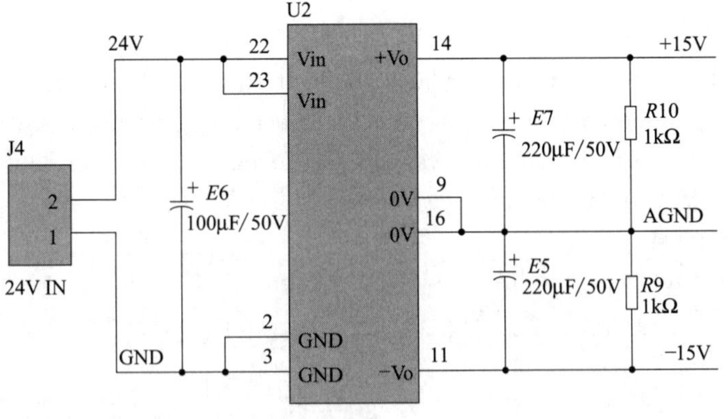

Figure 3. 3. 17 Schematic Diagram of DC Voltage and Current Sampling Module 2

3. Port Definition

The port definition of DC voltage and current sampling module is shown in Table 3. 3. 12.

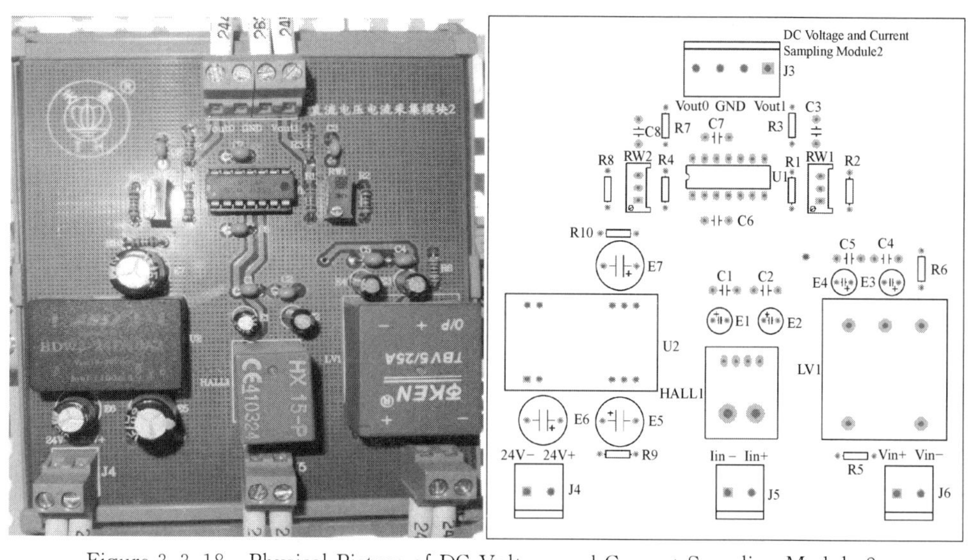

Figure 3. 3. 18 Physical Picture of DC Voltage and Current Sampling Module 2

Table 3. 3. 12 Port Definition Table

Number	Name	Description
1	J4:24V+	Isolated power supply DC/DC 24V input
2	J4:24V−	
3	J5:Iin−	Current sampling input
4	J5:Iin+	
5	J6:Vin−	Voltage sampling input
6	J6:Vin+	
7	J3:Vout0	Current sampling conditioning output
8	J3:GND	
9	J3:GND	Voltage sampling conditioning output
10	J3:Vout1	

Ⅹ. Main Circuit Module

The duty ratio of the main circuit can be adjusted by collecting the power of the fan array battery in real time, and the impedance of the load can be adjusted equivalently, so that the power taken by the load can be changed, which can always follow the maximum power output point of the fan array battery, equivalent to an impedance converter.

1. Schematic Diagram

The principle of main circuit module is shown in Figure 3. 3. 19.

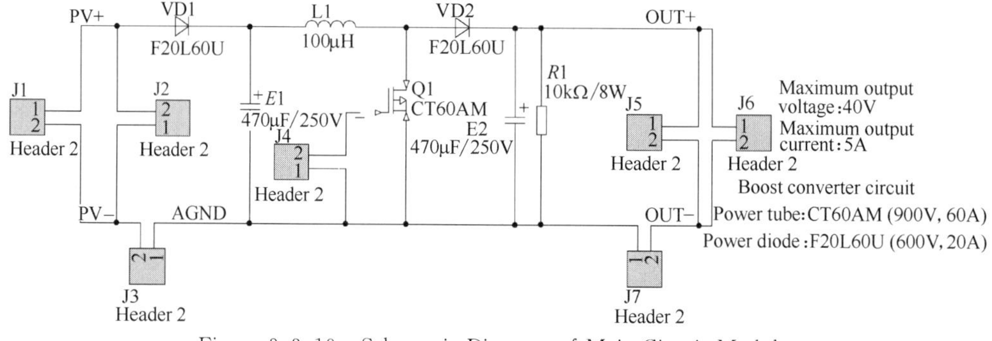

Figure 3. 3. 19 Schematic Diagram of Main Circuit Module

2. Physical Picture

Main circuit module is shown in Figure 3. 3. 20.

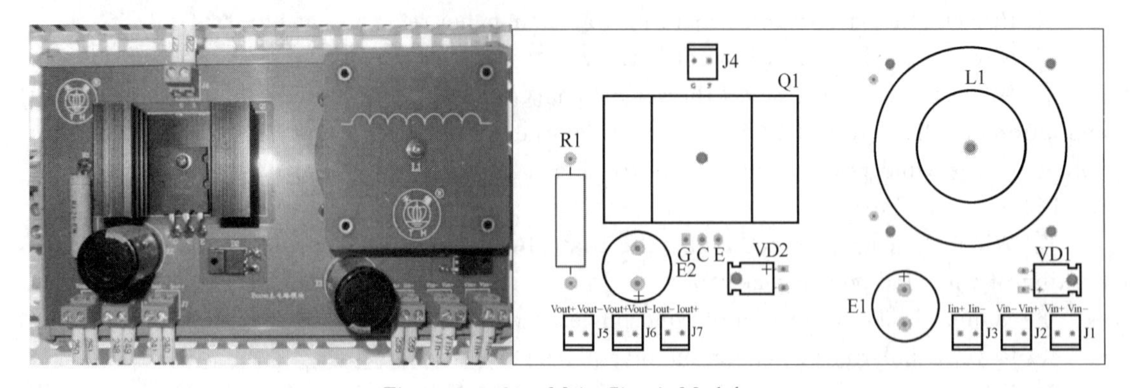

Figure 3. 3. 20 Main Circuit Module

3. Port Definition

The port definition of main circuit module is shown in Table 3. 3. 13.

Table 3. 3. 13 Port Definition Table

Number	Name	Description
1	J1:Vin+	Main circuit input voltage
2	J1:Vin−	
3	J2:Vin+	Input voltage sampling output
4	J2:Vin−	
5	J3:Iin−	Input current sampling output
6	J3:Iin+	
7	J5:Vin+	Main circuit output voltage
8	J5:Vin−	
9	J6:Vin+	Output voltage sampling output
10	J6:Vin−	
11	J7:Iin−	Output current sampling output
12	J7:Iin+	
13	J4:G	Isolated PWM driving signal input
14	J4:S	

Ⅺ. Intelligent Charge and Discharge Controller

Adjust the charging state and current according to the battery voltage, so as to prevent the battery from over-charging or over-discharging and prolong the service life of the battery. Intelligent charge and discharge controller can control wind power generator to charge battery intelligently. The equipment has elegant appearance, operating state indicator light, accurate charge and discharge control, intuitive LCD display and easy operation. It has perfect protection function. The core control component adopts American original microcontroller, and the control software adopts German advanced control technology. The current content voltage charge greatly prolongs the service life of battery, the charging efficiency of the equipment is high and the no-load loss is low. The system operates safely, stably and reliably with long service life. It has high cost performance.

1. Main Functions

① Always keep the battery fully charged.

② Prevent the battery from being over-charged.

③ Prevent the battery from over-discharged.

④ Prevent the battery from charging the solar panel reversely at night.

⑤ Reverse polarity protection of battery.

⑥ When the battery reaches the set off-charge voltage, disconnect the charging circuit and automatically detect the output voltage of the wind power generator. When the output voltage of the wind power generator is too high, automatically control the braking of the wind power generator.

⑦ When the current is too high, the controller automatically protects and controls the braking of the wind power generator.

⑧ Manual brake protection of wind power generator.

⑨ Reverse polarity protection of solar plate.

⑩ When the load current exceeds the rated current of the controller, the controller will automatically protect and lock and display "Load OverCurrent" .

⑪ The controller can protect against lightning strokes.

⑫ The controller accumulates and stores the charge ampere hours of the battery, the total generated kilowatt hours and the discharge ampere hours of generator and solar power.

⑬ Parameters such as off-charge voltage, load shutdown voltage, load recovery voltage, braking current and braking time will be automatically set according to battery voltage level (12V/24V) when the controller is started up.

⑭ The user can set the five parameters of off-charge voltage, load shutdown voltage, load recovery voltage, braking current and braking time according to their own needs.

⑮ In order to prevent the battery from being over-charged, the controller automatically controls the maximum off-charge voltage, which shall not be greater than 15V (for 12V battery) or 30V (for 24V battery), above this value, the "+" key will not work.

2. Operation Method

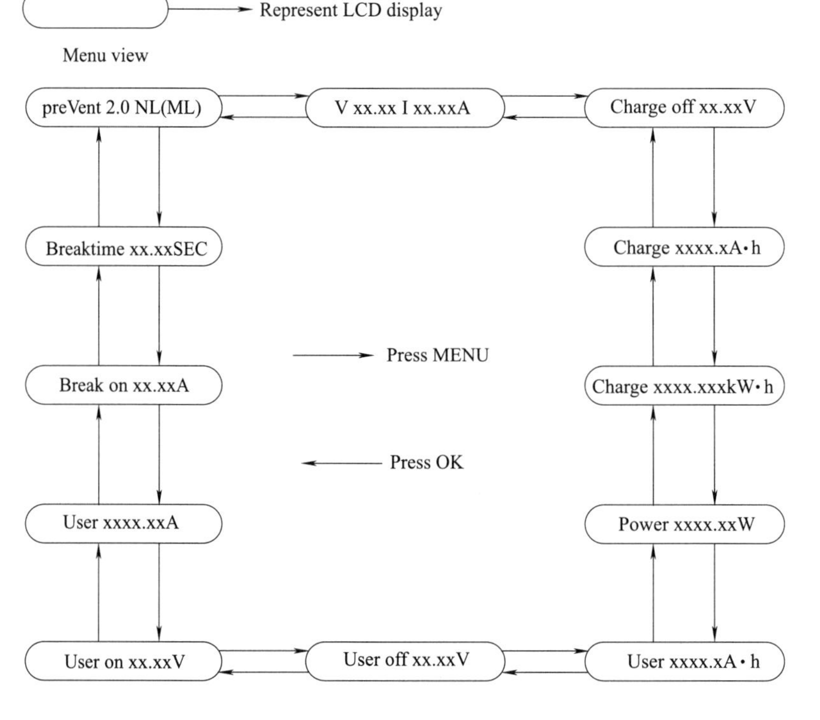

(1) Start-up interface display

preVent 2.0 NL Represent 12V battery, mode, preVent 2.0 ML Represent 24V battery mode.

(2) View battery voltage and charging current

V xx.xx I xx.xxA V (batterty voltage, unit: Volt), I (charging current, unit: Ampere).

(3) Set battery off-charge voltage

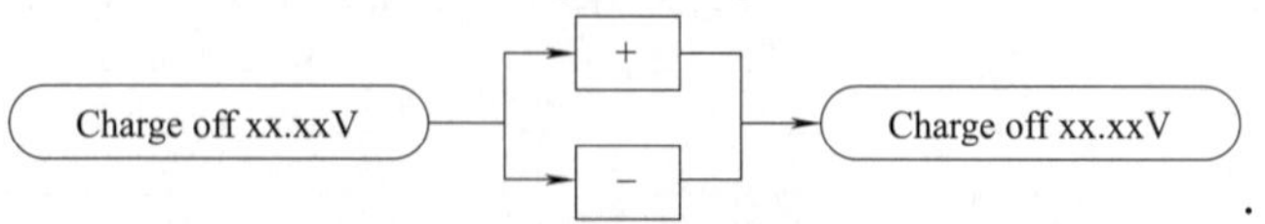

(4) View accumulated charge ampere hours

Charge xxxx.xA·h Charge (ampere hours, unit: Ampere hour).

(5) View accumulated kilowatt hours of generator and solar power

Charge xxxx.xxxkW·h Charge (generated kilowatt hours, unit: Kilowatt hour).

(6) View instantaneous power of generator and solar power

Power xxxx.xxW Power (generated instantaneous power, unit: Watt).

(7) View accumulated discharge ampere hours

User xxxx.xA·h User (discharge ampere hours, unit: Ampere hour).

(8) Set load shutdown voltage

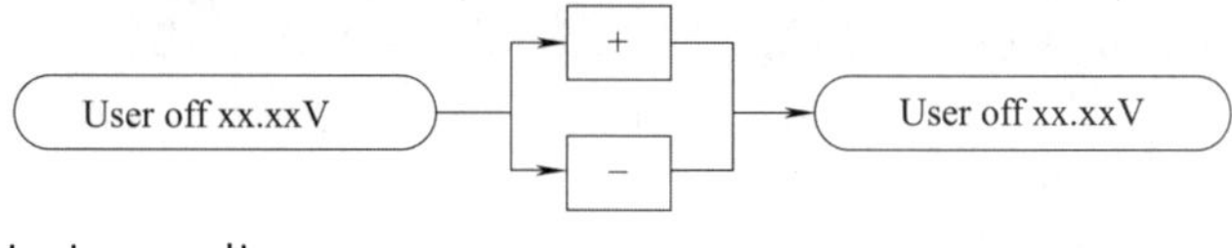

(9) Set load start-up voltage

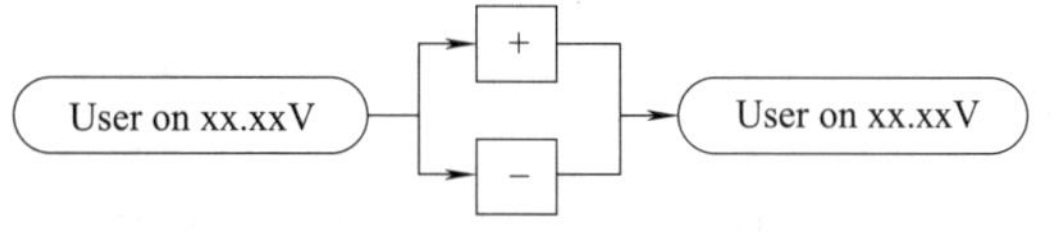

(10) View discharge current

User xxxx.xxA User (discharge current, unit: Ampere)

(11) Set braking current

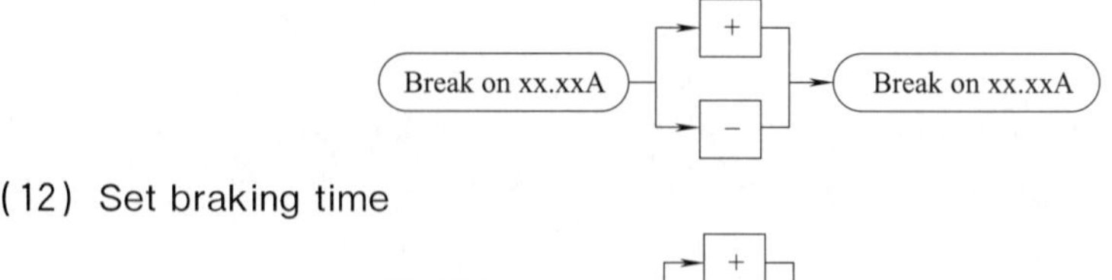

(12) Set braking time

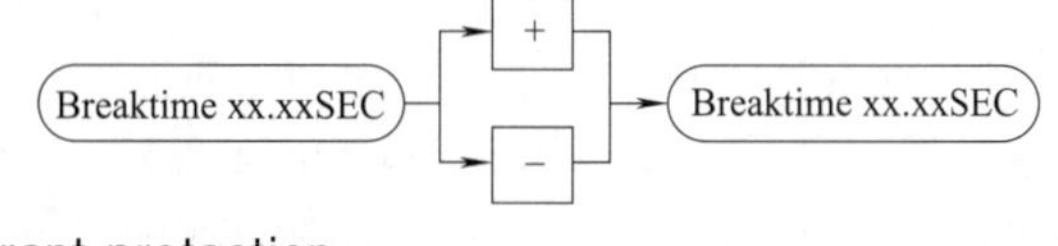

(13) Load overcurrent protection

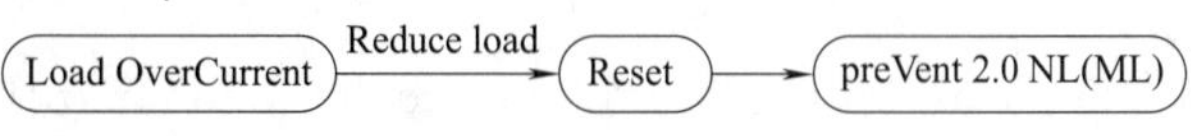

3. Technical Parameters

The technical parameters of intelligent charge and discharge controller is shown in Table 3.3.14.

Table 3. 3. 14 Technical Parameters Table

Rated current	12V,24V automatic voltage recognition
Maximum load current	35. 61A/17. 80A
Battery full charge off voltage	14. 32V/28. 64V(default,settable,range:0-15V/30V)
Load undervoltage off voltage	10. 98V/21. 97V(default,settable)
Load start-up recovery voltage	12. 44V/28. 88V(default,settable)
Braking current	15. 79A/7. 86A(default,settable,range:0-40A/20A)
Automatic recovery time after braking	10(default,settable)
Automatic braking voltage	17. 5V/32. 5V (rectified voltage of fan)
No-load loss	≤40mA
Overall dimension	228mm×133mm×75mm
Weight	1. 2kg
Working environment	Ambient temperature −10-+50℃,relative humidity 0-90%

4. Physical Picture

Intelligent charge and discharge controller is shown in Figure 3. 3. 21.

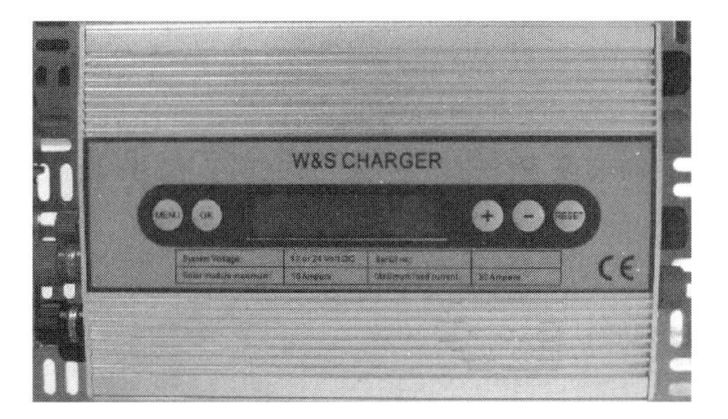

Figure 3. 3. 21 Intelligent Charge and Discharge Controller

Project Implementation

Ⅰ. Tool List

① Fan object of THWPWG-3B large-scale wind power generation system operational training platform;

② One wire stripper;

③ One crimping pliers;

④ One needle nose pliers;

⑤ One cross screwdriver and one flat screwdriver;

⑥ One multi-functional multimeter.

Ⅱ. Safety Notes

In order to successfully complete the operational training project and ensure the safe, reliable and long-term operation of the equipment during training, the operational training personnel shall strictly abide by the following safety regulations:

1. Preparation before Training

① Read the instruction carefully before training to get familiar with relevant parts of the system;

② Read the system operation instructions and precautions carefully before training;

③ Read carefully the operating instructions of relevant devices of energy storage system

before training;

④ Ensure that the main power supply of the system is disconnected before training;

⑤ Get familiar with the operation steps of this training according to the relevant contents in the training instruction before training.

2. Notes during Operational Training

① Check whether each power supply is normal before use.

② Before wiring, be sure to be familiar with the function and wiring position of each unit module of the device.

③ The main power supply must be disconnected before the training wiring, and live wiring is strictly prohibited.

④ Power on after wiring is completed and checked without error.

⑤ Proficient in wiring diagram of each module of the energy storage system and the parameter setting method of relevant modules.

⑥ There is AC 220V access point in the control cabinet. Pay attention to safety during training.

⑦ The training platform shall be kept clean and tidy all the time, and sundries shall not be placed randomly to avoid faults such as short circuit.

⑧ After the training is completed, the power switch shall be turned off in time and the training platform shall be cleaned timely.

⑨ Power on and off the system in strict accordance with correct operation steps to avoid damage to the system caused by misoperation.

⑩ In the process of training, prevent falling from high place and squeezing injury during equipment installation.

⑪ In the process of training, pay attention to the safety of strong current where there is "danger" sign.

3. Steps of Training

The following steps shall be achieved during training:

(1) Preview the report thoroughly get familiar with the equipment

Before the commencement of operational training, the instructor shall check the students' preview report and require students to understand the purpose, content and safe operation steps of this training. Only after meeting this requirement can the training be started.

The instructor shall make a detailed introduction to the training device. Students must be familiar with various equipment used for this training, and make clear the functions and using methods of these equipment.

(2) Establish teams and divide labor reasonably

Each training is carried out by a team consisting of 2-3 persons.

(3) Trial operation

Before formal training, get familiar with the operation of the device, then connect the power supply according to certain safe operation specifications, and observe whether the equipment is normal. If the equipment is abnormal, the power supply shall be cut off immediately and the fault shall be eliminated; if everything is normal, the training can be started formally.

(4) Be conscientious and responsible, make the training stead fast to the end

After the training is completed, the instructor shall check the training materials. After

being approved by the instructor, turn off all power supplies according to safe operation steps, arrange the articles used in training and put them back to their original position.

4. Summary of Training

This is the last and most important stage of operational training. We should analyze the training phenomenon and write the training report. Each training participant shall independently complete a training report, which shall be prepared with a serious and practical attitude.

The training report is a summary and feelings and experiences of operational training based on the observed problems found in the training through self-analysis and research or analysis and discussion among the team members. It shall be concise, clear in writing and clear in conclusion.

The training report shall include the following contents:

① Training name, major, class, student number, student name, name of the team member, etc.

② Training purpose, contents and steps.

③ Model and specification of training equipment.

④ Arrangement of training materials.

⑤ Analyze and summarize the training results with theoretical knowledge and draw correct conclusions.

⑥ Analyze and discuss the phenomena and problems encountered in the training, write out the experience, and put forward personal suggestions and improvement measures.

⑦ The training report should be written on the report paper of certain specification and kept neat and tidy.

⑧ Each training person shall independently complete a report and submit it to the instructor for review on time.

Ⅲ. Training Content

1. Installation: Device Layout

The aluminum guide rails, trunking, various function modules, terminals, connection modules, etc. shall be fixed to the appropriate positions of the mesh plate according to the device layout of the control cabinet of energy conversion storage control system in Figure 3. 3. 1 to prepare for wiring.

2. Connection

(1) Definition of Terminal Line Port

The definition of terminal line is shown in Figure 3. 3. 22.

Definition of terminal line number:

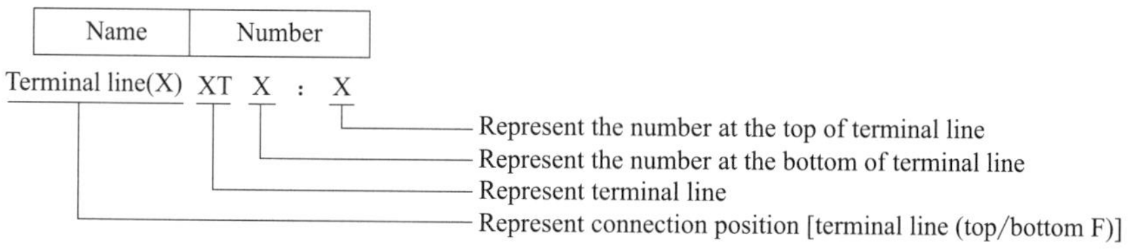

(2) Wiring of Fan Output Voltmeter

The wiring list of fan output voltmeter is shown in Table 3. 3. 15.

(3) Wiring of Fan Output Voltmeter Line Bank

The wiring list of fan output voltmeter line bank is shown in Table 3. 3. 16.

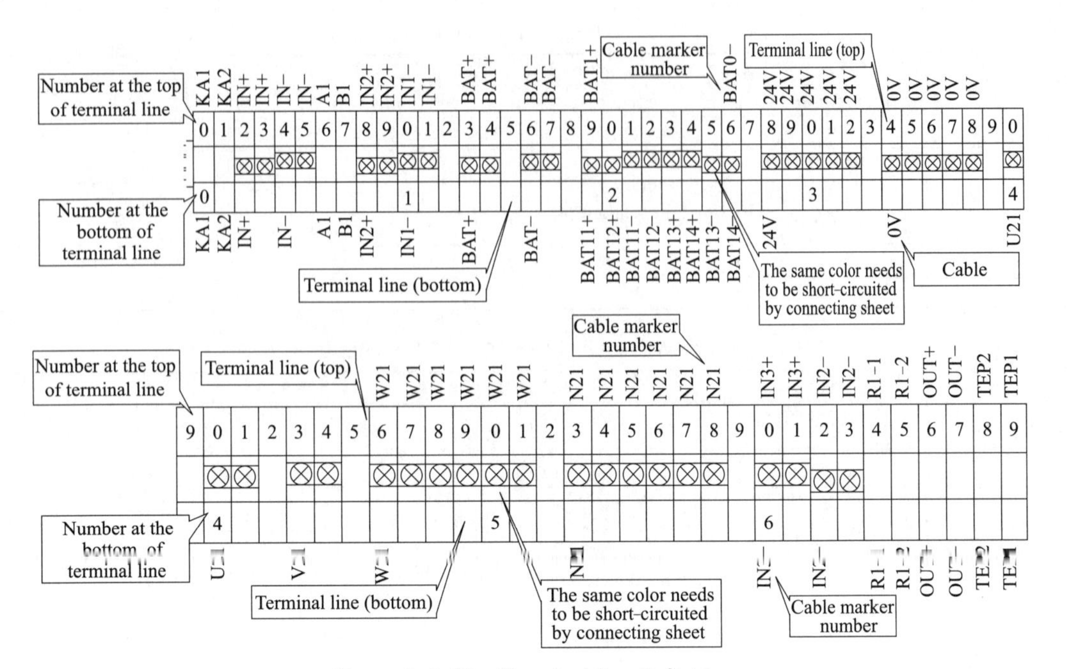

Figure 3. 3. 22 Terminal line Definition

Table 3. 3. 15 Wiring List of Fan Output Voltmeter

Number	Start Port Position	End Port Position		Cable marker number	Line Type
	Fan Output Voltmeter	Name	Number		
1	INPUT+	Line bank(top)	1	IN+	42 red
2	INPUT−		2	IN−	42 black
3	POWER18		3	W21	23 red
4	POWER17		4	N21	23 black
5	RS485A		5	A1	12 blue
6	RS485B		6	B1	12 blue

Table 3. 3. 16 Wiring List of Fan Output Voltmeter Line Bank

Number	Start Port Position	End Port Position		Cable marker number	Line Type
	Line Bank of Fan Output Voltmeter (bottom)	Name	Number		
1	1	Terminal line(top)	XT0:2	IN+	42 red
2	2		XT0:4	IN−	42 black
3	3	Terminal line(top)	XT4:8	W21	23 red
4	4		XT5:5	N21	23 black
5	5	Fan output ammeter Terminal line(top)	Serial No. 5; XT0:6	A1	12 blue
6	6		Serial No. 6; XT0:7	B1	12 blue

(4) Wiring of Fan Output Ammeter

The wiring list of fan output ammeter is shown in Table 3. 3. 17.

(5) Wiring of Fan Output Ammeter Line Bank

The wiring list of fan output ammeter line bank is shown in Table 3. 3. 18.

(6) Wiring of Fan Output Circuit Breaker

The wiring list of fan output circuit breaker is shown in Table 3. 3. 19.

Table 3. 3. 17 Wiring List of Fan Output Ammeter

Number	Start Port Position	End Port Position		Cable marker number	Line Type
	Fan output ammeter	Name	Number		
1	INPUT+	Line bank(top)	1	IN+	42 red
2	INPUT−		2	IN1+	42 red
3	POWER18		3	W21	23 red
4	POWER17		4	N21	23 black
5	RS485A		5	A1	12 blue
6	RS485B		6	B1	12 blue

Table 3. 3. 18 Wiring List of Fan Output Ammeter Line Bank

Number	Start Port Position	End Port Position		Cable marker number	Line Type
	Line Bank of Fan Output Ammeter (bottom)	Name	Number		
1	1	Terminal line(top)	XT0:3	IN+	42 red
2	2	Fan output circuit breaker K1	Bottom left	IN1+	42 red
3	3	Terminal line(top)	XT4:9	W21	23 red
4	4		XT5:6	N21	23 black
5	5	Battery voltmeter Fan output voltmeter	Serial No. 5	A1	12 blue
6	6		Serial No. 6	B1	12 blue

Table 3. 3. 19 Wiring List of Fan Output Circuit Breaker

Number	Start Port Position	End Port Position		Cable marker number	Line Type
	Fan Output Circuit Breaker K1	Name	Number		
1	Top left	Terminal line (bottom)	XT0:8	IN2+	42 red
2	Top right		XT1:0	IN1−	42 black
3	Bottom left	Fan output ammeter	Serial No. 2	IN1+	42 red
4	Bottom right	Terminal line(top)	XT0:5	IN−	42 black

(7) MPPT Circuit Breaker Wiring

The wiring list of MPPT circuit breaker is shown in Table 3. 3. 20.

Table 3. 3. 20 Wiring List of MPPT Circuit Breaker

Number	Start Port Position	End Port Position		Cable marker number	Line Type
	MPPT Circuit Breaker K2	Name	Number		
1	Top left	Terminal line(top)	XT0:8	IN2+	42 red
2	Top right		XT1:0	IN1−	42 black
3	Bottom left	main circuit module	J1:Vin+	Vin+	42 red
4	Bottom right		J1:Vin−	Vin−	42 black

(8) Wiring of Charge and Discharge Controller Circuit Breaker

The wiring list of charge and discharge controller circuit break is shown in Table 3. 3. 21.

(9) Wiring of Battery Circuit Breaker

The wiring list of battery circuit breaker is shown in Table 3. 3. 22.

(10) Wiring of Battery Voltmeter

The wiring list of battery voltmeter is shown in Table 3. 3. 23.

Table 3. 3. 21 Wiring List of Charge and Discharge Controller Circuit Breaker

Number	Start Port Position	End Port Position		Cable marker number	Line Type
	Charge and Discharge Controller Circuit Breaker K3	Name	Number		
1	Top left	Terminal line(top)	XT0:9	IN2+	42 red
2	Top right		XT1:1	IN1−	42 black
3	Bottom left	Terminal line (bottom)	XT6:0	IN3+	42 red
4	Bottom right		XT6:2	IN2−	42 black

Table 3. 3. 22 Wiring List of Battery Circuit Breaker

Number	Start Port Position	End Port Position		Cable marker number	Line Type
	Battery Circuit Breaker K4	Name	Number		
1	Top left	Battery ammeter	Serial No. 2	222	42 red
2	Top right	Terminal line(bottom)	XT1:6	BAT−	42 black
3	Bottom left	Fuse	Terminal of fuse	BAT0+	42 red
4	Bottom right	Terminal line(top)	XT2:6	BAT0−	42 black

Table 3. 3. 23 Wiring List of Battery Voltmeter

Number	Start Port Position	End Port Position		Cable marker number	Line Type
	Battery Voltmeter	Name	Number		
1	INPUT+	Line bank(top)	1	BAT+	42 red
2	INPUT−		2	BAT−	42 black
3	POWER18		3	W21	23 red
4	POWER17		4	N21	23 black
5	RS485A		5	A1	12 blue
6	RS485B		6	B1	12 blue

(11) Wiring of Battery Voltmeter Line Bank

The wiring list of battery voltmeter line bank is shown in Table 3. 3. 24.

Table 3. 3. 24 Wiring List of Battery Voltmeter Line Bank

Number	Start Port Position	End Port Position		Cable marker number	Line Type
	Battery Voltmeter Line Bank(bottom)	Name	Number		
1	1	Line bank(top)	XT1:4	BAT+	42 red
2	2		XT1:7	BAT−	42 black
3	3	Line bank(top)	XT4:6	W21	23 red
4	4		XT5:3	N21	23 black
5	5	Battery ammeter	Serial No. 5	A1	12 blue
6	6	Fan output ammeter	Serial No. 6	B1	12 blue

(12) Wiring of Battery Ammeter

The wiring list of battery ammeter is shown in Table 3. 3. 25.

(13) Wiring of Battery Ammeter Line Bank

The wiring list of battery ammeter line bank is shown in Table 3. 3. 26.

(14) Wiring of Charge and Discharge Controller

The wiring list of charge and discharge controller is shown in Table 3. 3. 27.

(15) Wiring of Handle

The wiring list of handle is shown in Table 3. 3. 28.

Table 3. 3. 25 Wiring List of Battery Ammeter

Number	Start Port Position	End Port Position		Cable marker number	Line Type
	Battery ammeter	Name	Number		
1	INPUT+	Line bank(top)	1	BAT+	42 red
2	INPUT−		2	222	42 red
3	POWER18		3	W21	23 red
4	POWER17		4	N21	23 black
5	RS485A		5	A1	12 blue
6	RS485B		6	B1	12 blue

Table 3. 3. 26 Wiring List of Battery Ammeter Line Bank

Number	Start Port Position	End Port Position		Cable marker number	Line Type
	Battery Ammeter Line Bank(bottom)	Name	Number		
1	1	Line bank(bottom)	XT1:3	BAT+	42 red
2	2	Battery Circuit Breaker K4	Top left	222	42 red
3	3	Terminal line(top)	XT4:7	W21	23 red
4	4		XT5:4	N21	23 black
5	5	Battery Voltmeter	Serial No. 5	A1	12 blue
6	6		Serial No. 6	B1	12 blue

Table 3. 3. 27 Wiring List of Charge and Discharge Controller

Number	Start Port Position	End Port Position		Cable marker number	Line Type
	Charge and Discharge Controller	Name	Number		
1	WINDGENERATOR+	Terminal line(top)	XT6:0	IN3+	42 red
2	WINDGENERATOR−		XT6:2	IN2−	42 black
3	BATTERY+	Terminal line(top)	XT1:3	BAT+	42 red
4	BATTERY−		XT1:6	BAT−	42 black
5	OUTPUT+		XT6:6	OUT+	42 red
6	OUTPUT−		XT6:7	OUT−	42 black
7	BRAKE+	Temperature alarm module	J3:IN1	BK1	12 blue
8	BRAKE−		J3:GND	BK2	12 blue

Table 3. 3. 28 Wiring List of Handle

Number	Start Port Position	End Port Position		Cable marker number	Line Type
	Handle(two-way tumbler switch)	Name	Number		
1	Bottom(with COM2)	Terminal line(top)	XT6:1	IN3+	42 red
2	Bottom(with COM1)		XT6:3	IN2−	42 black
3	Top(with COM1)		XT6:4	R1−1	42 red
4	Top(with COM2)		XT6:5	R1−2	42 black
5	COM1	main circuit module	J5:Vout+	218	42 red
6	COM2		J5:Vout−	219	42 black

(16) Wiring of Temperature Alarm Module

The wiring list of temperature alarm module is shown in Table 3. 3. 29.

(17) Wiring of Fuse

The wiring list of fuse is shown in Table 3. 3. 30.

Table 3. 3. 29 Temperature Alarm Module

Number	Start Port Position	End Port Position		Cable marker number	Line Type
	Temperature alarm module	Name	Number		
1	J2:KA1	Terminal line(top)	XT0:0	KA1	23 red
2	J2:KA2		XT0:1	KA2	23 red
3	J1:24V+		XT3:0	24V	23 red
4	J1:24V−		XT3:6	0V	23 black
5	J3:IN1	Charge and Discharge Controller	BRAKE+	BK1	12 blue
6	J3:GND		BRAKE−	BK2	12 blue
7	J4:GND	CPU core module	J9:GND	GND	23 black
8	J4:+5V		J9:+5V	+5V	23 red
9	J4:TM		J4:IN4	TM	12 blue
10	J5:+5V	Temperature sensor	XT6:8	TEP1	12 blue
11	J5:TM		XT6:9	TEP2	12 blue

Table 3. 3. 30 Wiring List of Fuse

Number	Start Port Position	End Port Position		Cable marker number	Line Type
	Fuse	Name	Number		
1	Fuse(top)	Battery Circuit Breaker K4	Bottom left	BAT0+	42 red
2	Fuse(bottom)	Terminal line(top)	XT1:9	BAT1+	42 red

(18) Wiring of Switching Power Supply

The wiring list of switching power supply is shown in Table 3. 3. 31.

Table 3. 3. 31 Wiring List of Switching Power Supply

Number	Start Port Position	End Port Position		Cable marker number	Line Type
	Switching power supply	Name	Number		
1	L	Terminal line(top)	XT5:0	W21	23 red
2	N		XT5:7	N21	23 black
3	EARTH	Cabinet enclosure		EARTH	23 yellow green
4	−V	Terminal line (bottom)	XT3:4	0V	42 black
5	+V		XT2:8	24V	42 red

(19) Wiring of Human-Computer Interaction Module

The wiring list of human-computer interaction module is shown in Table 3. 3. 32.

Table 3. 3. 32 Wiring List of Human-Computer Interaction Module

Number	Start Port Position	End Port Position		Cable marker number	Line Type
	Human-computer interaction module	Name	Number		
1	J1	CPU core module	J1	None	20P double-end cable

(20) Wiring of CPU Core Module

The wiring list of CPU core module is shown in Table 3. 3. 33.

(21) Wiring of PWM Driving Module

The wiring list of driving module is shown in Table 3. 3. 34.

(22) Wiring of DC Voltage and Current Sampling Module 1

The wiring list of DC voltage and current sampling module 1 is shown in Table 3. 3. 35.

Table 3. 3. 33 Wiring List of CPU Core Module

Number	Start Port Position	End Port Position		Cable marker number	Line Type
	CPU core module	Name	Number		
1	J2:IN0	DC voltage and current sampling module 1	J2:Vout1	211	12 blue
2	J2:GND		J2:GND	210	23 black
3	J2:IN1		J2:Vout0	209	12 blue
4	J3:IN2	DC voltage and current sampling module 2	J3:Vout1	208	12 blue
5	J3:GND		J3:GND	207	23 black
6	J3:IN3		J3:Vout0	206	12 blue
7	J4:IN4	Temperature alarm module	J4:TM	TM	12 blue
9	J8:24V+	Terminal line(top)	XT3:1	24V	23 red
8	J8:24V−		XT3:7	0V	23 black
10	J9:+5V	Temperature alarm module	J4:+5V	+5V	23 red
11	J9:GND		J4:GND	GND	23 black
12	J11:GND	PWM driving module	J1:GND	200	23 black
13	J11:RXD		J1:TXD	201	12 blue
14	J11:TXD		J1:RXD	202	12 blue
15	J11:+5V		J1:+5V	203	23 red

Table 3. 3. 34 Wiring List of Driving Module

Number	Start Port Position	End Port Position		Cable marker number	Line Type
	PWM driving module	Name	Number		
1	J6:S2	Main circuit module	J4:S	204	12 blue
2	J6:G2		J4:G	205	12 blue
3	J1:+5V	CPU core module	J11:+5V	203	23 red
4	J1:GND		J11:GND	200	23 black
5	J1:RXD		J11:TXD	202	12 blue
6	J1:TXD		J11:RXD	201	12 blue
7	J5:24V+	Terminal line(top)	XT3:2	24V	23 red
8	J5:24V−		XT3:8	0V	23 black

Table 3. 3. 35 Wiring List of DC Voltage and Current Sampling Module 1

Number	Start Port Position	End Port Position		Cable marker number	Line Type
	DC Voltage and Current Sampling Module 1	Name	Number		
1	J3:Vin+	main circuit module	J2:Vin+	223	42 red
2	J3:Vin−		J2:Vin−	224	42 black
3	J1:Iin+		J3:Iin+	212	42 red
4	J1:Iin−		J3:Iin−	213	42 black
5	J4:24+	Terminal line(top)	XT2:8	24V	23 red
6	J4:24−		XT3:4	0V	23 black
7	J2:Vout0	CPU core module	J2:IN1	209	12 blue
8	J2:GND		J2:GND	210	23 black
9	J2:Vout1		J2:IN0	211	12 blue

(23) Wiring of DC Voltage and Current Sampling Module 2

The wiring list of DC voltage and current sampling module 2 is shown in Table 3. 3. 36.

(24) Wiring of Main Circuit Module

The wiring list of main circuit module is shown in Table 3. 3. 37.

(25) Wiring of Connection Module XS4

The wiring list of connection module XS4 is shown in Table 3. 3. 38.

Table 3. 3. 36 Wiring List of DC Voltage and Current Sampling Module 2

Number	Start Port Position	End Port Position		Cable marker number	Line Type
	DC Voltage and Current Sampling Module 2	Name	Number		
1	J6:Vin+	main circuit module	J6:Vout+	216	42 red
2	J6:Vin−		J6:Vout−	217	42 black
3	J5:Iin+		J7:Iout+	214	42 red
4	J5:Iin−		J7:Iout−	215	42 black
5	J4:24+	Terminal line(top)	XT2:9	24V	23 red
6	J4:24−		XT3:5	0V	23 black
7	J3:Vout0	CPU core module	J3:IN3	206	12 blue
8	J3:GND		J3:GND	207	23 black
9	J3:Vout1		J3:IN2	208	12 blue

Table 3. 3. 37 Wiring List of Main Circuit Module

Number	Start Port Position	End Port Position		Cable marker number	Line Type
	Main circuit module	Name	Number		
1	J1:Vin+	MPPT circuit breaker	Bottom left	Vin+	42 red
2	J1:Vin−		Bottom right	Vin−	42 black
3	J2:Vin+	DC voltage and current sampling module 1	J3:Vin+	223	42 red
4	J2:Vin−		J3:Vin−	224	42 black
5	J3:Iin+		J1:Iin+	212	42 red
6	J3:Iin−		J1:Iin−	213	42 black
7	J7:Iout+	DC voltage and current sampling module 2	J5:Iin+	214	42 red
8	J7:Iout−		J5:Iin−	215	42 black
9	J6:Vout+		J6:Vin+	216	42 red
10	J6:Vout−		J6 Vin−	217	42 black
11	J5:Vout+	Handle(two-way tumbler switch)	COM1	218	42 red
12	J5:Vout−		COM2	219	42 black
13	J4:S	PWM driving module	J6:S2	204	12 blue
14	J4:G		J6:G2	205	12 blue

Table 3. 3. 38 Wiring List of Connection Module XS4

Number	Start Port Position	End Port Position		Cable marker number	Line Type
	Connection Module XS4	Name	Number		
1	A1	Terminal line (bottom)	XT0:0	KA1	23 red
2	A2		XT0:1	KA2	23 black
3	B1		XT0:2	IN+	23 red
4	B2		XT0:4	IN−	23 black

(26) Wiring of Connection Module XS5

The wiring list of connection module XS5 is shown in Table 3. 3. 39.

Table 3. 3. 39 Wiring List of Connection Module XS5

Number	Start Port Position	End Port Position		Cable marker number	Line Type
	Connection Module XS5	Name	Number		
1	A7	Terminal line (bottom)	XT0:6	A1	42 red
2	A8		XT0:7	B1	42 black
3	A9		XT6:6	OUT+	42 red
4	A10		XT6:7	OUT−	42 black

(27) Wiring of Battery Pack

The wiring list of battery pack is shown in Table 3. 3. 40.

Table 3. 3. 40 Wiring List of Battery Pack

Number	Start Port Position	End Port Position		Cable marker number	Line Type
	Battery pack	Name	Number		
1	BAT11+	Terminal line (bottom)	XT1:9	BAT11+	42 red
2	BAT11−		XT2:1	BAT11−	42 black
3	BAT12+		XT2:0	BAT12+	42 red
4	BAT12−		XT2:2	BAT12−	42 black
5	BAT13+		XT2:3	BAT13+	42 red
6	BAT13−		XT2:5	BAT13−	42 black
7	BAT14+		XT2:4	BAT14+	42 red
8	BAT14−		XT2:6	BAT14−	42 black

(28) Wiring of Fixed Load

The wiring list of fixed load is shown in Table 3. 3. 41.

Table 3. 3. 41 Wiring List of Fixed Load

Number	Start Port Position	End Port Position		Cable marker number	Line Type
	Fixed Load	Name	Number		
1	1	Terminal line (bottom)	XT6:4	R1−1	42 red
2	2		XT6:5	R1−2	42 black

(29) Wiring of Temperature Sensor

The wiring list of temperature sensor is shown in Table 3. 3. 42.

Table 3. 3. 42 Wiring List of Temperature Sensor

Number	Start Port Position	End Port Position		Cable marker number	Line Type
	Temperature sensor	Name	Number		
1	1	Terminal line (bottom)	XT6:8	TEP2	12 blue
2	2		XT6:9	TEP1	12 blue

Ⅳ. Operation Instruction

1. Working Mode Switching Setting

Working mode switching setting is shown in Table 3. 3. 43.

Table 3. 3. 43 Working Mode Switch Setting Table

Number	Function	MPPT(fixed load)	MPPT(battery)	Charge and Discharge Controller
1	Fan output(K1)	On	On	On
2	MPPT(K2)	On	On	Off
3	Charge and discharge controller(K3)	Off	Off	On
4	Battery(K4)	On	On	On
5	Handle	Up	Down	Medium

2. Operation Steps of Maximum Power Tracking Test

① Close the "main power" switch on the "energy storage control unit", the system is

powered on and the three-phase power indicator light is on.

② Set the brake of charge and discharge controller to "RELEASE" state, close the "battery" circuit breaker of the "energy storage control unit", connect the battery to the MPPT controller and supply power to charge and discharge controller at the same time. At this time, the charge and discharge controller is initialized and the red indicator light is on (working in braking state). Do not proceed until the red indicator goes out (exit the braking state).

③ Close the "fan output" and "MPPT" circuit breaker on the "energy storage control unit".

④ Press reset key K1 on "CPU core module" to reset the system.

⑤ Press the "ENTER" key on the "human-computer interaction module" to enter the manual power tracking interface, and then press "UP" and "DOWN" key to manually adjust the duty ratio, measure the output current and voltage of multiple groups of fans, record it in the following table.

Number	Voltage/V	Current/A	Power/W
1			
2			
3			
4			
5			
6			
7			
8			
9			
10			

⑥ Based on the recorded voltage and current data, calculate the power corresponding to each voltage and current.

⑦ Select "power tracking PQ" through "UP" and "DOWN" on "human-computer interaction module", and press "ENTER", enter automatic power tracking, record voltage and current data respectively after a period of time, and calculate corresponding power.

⑧ Turn off the "energy storage control unit" "fan output" "MPPT", circuit breaker of "battery" and "main power" switch in turn.

3. Driver Installation of ISP Downloader

① Connect the ISP downloader to the computer using a USB cable. When you use the downloader for the first time, the computer will pop up a prompt interface for discovering new hardware, as shown in Figure 3. 3. 23.

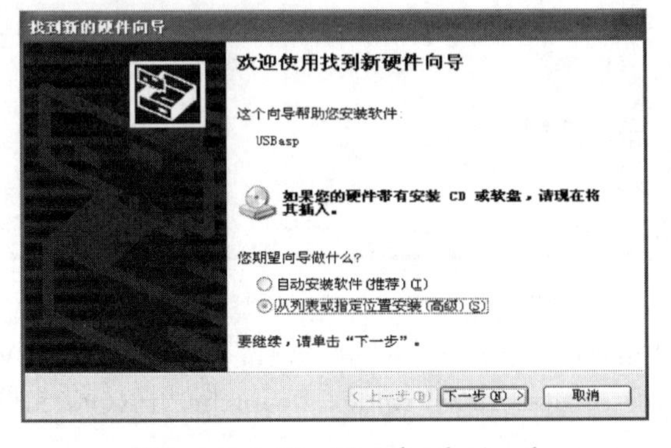

Figure 3. 3. 23 New hard wizard

② Select "install from list or specified location (advanced) (S)" and click "next". In the displayed interface, select the path where the USB ISP driver is located. Then click "next" to start the installation of the driver, as shown in Figure 3. 3. 24.

③ Click "finish" after installation, as shown in Figure 3. 3. 25.

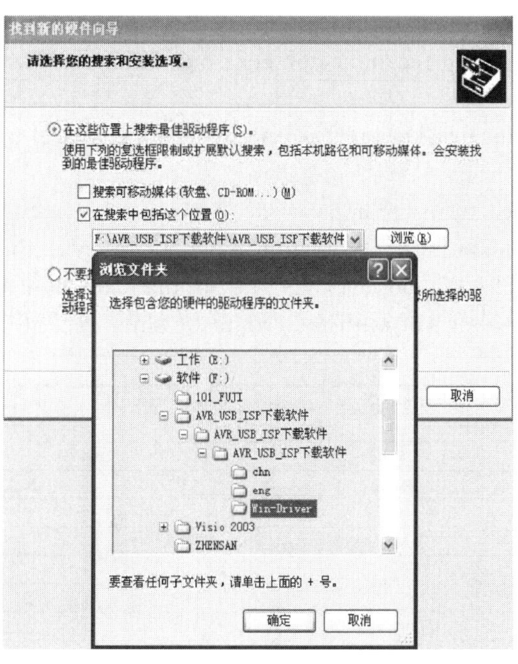

Figure 3. 3. 24 Select a Path

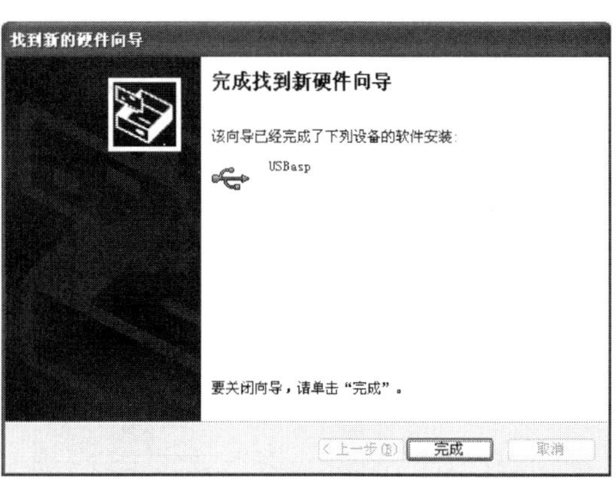

Figure 3. 3. 25 complete the installation

④ Right click on "my computer", open "device manager", connect the ISP downloader to the computer, and this item will appear as shown in the figure below, which indicates that the driver installation is successful as shown in Figure 3. 3. 26.

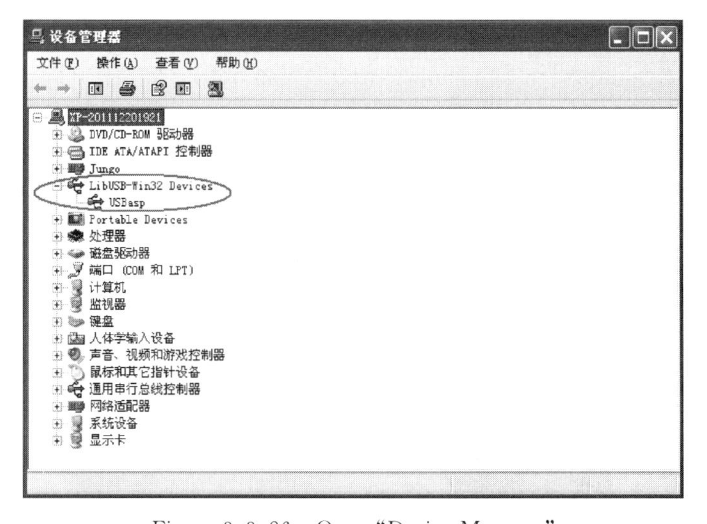

Figure 3. 3. 26 Open "Device Manager"

⑤ Connect one end of the ISP downloader to the J10 of the THWPG020. PCB, and the other end to the computer through USB cable, open the "PROGISP" download software on the computer and set it as shown in Figure 3. 3. 27.

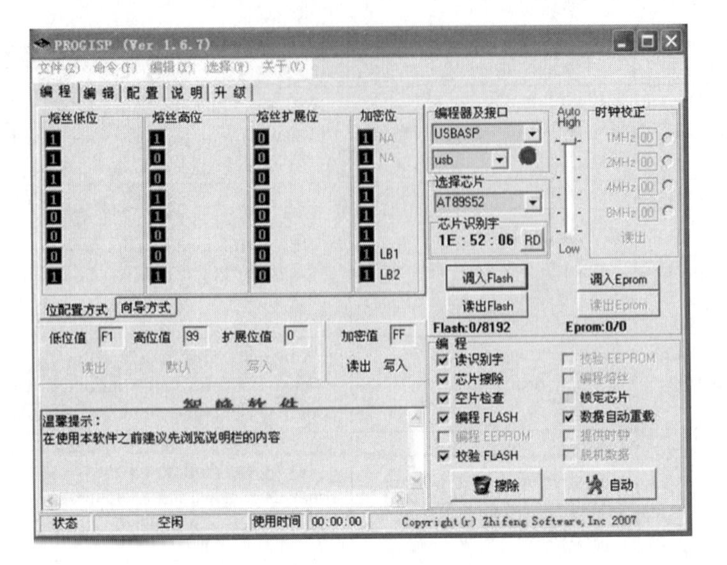

Figure 3. 3. 27 Open "PROGISP"

⑥ Click the "call Flash" key to pop up the "open" dialog box. Then click the "auto" key to start the burning program. After successful programming, there will be a prompt "1: Erase, Blank, WriteFlash, VerifyFlash, Successfully done", disconnect the power supply.

V. Troubleshooting of Common Fault

Common fault analysis and troubleshooting is shown in Table 3. 3. 44.

Table 3. 3. 44 Summary Table of Common Fault Analysis and Troubleshooting

Fault Phenomenon	Possible Cause of Fault	Troubleshooting Methods and Steps
Abnormal power supply	1. The fuse is damaged 2. There is short circuit or electric leakage	1. Check whether the fuse in the fuse holder is burnt. If it is burnt, replace it with a new one 2. Carefully check whether there is short circuit in the circuit, and use multimeter to measure whether the resistance between phases and the resistance between phases to ground is normal. If the resistance value is zero or very small, it indicates that there is a short circuit, and successive disconnection method shall be adopted for inspection
The fan object works normally, but cannot be charged to the battery	1. The circuit breaker is damaged 2. The circuit breaker is not closed 3. The fuse core of battery is damaged 4. The charge and discharge controller is damaged 5. The MPPT controller module is damaged 6. Open circuit or poor contact of circuit	1. Check whether the fan output, MPPT, battery circuit breaker input and output are connected 2. Close the fan output, MPPT and battery circuit breaker 3. Check whether 10A fuse core in fuse is damaged. If damaged, replace fuse core 4. If the circuit is normal and the fault still exists after the fuse core is replaced, disconnect the wiring at the fan input end of the charge and discharge controller and measure the output voltage of the fan with multimeter. If the voltage is normal, replace or repair the controller; if there is no voltage, replace or repair the generator and check whether each module of MPPT controller is damaged 5. Check whether MPPT controller module is damaged 6. Check system circuit for open circuit or poor contact with multimeter

Continued

Fault Phenomenon	Possible Cause of Fault	Troubleshooting Methods and Steps
The green light of charge and discharge controller is off, and the load is cut off	Insufficient battery voltage	Charge the battery or replace the battery
The green light of charge and discharge controller is off, the load is cut off, and the LCD screen displays Load Over Current	Overload or short circuit of load	Reduce the load or eliminate the fault of short circuit, press Reset key to restart the machine
The red light of charge and discharge controller is off	Uncharged	If it is at night, it is a normal phenomenon. If there is no charging for a long time, check whether the fan wiring is correct, whether there is looseness and open circuit
Maximum power tracking of fan fails to enter stable state	1. The voltage and current sampling module 1 is abnormal 2. ADC0809 damaged	1. Use multimeter to detect power supply and circuit connection of voltage and current sampling module 1 2. Replace ADC0809
Failure to track the maximum power of fan (normal fan output voltage, no current)	1. The circuit breaker is damaged 2. The circuit breaker is not closed 3. The CPU core module is abnormal 4. The communication interface module is abnormal 5. The PWM driving module is abnormal 6. The main circuit module is abnormal 7. Poor circuit contact of MPPT circuit breaker	1. Check whether the fan output, MPPT, battery circuit breaker input and output are connected 2. Close the fan output, MPPT and battery circuit breaker 3. Check whether the power supply of CPU core module is normal. If the power supply is normal, use multimeter to detect the circuit connection 4. Check whether the power supply of the communication interface module is normal. If the power supply is normal, use multimeter to detect the circuit connection 5. Check whether the power supply of PWM driving module is normal. If the power supply is normal, use multimeter to detect the circuit connection 6. Use multimeter to detect the circuit connection of main circuit module 7. Use multimeter to check the circuit connection of MPPT circuit breaker
Enter the protection state immediately after starting the fan	1. The circuit breaker is damaged 2. The circuit breaker is not closed 3. Brake protection of charge and discharge controller 4. Battery feeding	1. Check whether input and output of battery circuit breaker are connected 2. Close the battery circuit breaker 3. Check whether the manual brake key of charge and discharge controller is in "BRAKE" state. If it is "BRAKE", manually set it to "RELEASE". When the charge and discharge controller is just started, the system initialization needs a period of time. At this time, the charge and discharge controller is in the brake protection stage, so it is required to wait for the completion of system initialization before charging 4. Observe whether the battery voltage is lower than 18V. If it is lower than 18V, the charge and discharge controller will automatically enter into 12V mode. Once the input voltage is higher than 15V, the charge and discharge controller will send brake protection signal. Since the wind power generator is of three-phase 12V output, the charge and discharge controller will always be in the protection state, so it is necessary to charge the battery and start the wind power generator normally only after the normal voltage is restored

Project Work

(1) What are the main modules of the energy conversion storage control system? What are their functions?

(2) Describe the operation process of the maximum power tracking test project with words or flow chart.

Project Ⅳ Realization of Yaw Function

Project Description

Understand the working principle of yaw control system, implement manual yaw control and automatic yaw control by main control computer; complete program editing and realize yaw function of given angle.

Competency Objectives:

① Grasp the working principle of absolute value encoder and Gray code;

② Understand the yaw control process and realize it by PLC programming.

Project Environment

Yaw control system mainly consists of slave PLC, frequency converter, absolute value encoder, yaw motor (AC gear motor), control button and other devices. It can complete the installation of yaw control system, the programming and debugging of PLC control program for manual and automatic yaw, and the communication between slave PLC and master PLC. Yaw pitch control system control cabinet is shown in Figure 3. 4. 1.

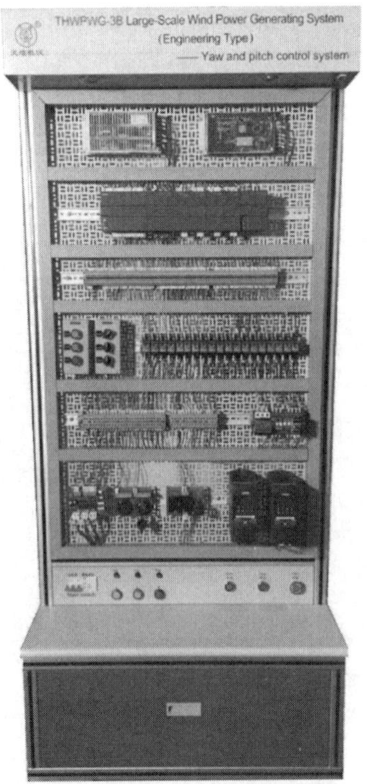

Figure 3. 4. 1 Yaw Pitch Control System Control Cabinet

Ⅰ. Main Equipment of Yaw Control System

See Table 3. 4. 1 for the list of main equipment of yaw control system.

Table 3. 4. 1 List of Main Equipment of Yaw Control System

Number	Name	Specification	Quantity
1	Control cabinet of yaw pitch control system	880mm × 620mm × 2120mm	1 set
2	Frequency converter	One MM420 and MM440 respectively; three-phase input; Power: 0. 75kW	2 sets
3	Button module	One 24V/6A switching power supply; one emergency stop button; one reset, start and stop button(yellow, green and red) respectively; one self-locking button(yellow, green and red) respectively; two change-over switches; two 24V indicator lights(yellow, green and red) respectively	1 group
4	PLC	Siemens CPU ST40 host	1 set
		EM DR32 digital quantity input/output module, 16×24V DC input/16-point relay output	1 set
		EM AQ02 module, 2-way analog quantity output(2AI)	1 set
		EM AR02 module, 2-way thermal resistance input module	3 sets
		EM DP01 communication module	1 set

Ⅱ. Main Parameter Setting of MM420 Frequency Converter

See Table 3. 4. 2 for the main parameter of MM420 frequency converter.

Table 3. 4. 2 Main Parameter of MM420 Frequency Converter

Number	Parameter of Frequency Converter	Set Value	Function Description
1	P0010	30	
2	P0970	1	Factory reset
3	P0010	1	Fast debugging
4	P0003	2	Allow access to extended parameters
5	P0304	380	Rated voltage of motor
6	P0305	3	Rated current of motor
7	P0307	0. 3	Rated power of motor
8	P0310	50. 00	Rated frequency of motor
9	P0311	1300	Rated speed of motor
10	P1000	2	Analog input
11	P1080	0	Minimum frequency of motor(0Hz)
12	P1082	50. 00	Maximum frequency of motor(50Hz)
13	P1120	5. 0	Ramp-up time
14	P1121	0. 1	Ramp-down time
15	P0003	3	
16	P0010	0	
17	P0700	2	Select command source(input from terminal line)
18	P0701	1	ON/OFF(turn on forward/stop command 1)
19	P0702	2	Turn on reverse

Ⅲ. Main Control Principle and Wiring Diagram

Electrical schematic diagram is shown in Figure 3. 4. 2, wiring dragram of 420 frequency converter is shown in Figure 3. 4. 3.

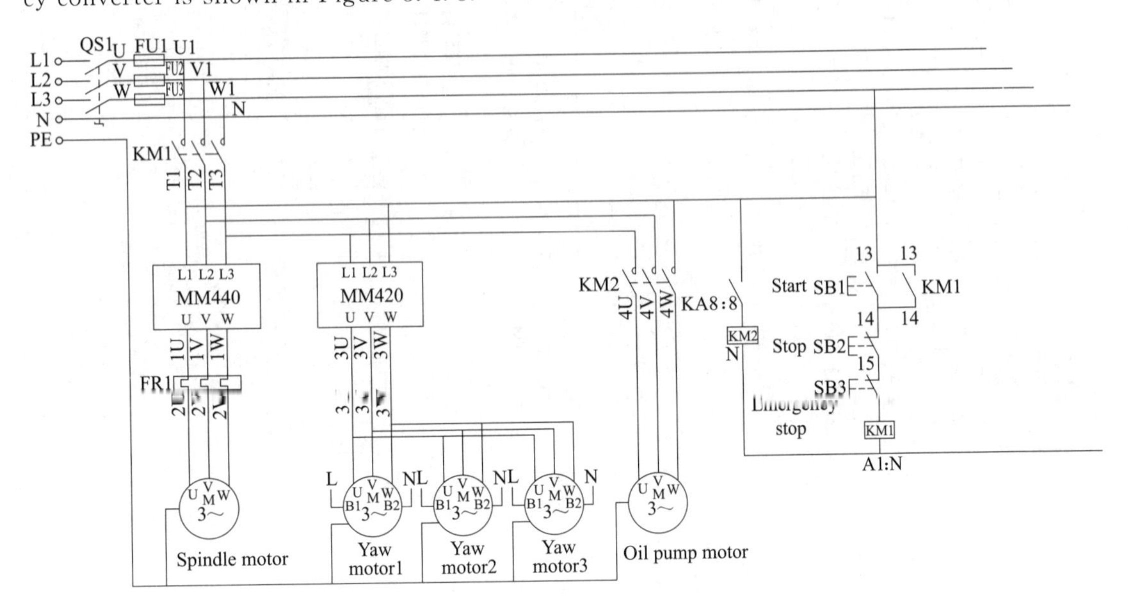

Figure 3. 4. 2 Electrical Schematic Diagram

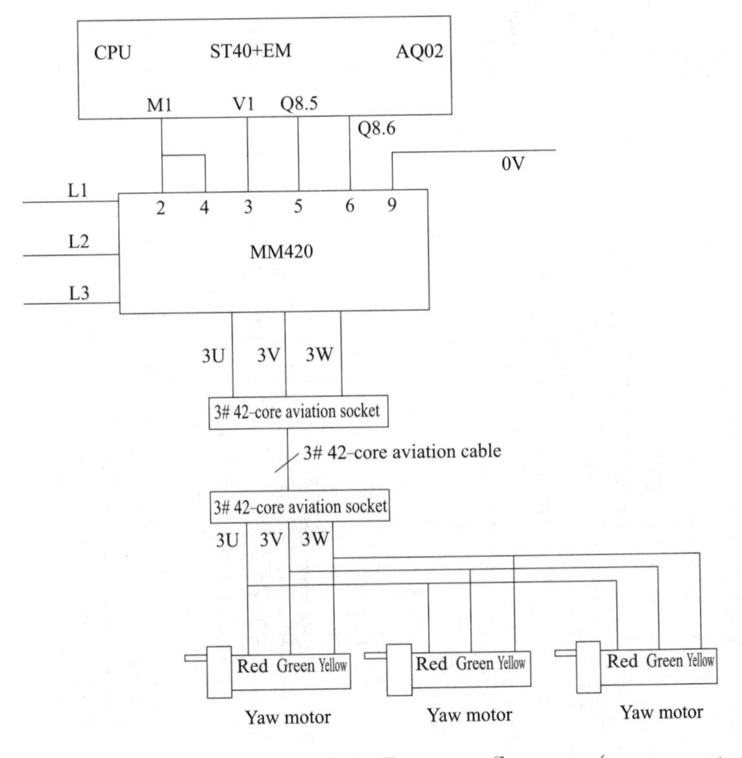

Figure 3. 4. 3 Wiring Diagram of 420 Frequency Converter (yaw converter)

Exploded view of control cabinet of yaw pitch control system is shown in Figure 3. 4. 4.

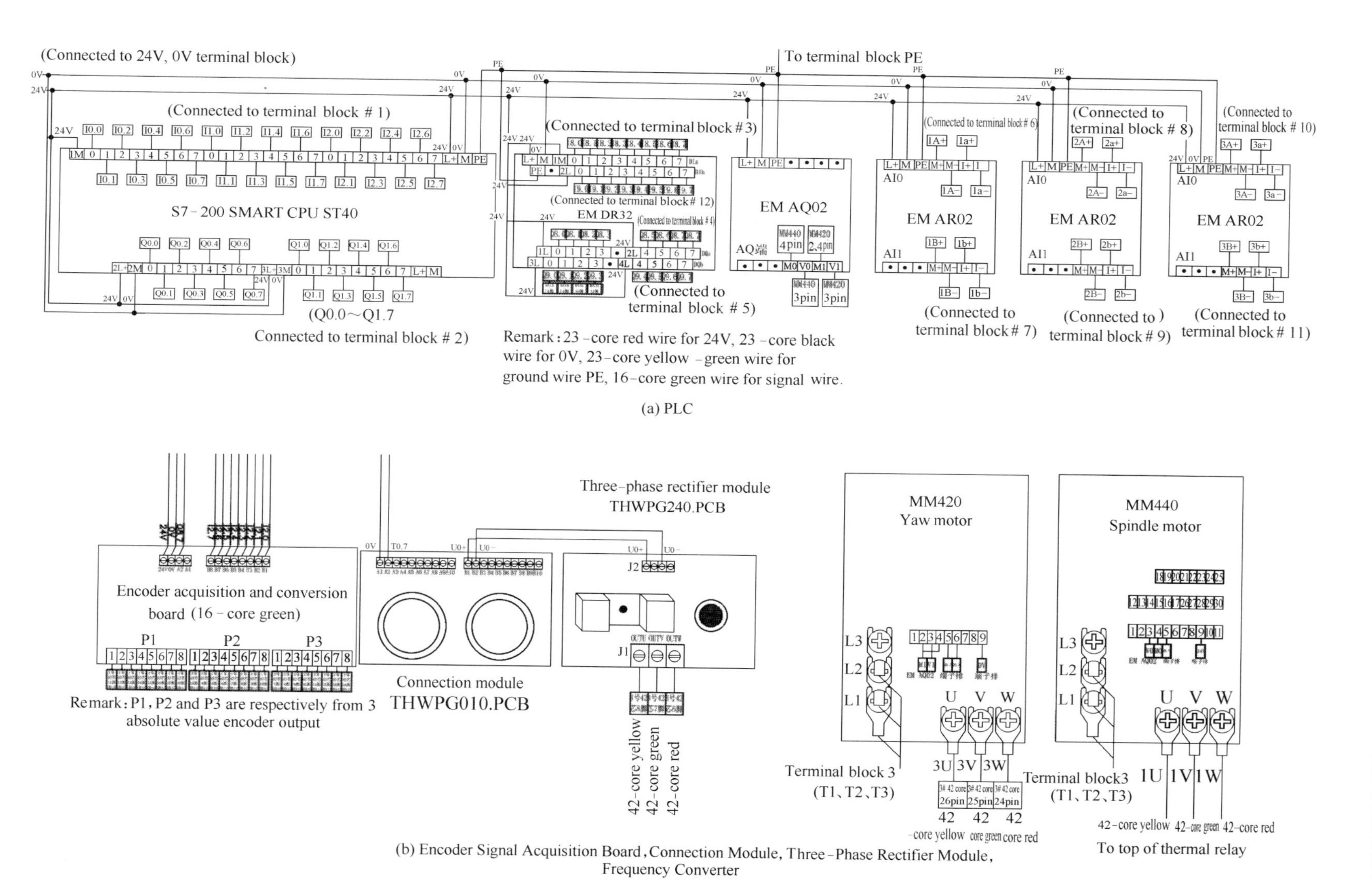

(a) PLC

(b) Encoder Signal Acquisition Board, Connection Module, Three-Phase Rectifier Module, Frequency Converter

Figure 3.4.4

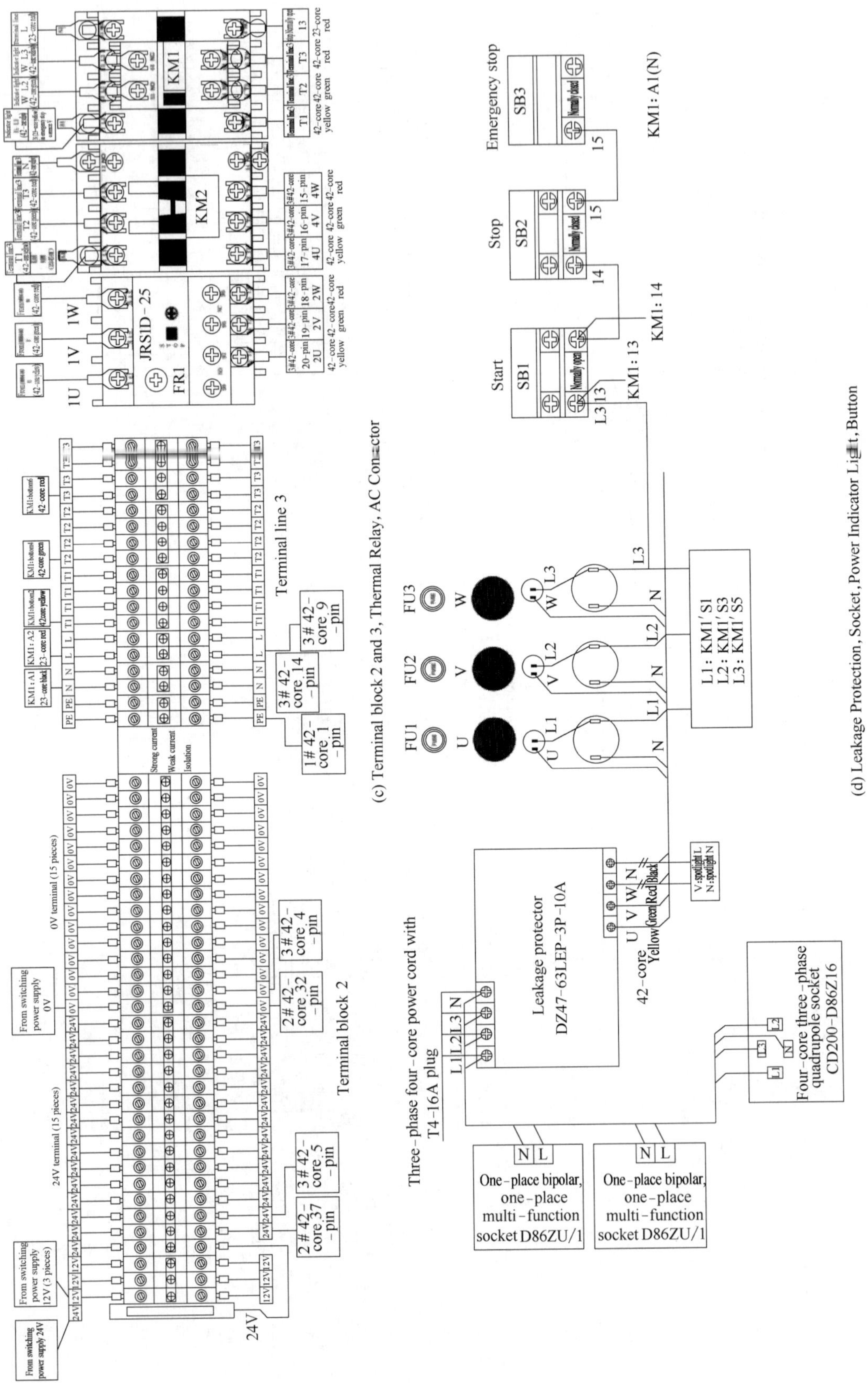

(c) Terminal block 2 and 3, Thermal Relay, AC Connector

(d) Leakage Protection, Socket, Power Indicator Light, Button

Remark: 42-core red wire for L; 42-core black wire for N
23-core yellow-green wire for ground wire PE

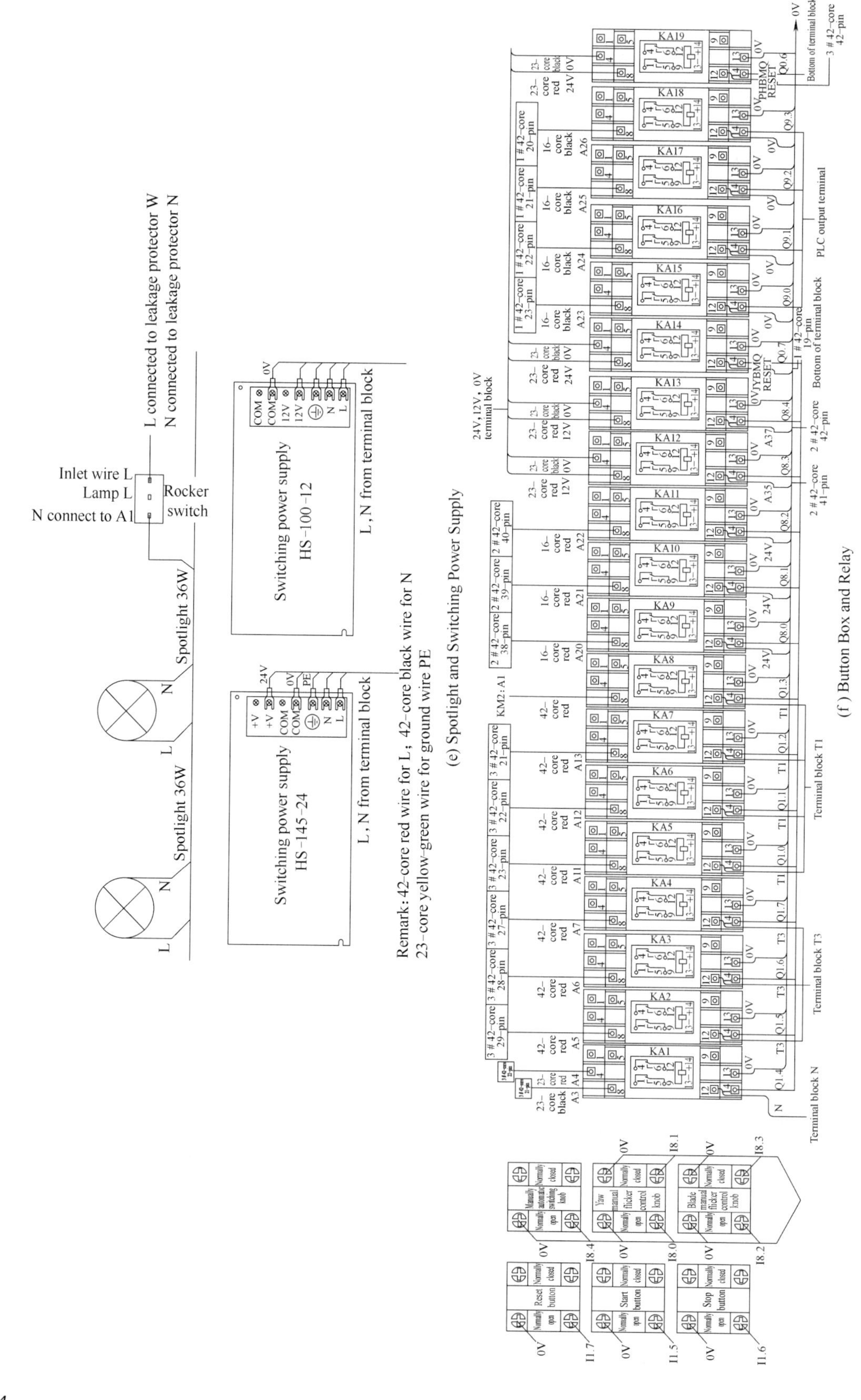

(e) Spotlight and Switching Power Supply

(f) Button Box and Relay

Figure 3. 4. 4

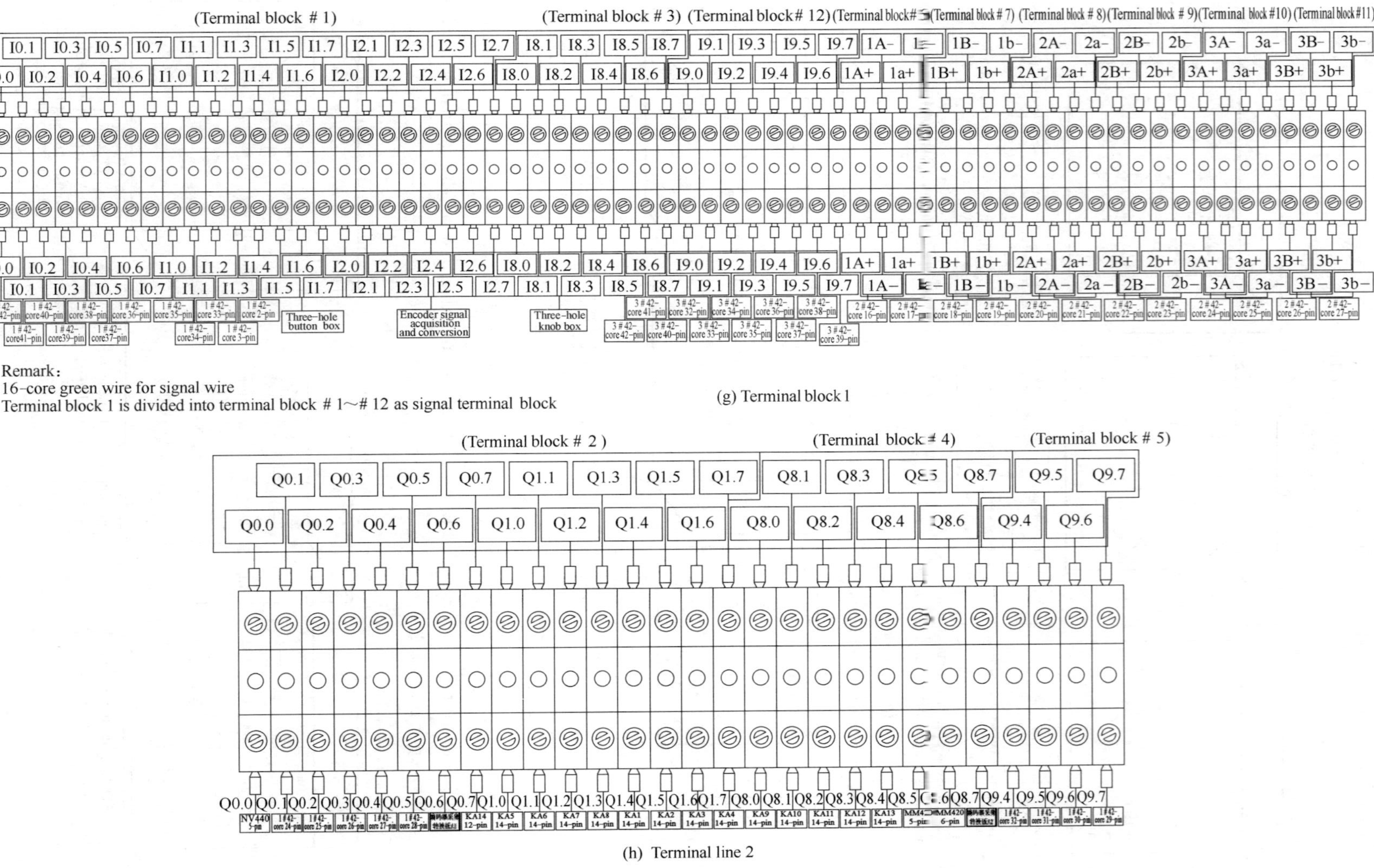

Remark:
16-core green wire for signal wire
Terminal block 1 is divided into terminal block # 1~# 12 as signal terminal block

(g) Terminal block 1

(h) Terminal line 2

Figure 3. 4. 4 Exploded View of Control Cabinet of Yaw Pitch Control System

Butt joint plug is shown in Figure 3. 4. 5.

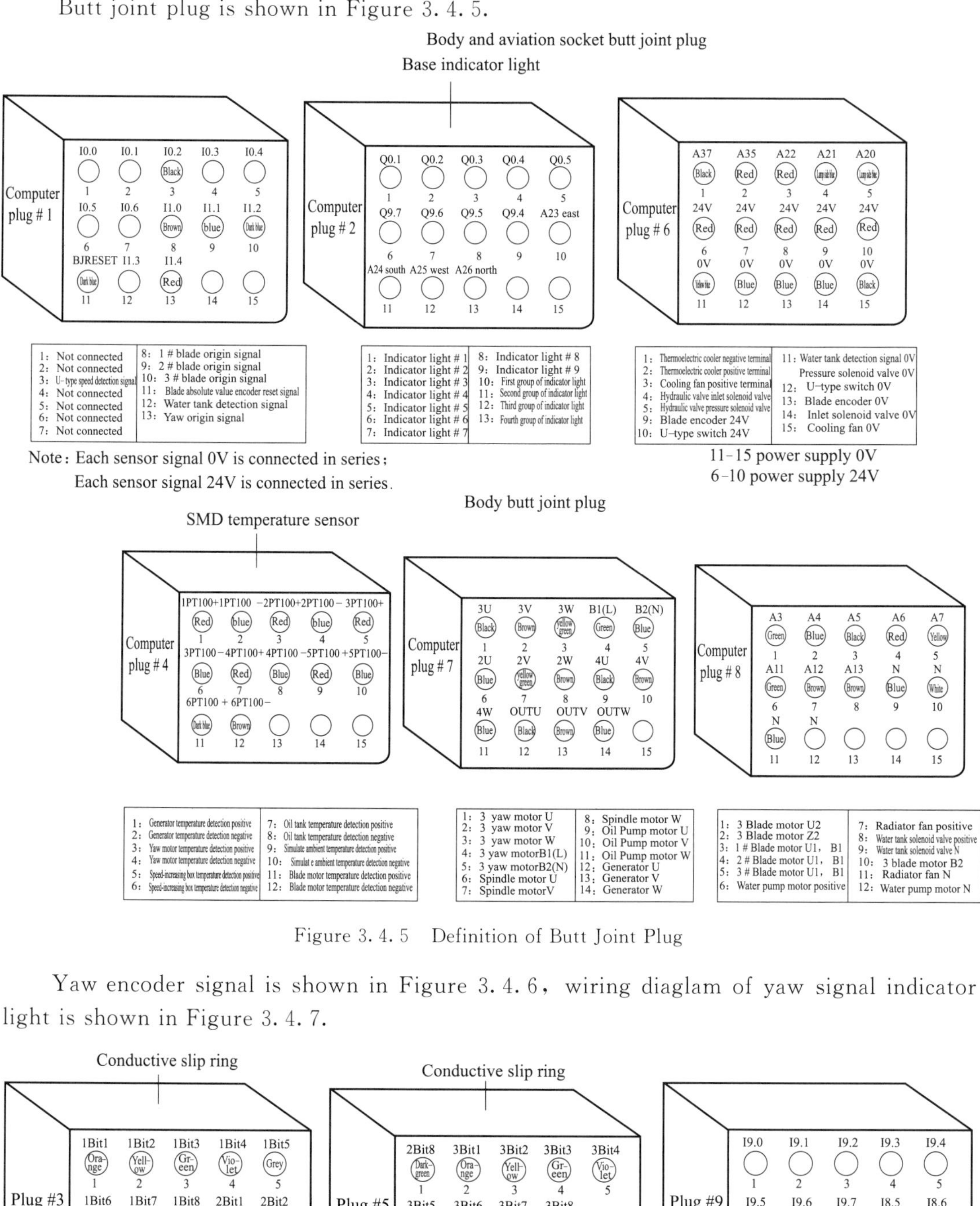
Body and aviation socket butt joint plug
Base indicator light
Computer plug # 1
I0.0 I0.1 I0.2 I0.3 I0.4
I0.5 I0.6 I1.0 I1.1 I1.2
BJRESET I1.3 I1.4
Computer plug # 2
Q0.1 Q0.2 Q0.3 Q0.4 Q0.5
Q9.7 Q9.6 Q9.5 Q9.4 A23 east
A24 south A25 west A26 north
Computer plug # 6
A37 A35 A22 A21 A20
24V 24V 24V 24V 24V
0V 0V 0V 0V 0V
1: Not connected
2: Not connected
3: U-type speed detection signal
4: Not connected
5: Not connected
6: Not connected
7: Not connected
8: 1 # blade origin signal
9: 2 # blade origin signal
10: 3 # blade origin signal
11: Blade absolute value encoder reset signal
12: Water tank detection signal
13: Yaw origin signal
1: Indicator light # 1
2: Indicator light # 2
3: Indicator light # 3
4: Indicator light # 4
5: Indicator light # 5
6: Indicator light # 6
7: Indicator light # 7
8: Indicator light # 8
9: Indicator light # 9
10: First group of indicator light
11: Second group of indicator light
12: Third group of indicator light
13: Fourth group of indicator light
1: Thermoelectric cooler negative terminal
2: Thermoelectric cooler positive terminal
3: Cooling fan positive terminal
4: Hydraulic valve inlet solenoid valve
5: Hydraulic valve pressure solenoid valve
9: Blade encoder 24V
10: U-type switch 24V
11: Water tank detection signal 0V
Pressure solenoid valve 0V
12: U-type switch 0V
13: Blade encoder 0V
14: Inlet solenoid valve 0V
15: Cooling fan 0V
Note: Each sensor signal 0V is connected in series;
Each sensor signal 24V is connected in series.
11-15 power supply 0V
6-10 power supply 24V
Body butt joint plug
SMD temperature sensor
Computer plug # 4
1PT100+1PT100 -2PT100+2PT100- 3PT100+
3PT100-4PT100+ 4PT100 -5PT100 +5PT100-
6PT100 + 6PT100-
Computer plug # 7
3U 3V 3W B1(L) B2(N)
2U 2V 2W 4U 4V
4W OUTU OUTV OUTW
Computer plug # 8
A3 A4 A5 A6 A7
A11 A12 A13 N N
N N
1: Generator temperature detection positive
2: Generator temperature detection negative
3: Yaw motor temperature detection positive
4: Yaw motor temperature detection negative
5: Speed-increasing box temperature detection positive
6: Speed-increasing box temperature detection negative
7: Oil tank temperature detection positive
8: Oil tank temperature detection negative
9: Simulate ambient temperature detection positive
10: Simulat e ambient temperature detection negative
11: Blade motor temperature detection positive
12: Blade motor temperature detection negative
1: 3 yaw motor U
2: 3 yaw motor V
3: 3 yaw motor W
4: 3 yaw motorB1(L)
5: 3 yaw motorB2(N)
6: Spindle motor U
7: Spindle motorV
8: Spindle motor W
9: Oil Pump motor U
10: Oil Pump motor V
11: Oil Pump motor W
12: Generator U
13: Generator V
14: Generator W
1: 3 Blade motor U2
2: 3 Blade motor Z2
3: 1 # Blade motor U1, B1
4: 2 # Blade motor U1, B1
5: 3 # Blade motor U1, B1
6: Water pump motor positive
7: Radiator fan positive
8: Water tank solenoid valve positive
9: Water tank solenoid valve N
10: 3 blade motor B2
11: Radiator fan N
12: Water pump motor N

Figure 3. 4. 5 Definition of Butt Joint Plug

Yaw encoder signal is shown in Figure 3. 4. 6, wiring diaglam of yaw signal indicator light is shown in Figure 3. 4. 7.

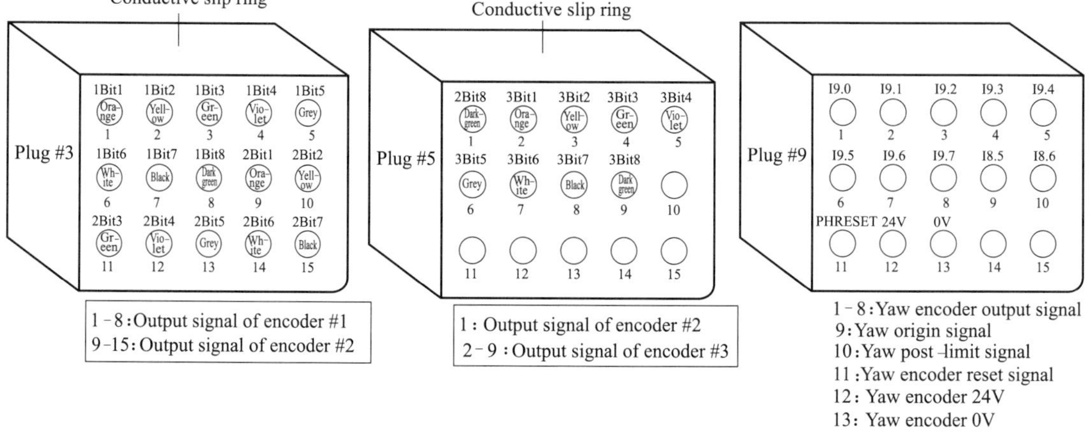
Conductive slip ring
Plug #3
1Bit1 1Bit2 1Bit3 1Bit4 1Bit5
1Bit6 1Bit7 1Bit8 2Bit1 2Bit2
2Bit3 2Bit4 2Bit5 2Bit6 2Bit7
1-8: Output signal of encoder #1
9-15: Output signal of encoder #2
Conductive slip ring
Plug #5
2Bit8 3Bit1 3Bit2 3Bit3 3Bit4
3Bit5 3Bit6 3Bit7 3Bit8
1: Output signal of encoder #2
2-9: Output signal of encoder #3
Plug #9
I9.0 I9.1 I9.2 I9.3 I9.4
I9.5 I9.6 I9.7 I8.5 I8.6
PHRESET 24V 0V
1-8: Yaw encoder output signal
9: Yaw origin signal
10: Yaw post -limit signal
11: Yaw encoder reset signal
12: Yaw encoder 24V
13: Yaw encoder 0V

Figure 3. 4. 6 Yaw Encoder Signal

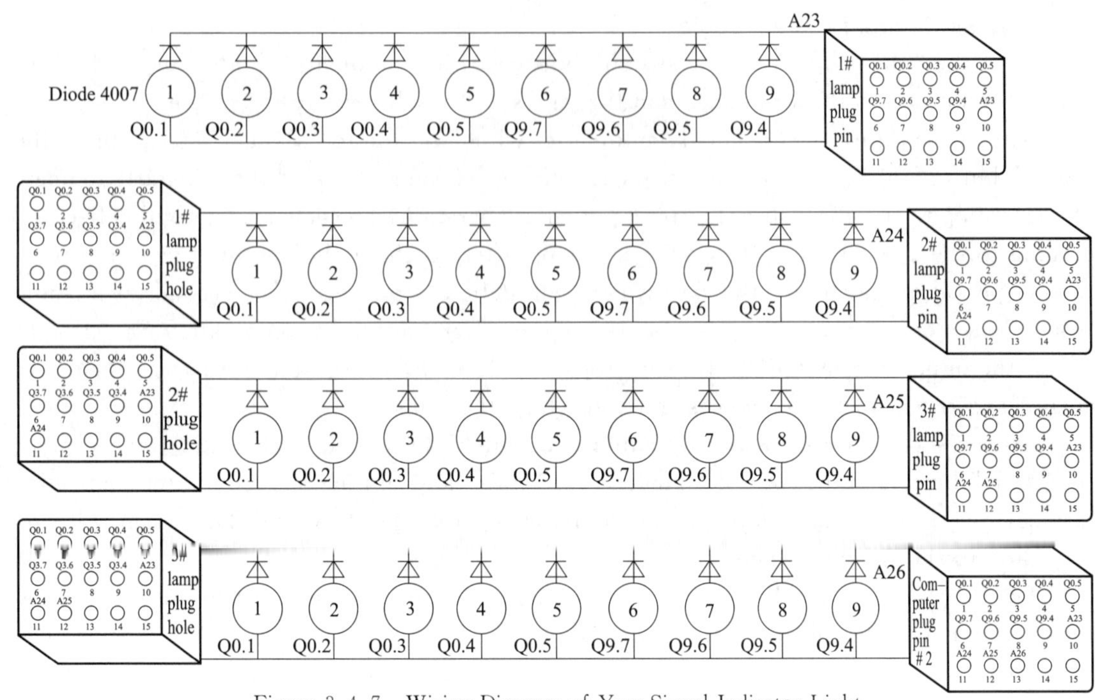

Figure 3. 4. 7 Wiring Diagram of Yaw Signal Indicator Light

Ⅳ. Yaw Control

1. I/O Distribution Related to Yaw Control

See Table 3. 4. 3 for yaw control related I/O distribution.

Table 3. 4. 3 Yaw Control Related I/O Distribution Table

Input Signal/Equipment	Input Terminal	Output Signal/Equipment	Output Terminal
Manual yaw forward	I8. 0	Yaw position indicator light # 1	Q0. 1
Manual yaw reverse	I8. 1	Yaw position indicator light # 2	Q0. 2
Yaw origin	I8. 5	Yaw position indicator light # 3	Q0. 3
Yaw post-limit	I8. 6	Yaw position indicator light # 4	Q0. 4
Yaw bit 1	I9. 0	Yaw position indicator light # 5	Q0. 5
Yaw bit 2	I9. 1	Yaw encoder RESET	Q0. 6
Yaw bit 3	I9. 2	Yaw frequency converter forward	Q8. 6
Yaw bit 4	I9. 3	Yaw frequency converter reverse	Q8. 5
Yaw bit 5	I9. 4	Yaw position indicator light group I(east)	Q9. 0
Yaw bit 6	I9. 5	Yaw position indicator light group II(south)	Q9. 1
Yaw bit 7	I9. 6	Yaw position indicator light group III(west)	Q9. 2
Yaw bit 8	I9. 7	Yaw position indicator group 4(north)	Q9. 3
Read yaw motor temperature	AIW50	Yaw position indicator light # 9	Q9. 4
		Yaw position indicator light # 8	Q9. 5
		Yaw position indicator light # 7	Q9. 6
		Yaw position indicator light # 6	Q9. 7
		Yaw frequency converter analog value	AQW34

2. Main Variables Related to Yaw Control

See Table 3. 4. 4 for yaw control related variables.

Table 3. 4. 4 Yaw Control Related Variables Table

Variable Name	Address	Variable Name	Address
Yaw angle input	VW102	Current yaw angle display	VW112
		Yaw motor temperature display	VW120

3. Yaw Control Program Logic

Turn the "manually automatic switch" button to the "automatic" state, press the "reset" button on the control cabinet, and the yaw system will reset to the origin.

After reset, turn the "manually automatic switch" button to "Manual" state, press the "Start" button on the control cabinet to enter the operation state, and the fan starts to operate according to the given speed; after a period of operation, the yaw system will deflect according to the first given angle value, and stop after reaching the specified angle; after a further period of operation, the yaw system will deflect according to the second given angle value, and stop after reaching the specified angle; after continuing to operate for a period of time, the brake system works, the fan stops running, the yaw system returns to the origin again, and repeat the above process after reaching the origin.

After reset, turn the "manually automatic switch" button to "automatic" state, press the "start" button on the control cabinet or the "start" key on the upper computer control interface, set the yaw angle on the upper computer control interface, and the system will yaw according to the set angle value.

The yaw speed is small when yaw is initiated and near the yaw given angle; the yaw speed is higher in the middle of the yaw process.

There are a total of 36 (9×4) yaw angle indicator lights on the fan object base. When the yaw runs to the given angle value, it stops, the corresponding position indicator lights up, and the LED dot matrix screen interface displays the current yaw angle.

Project Principle and Basic Knowledge

Ⅰ. Encoder

There are many scribed lines on the optical coded disk of absolute encoder. Each scribed line is arranged in sequence of 2 lines, 4 lines, 8 lines, 16 lines..., each position of encoder obtains a group of unique binary code from 2^0 to 2^{n-1} by reading the open and dark of each scribed line. This is called n-bit absolute encoder. Because it is not affected by power failure and interference, there is no need to remember, find reference points, and count all the time, the anti-inference characteristics of the encoder and the reliability of the date are greatly improved. The absolute value encoder is mostly used for angle measurement.

Ⅱ. Gray Code

Absolute encoder is a sensor directly output the digital quantity, using natural binary or cyclic binary (Gray code) mode for photoelectric conversion, coding design generally adopts natural binary code, cyclic binary code, two′s complement and so on. Strong anti-interference ability, no accumulated error; position information will not be lost after the power supply is cut off, but the resolution is determined by the number of bits in binary system. According to different accuracy requirements, different resolutions, i. e. the number of bits, can be selected. Absolute encoder using cyclic binary coding, the output signal is a sort of digital sequence, not a weight code. Each bit has no definite size, so it cannot be directly compared and carried out arithmetic operations, nor can it be directly converted into other signals. After a code conversion, it becomes a natural binary code, which is read by the upper computer to realize the corresponding control, see Table 3. 4. 5.

Table 3.4.5 Comparison Table of Several Natural Binary Code and Gray Code

Decimal	Binary	Gray Code	Decimal	Binary	Gray Code
0	0000	0000	8	1000	1100
1	0001	0001	9	1001	1101
2	0010	0011	10	1010	1111
3	0011	0010	11	1011	1110
4	0100	0110	12	1100	1010
5	0101	0111	13	1101	1011
6	0110	0101	14	1110	1001
7	0111	0100	15	1111	1000

1. Conversion of natural binary code into binary Gray code

The natural binary code is converted into binary Gray code. The rule is to keep the highest bit of the natural binary code as the highest bit of the Gray code, while the next high bit of the Gray code is the XOR of the higher bit and the next high bit of the binary code, and the rest bits of Gray code are similar to those of the next high bit.

2. Conversion of binary Gray code into natural binary code

The binary Gray code is converted into natural binary code, its rule is to keep the highest bit of Gray code as the highest bit of natural binary code, while the next high bit of the natural binary code is the XOR of the higher bit of the natural binary code and the next high bit of gray code, and the rest bits of natural binary code are similar to those of the next high bit.

Siemens S7-200 SMART does not have Gray code conversion library, you need to download the Gray code library on the official website and add the library, in the programming only need to call to complete the Gray code to binary conversion, thus intuitively see the value of absolute value encoder.

Ⅲ. Yaw Principle

The absolute value encoder of this set of equipment adopts eight-wire connection, and the conversion value range is 0-255. Since the absolute value encoder reflects the rotation angle through the absolute position, the encoder output value executes a cycle for each revolution of the encoder shaft. Therefore, the yaw angle range reflected by yaw system is closely related to the ratio of the number of teeth on the pinions to the number of teeth on the bull gears where the yaw motor is located.

The number of teeth on the yaw pinions of this set of equipment is 21 and the number of teeth on the yaw bull gear is 94. Therefore, the rotation angle value of fan object is $21/94\times 360° = 80.43°$ for each rotation (360°) of yaw motor shaft, i. e. the corresponding angle range of 0-255 converted by absolute value encoder is 0-80.43°.

Yaw angle given by upper computer is stored in VW102 of PLC, which is an angle value. The correspondence between the angle value and the output value of absolute value encoder is related to the number of teeth on the yaw pinions/yaw bull gears and is calculated as follows:

Pitch angle corresponding to encoder output value 255 $21/94\times 360\ (°)\approx 80°$

Angle corresponding to 1 encoder output value $80°/255=0.3137\ (°)$/piece

Code value corresponding to given yaw angle = given angle value/0.3137

Therefore, the main work of this training project is to convert the numerical values sent by absolute value encoder into intuitionistic values after sorting, and make comparison, so as to determine the yaw direction and angle value.

Project Implementation

Ⅰ. List of Instruments, Equipment and Tools

THWPWG-3B large-scale wind power generation system operational training platform fan object model, control cabinet of yaw pitch control system, control cabinet of energy control. monitoring management. weather station.

Ⅱ. Safe Operation Specification

① Proficient in control principle and program control method of yaw motor;

② Before training, read the instruction manual carefully to get familiar with the parts related to yaw and relevant operation instructions to ensure that the power supply of each system control cabinet is disconnected, and get familiar with the operation steps of this training according to relevant contents in the training instruction;

③ Power on and off the system in strict accordance with correct operation steps, so as to avoid damage to the system caused by misoperation;

④ Check whether each power supply and equipment are normal first, and then formally start the training task after ensuring they are normal;

⑤ In the process of training, always keep the training platform clean and tidy, do not place sundries at will, so as to avoid short circuit and other faults, and pay attention to safety when there is " danger" sign;

⑥ After the training is completed, the power switch shall be turned off in time and the training platform shall be cleaned in time.

Ⅲ. Operational Training Steps

1. Yaw back to origin

① Turn on the main power supply of the control system, the system is powered on, and the power indicator light is on.

② Turn the three-position knob "M/A" knob to "A" state.

③ Press "RESET" button.

④ Fan object yaw back to origin.

⑤ When the fan object yaws back to the origin, the origin sensor indicator light is on.

⑥ KA19 relay acts, yaw absolute value encoder resets, blade acquisition signal I9. 0-I9. 7 signal is cleared, corresponding indicator light on PLC goes out, yaw reset is completed.

2. Yaw manual forward and reverse control

① Turn on the main power supply of the control system, the system is powered on, and the power indicator light is on.

② Turn the three-position knob "M/A" knob to "M" state.

③ Turn the three-position knob "Y: FWD/REV" knob to "FWD" and rotate the yaw clockwise.

④ Turn the three-position knob "Y: FWD/REV" knob to "REV" and rotate the yaw anticlockwise.

3. Forward and reverse control of main control computer

① Turn the three-position knob "M/A" knob to "A" state.

② Open the control software of main control computer and connect step by step.

③ First, reset the three blades to the origin.

④ Enter the yaw angle on the main control computer, press the "confirm" button, and then press the "Start" button.

⑤ Yaw rotates to set angle.

⑥ The indicator light goes on or goes off depending on the deflection angle.

⑦ At the next yaw, reset before yawing.

4. Program reading——yaw control

① Master yaw control logic of wind power generating unit.

② List the main I/O distribution related to yaw control and the meaning of the main intermediate relay and variable.

③ Describe the yaw control logic of the network 5, HSC_INIT, network 61, 63, 30, 31, 32, 19 with flow chart or natural language.

5. Programming——yaw control

① Compile the fan yaw control program of the wind power generating unit.

② The user can adjust the analog quantity input into frequency converter MM420 according to different phases of yaw through the given yaw angle by upper computer software.

At the same wind speed, when the yaw angle is within the range of $\pm 6°$ from the origin, the analog quantity output value is subtracted by 500. When the blade angle is within the range of $\pm 3°$, the analog quantity output value is subtracted by 1000.

Press the start button to start the spindle frequency converter; press the stop button or the wind speed exceeds the safe speed to stop the spindle frequency converter.

a. Measure the rotating speed of spindle;

b. The analog quantity of spindle frequency converter is given;

c. The analog quantity of yaw frequency converter is given;

6. Write training report

Project Work

Describe the realization process of yaw function with words or flow chart.

Project Ⅴ Realization of Pitch Function

Project Description

Understand the working principle of pitch control system, implement manual pitch control and automatic pitch control of upper computer; complete program editing and realize pitch function of given angle.

Competency Objectives:

① Grasp the working principle of absolute value encoder and Gray code;

② Understand the pitch control process and realize it by PLC programming.

Project Environment

The pitch control system is mainly composed of slave PLC, frequency converter, absolute value encoder, pitch motor (AC gear motor), control button and other devices. It can complete the installation of pitch control system, the programming and debugging of PLC control program for manual and automatic pitch, and the communication between slave PLC and master PLC.

Ⅰ. Main Equipment of Pitch Control System

See Table 3. 5. 1 for the list of main equipment of pitch control system.

Table 3. 5. 1 List of Main Equipment of Pitch Control System

Number	Name	Specification	Quantity
1	Control cabinet of yaw pitch control system	880mm×600mm×2100mm	1 set
2	Button module	One 24V/6A switching power supply; one emergency stop button; one reset, start and stop buttons(yellow, green and red) respectively; one self-locking button(yellow, green and red) respectively; one change-over switch; two 24V indicator lights(yellow, green and red) respectively	1 group
3	PLC	Siemens CPU ST40 host	1 set
		EM DR32 digital quantity input/output module, 16×24V DC input/16-point relay output	1 set
		EM AQ02 module, 2-way analog quantity output(2AI)	1 set
		EM AR02 module, 2-way thermal resistance input module (PT100)	3 sets
		EM DP01 communication module	1 set
4	Switching power supply	HS-145-24, 24V output	1 piece
		HS-100-12, 12V output	1 piece
5	AC contactor	LC1-D0610M5N 220V	2 pieces

Ⅱ. Main Control Principle and Wiring Diagram

See Figure 3. 5. 1 for wiring diagram of pitch motor, see Figure 3. 5. 2 for wiring dragram of conductive slip ring, the operating capacitance is connected between each blade motor U2 and Z2.

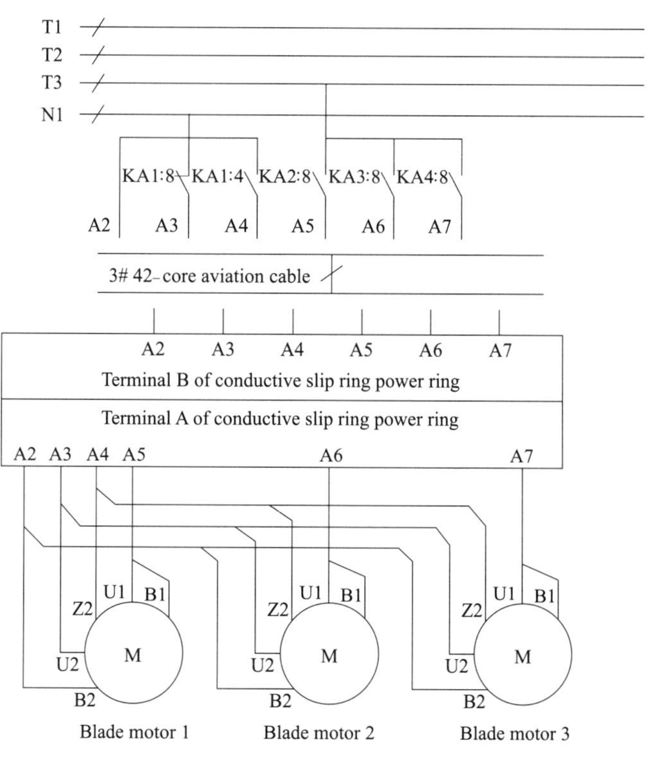

Figure 3. 5. 1 Wiring Diagram of Pitch Motor

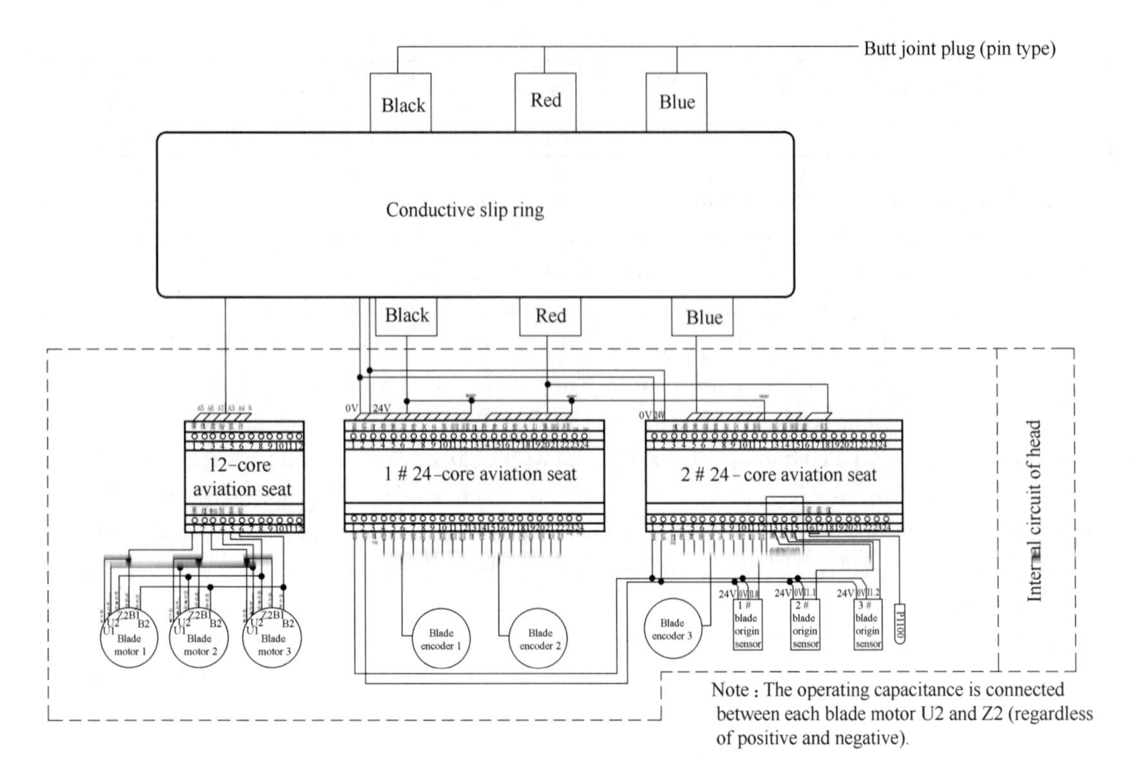

Figure 3. 5. 2 Wiring Diagram of Conductive Slip Ring

See Figure 3. 4. 4-Figure 3. 4. 6 for wiring diagram and exploded view of yaw pitch control system control cabinet, and the definition of butt joint plug.

Ⅲ. Pitch Control

1. I/O Distribution Related to Pitch Control

See Table 3. 5. 2 for pitch control related I/O distribution.

Table 3. 5. 2 Pitch Related I/O Distribution Table

Input Signal/Equipment	Input Terminal	Output Signal/Equipment	Output Terminal
Blade 1 origin	I1. 0	Blade encoder RESET	Q0. 7
Blade 2 origin	I1. 1	Blade motor rotate forward	Q1. 4
Blade 3 origin	I1. 2	Blade 1 start	Q1. 5
Blade Bit 1	I2. 0	Blade 2 start	Q1. 6
Blade Bit 2	I2. 1	Blade 3 start	Q1. 7
Blade Bit 3	I2. 2		
Blade Bit 4	I2. 3		
Blade Bit 5	I2. 4		
Blade Bit 6	I2. 5		
Blade Bit 7	I2. 6		
Blade Bit 8	I2. 7		
Manual blade reverse	I8. 2		
Manual blade rotate forward	I8. 3		
Read blade motor temperature	AIW82		

2. Main Variables Related to Pitch Control

See Table 3. 5. 3 for pitch control related variables.

Table 3. 5. 3 Pitch Related Variables Table

Variable Name	Address	Variable Name	Address
Pitch angle input	VW104	Current blade angle display	VW114
		Blade motor temperature display	VW128

3. Pitch Control Program Logic

Turn the "manually automatic switch" button to the "automatic" state, press the "reset" button on the control cabinet to reset the blade system to the origin.

After reset, turn the "manually automatic switch" button to "manual" state, press the "start" button on the control cabinet to enter the operation state, and the fan starts running according to the given speed; after running for a period of time and completing the first yaw, the blade system rotates according to the first given angle, and stops after reaching the specified angle; after running for another period of time and completing the second yaw, the blade system rotates according to the second given angle and stops after reaching the specified angle; after continuing to operate for a period of time, the brake system works, the fan stops running, the blade system returns to the origin, and repeat the above process after reaching the origin.

After reset, rotate the "manually automatic switch" button to "automatic" state, press the "start" button on the control cabinet or the "start" key on the upper computer control interface, and set the blade angle on the upper computer control interface. The system will pitch according to the set angle value.

The pitch control system adjusts the pitch based on the pitch angle value given in the upper computer software, and the current pitch angle is displayed in the upper computer software interface. When the pitch runs to the given angle value, it stops, and the LED dot matrix screen interface displays the current yaw angle.

Project Principle and Basic Knowledge

Ⅰ. Encoder (See Project Ⅳ)

Ⅱ. Gray Code (See Project Ⅳ)

Ⅲ. Pitch Principle

The absolute value encoder of this set of equipment adopts eight-wire connection, and the conversion value range is 0-255. Since the absolute value encoder reflects the rotation angle through the absolute position, the encoder output value executes a cycle for each revolution of the encoder shaft. Therefore, the pitch angle range reflected by pitch system is closely related to the ratio of the number of pinions to the number of pitch bull gears where the pitch motor is located.

The number of pitch pinions of this set of equipment is 20, and the number of internal gears of pitch bearing is 37. Therefore, the rotation angle value of blade is $20/37 \times 360° = 194.594595°$ for each rotation (360°) of pitch motor shaft, i. e. the corresponding angle range of 0-255 converted by absolute value encoder is 0-194. 594595°.

Therefore, the main work of this training project is to convert the numerical values sent by absolute value encoder into intuitionistic values after sorting, and make comparison, so as to determine the rotation direction and angle value of pitch motor.

Project Implementation

Ⅰ. List of Instruments, Equipment and Tools

THWPWG-3B large-scale wind power generation system operational training platform fan object model, control cabinet of yaw pitch control system, control cabinet of energy control. monitoring management. weather station.

Ⅱ. Safe Operation Specification

① Proficient in control principle and program control method of pitch motor;

② Before training, read the instruction manual carefully to get familiar with the parts related to the pitch and relevant operation instructions to ensure that the power supply of each system control cabinet is disconnected, and get familiar with the operation steps of this training according to the relevant contents in the training instruction;

③ Power on and off the system in strict accordance with correct operation steps, so as to avoid damage to the system caused by misoperation;

④ Check whether each power supply and equipment are normal first, and then formally start the training task after ensuring they are normal;

⑤ In the process of training, always keep the training platform clean and tidy, do not place sundries at will, so as to avoid short circuit and other faults, and pay attention to safety when there is " danger" sign;

⑥ After the training is completed, the power switch shall be turned off in time and the training platform shall be cleaned in time.

Ⅲ. Operational Training Steps

1. Program reading and understanding

See Table 3. 5. 4 for pitch absolute value encoder pin information.

Table 3. 5. 4 Pin Information Table of Pitch Absolute Value Encoder

Pin Number, Definition	Wire Color	Resolution		
		512	256/180	128/90
1	Blue	0V	←	←
2	Brown	10. 8-26. 4V	←	←
3	Black	Not connected	←	←
4	Red	Bit 1 (2^0)	Not connected	←
5	Orange	Bit 2 (2^1)	Bit 1 (2^0)	Not connected
6	Yellow	Bit 3 (2^2)	Bit 2 (2^1)	Bit 1 (2^0)
7	Green	Bit 4 (2^3)	Bit 3 (2^2)	Bit 2 (2^1)
8	Violet	Bit 5 (2^4)	Bit 4 (2^3)	Bit 3 (2^2)
9	Grey	Bit 6 (2^5)	Bit 5 (2^4)	Bit 4 (2^3)
10	White	Bit 7 (2^6)	Bit 6 (2^5)	Bit 5 (2^4)
11	Black white	Bit 8 (2^7)	Bit 7 (2^6)	Bit 6 (2^5)
12	Red white	Bit 9 (2^8)	Bit 8 (2^7)	Bit 7 (2^6)
13	Blue white	Reset	←	←

As shown in Table 3. 5. 4, the weights of various output bits corresponding to various wiring methods of absolute value encoder have been given. This set of equipment adopts 256/180 wiring method. Orange-red/white wires are respectively connected to PLC input points I2. 0-I2. 7. These signals need to be processed before they can be transformed into intuitionistic data. The specific method is to treat these data as binary numbers first, convert them into decimal numbers through different weights, and then call Gray code conversion library to convert them into data that can be used directly.

Pitch absolute value encoder reset program:

When the voltage is between 0-24 V DC (high level), the position of the current shaft is set to the "0" position of the encoder. During normal operation, the encoder needs to maintain a voltage between 0-0. 8 V DC (low level).

Q0. 7 is used to drive relay KA14 to switch on 24V, switch on 0V after 1s, and the encoder is reset. The procedure is shown in Figure 3. 5. 3.

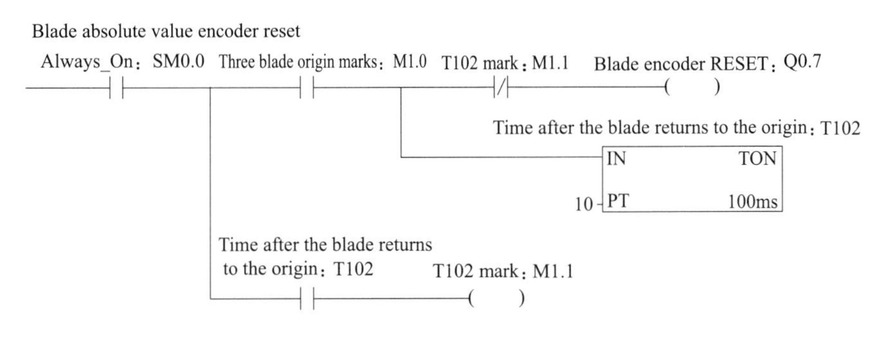

Figure 3. 5. 3 Pitch Encoder Reset Program

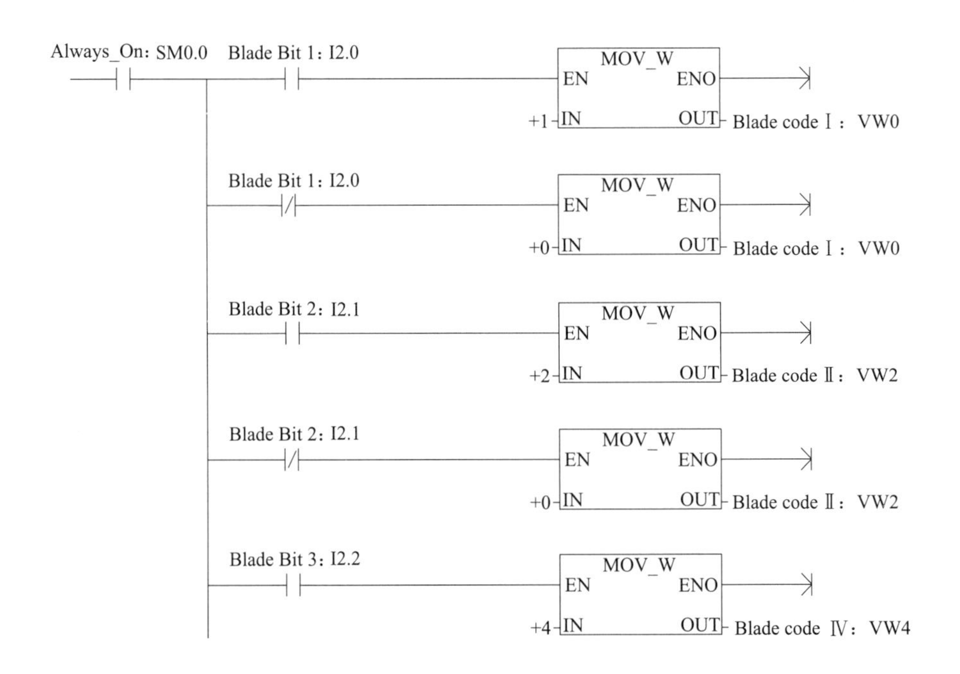

Figure 3. 5. 4

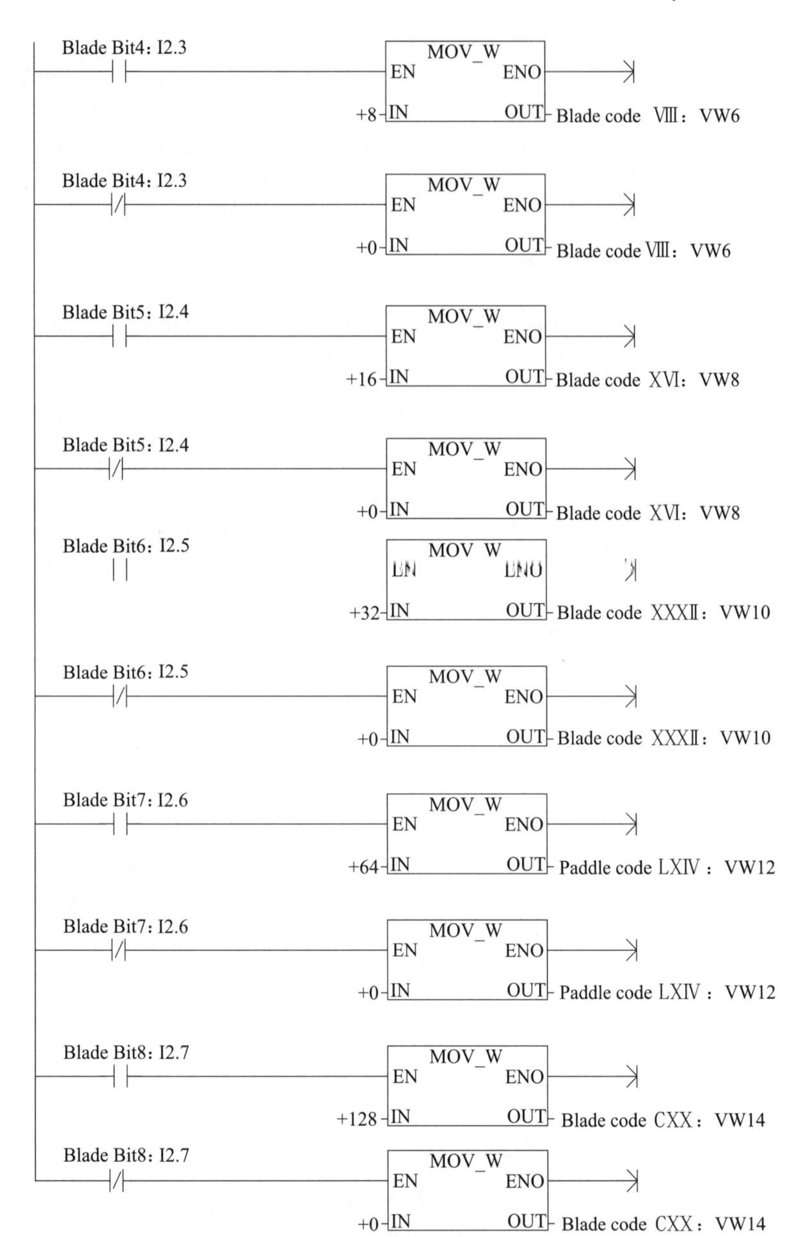

Figure 3.5.4 Signal Acquisition Program for Pitch Encoder

Pitch encoder signal acquisition program is shown in Figure 3.5.4.

The program shown in Figure 3.5.4 treats the data from the absolute value encoder as binary processing, that is, when there is no signal of 8-bit signals of the encoder, the corresponding signal storage area is all 0.

When I2.0=1, $2^0=1$ is transferred to VW0.

When I2.1=1, $2^1=2$ is transferred to VW2.

When I2.2=1, $2^2=4$ is transferred to VW4.

When I2.3=1, $2^3=8$ is transferred to VW6.

When I2.4=1, $2^4=16$ is transferred to VW8.

When I2.5=1, $2^5=32$ is transferred to VW10.

When I2.6=1, $2^6=64$ is transferred to VW12.

When I2.7=1, $2^7=128$ is transferred to VW14.

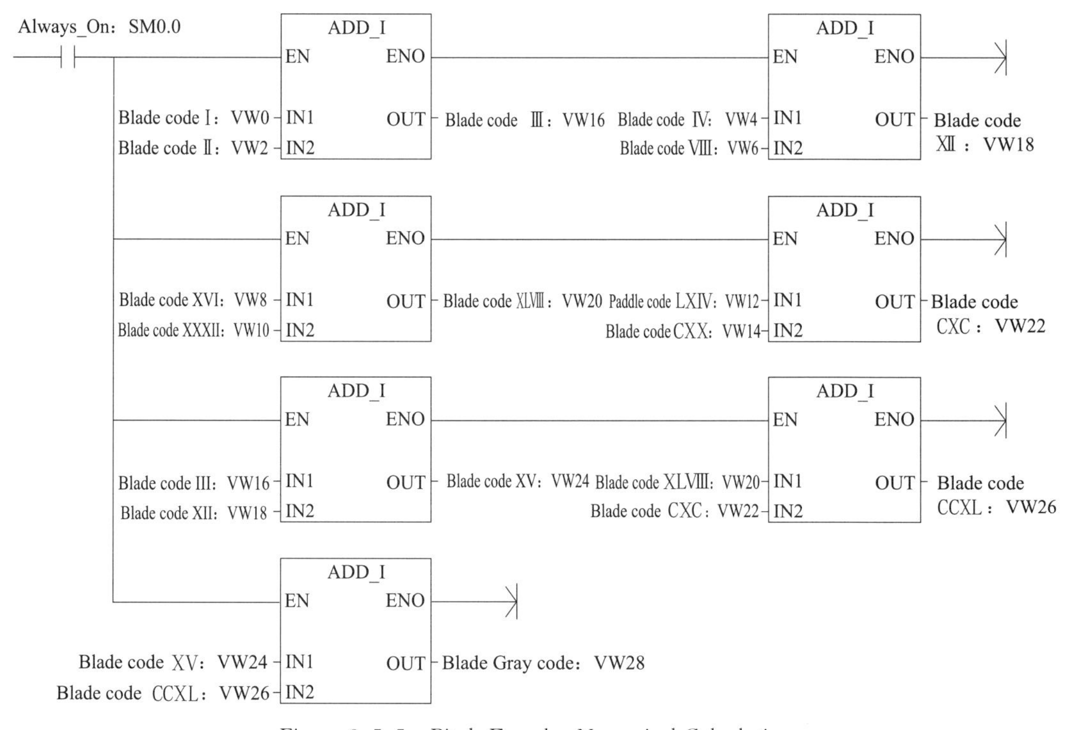

Figure 3. 5. 5 Pitch Encoder Numerical Calculation

Pitch encoder numerical calculation is shown in Figure 3. 5. 5.

The program shown in Figure 3. 5. 5 converts the resulting binary numbers into decimal numbers. Add the values of VW0 and VW2 and put in VW16, add the values of VW4 and VW6 and put in VW18; add the values of VW8 and VW10 and put in VW20; add the values of VW12 and VW14 and put in VW22; add the values of VW16 and VW18 and put in VW24; add the values of VW20 and VW22 and put in VW26; add the values of VW24 and VW26 and put in VW28.

The value obtained in VW28 is the Gray code value generated by the encoder, as shown in Figure 3. 5. 6.

Gray code conversion:

Always_On: SM0.0
GRAY_BIN_W
EN
Blade Gray code: VW28 IN
OUT Actual code value of blade: VW30

Figure 3. 5. 6 Gray Code Conversion of Pitch Encoder

The program shown in Figure 3. 5. 6 converts the calculated Gray code decimal number into a directly usable decimal number by calling the Gray code conversion library and stores it in VW30.

Conversion of values between pitch angle and absolute value encoder:

The pitch angle given by the main control computer is stored in the VW104 of PLC and is an angle value. The correspondence between the angle value and the absolute value encoder output value is related to the number of teeth on the pitch pinions (20) /the number of teeth on the pitch bearing internal gears (37) and is calculated as follows:

Pitch angle corresponding to encoder output value 255: $\frac{20}{37}\times 360^\circ \approx 194^\circ$;

Angle corresponding to 1 encoder output value: $\frac{194°}{255}=0.763116(°)/\text{piece}$;

Code value corresponding to given pitch angle $=\frac{\text{given code value}}{0.763116}$. The implementation program is shown in Figure 3. 5. 7.

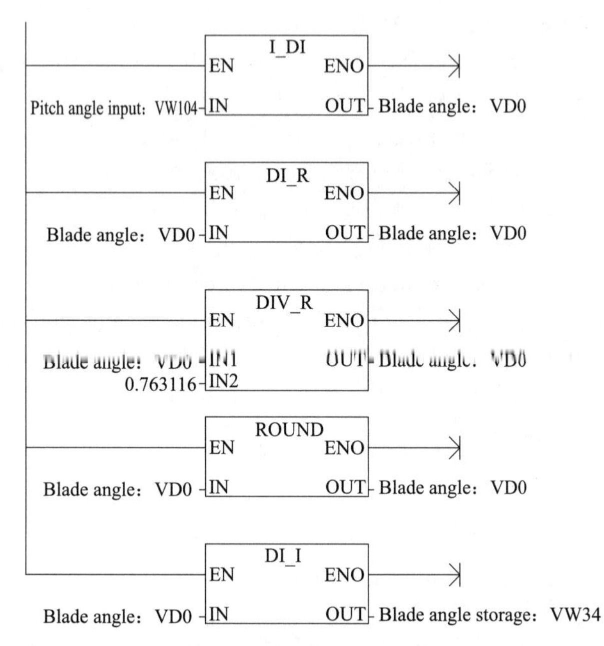

Figure 3. 5. 7 Conversion of Values between Pitch Angle and Absolute Value Encoder

Convert the angle input in VW104 from integer to double integer, and store it in VD0; convert the value in VD0 into real number to facilitate subsequent calculation;

Divide the value in VD0 converted into real number by 0. 763116 to obtain the blade angle of real number to be stored in VD0 again;

Rounding the value in VD0;

Converte the integer value in VD0 from double integer to integer, and store in VW34;

In this way, the obtained code value VW34 is compared with the actual code value VW30 and applied to the program.

2. Blade control

(1) Blade return to origin

Turn on the main power supply of the control system, the system is powered on, and the power indicator light is on. If the three blades do not return to the origin, that is, the indicator light of the origin sensor is not on, the blades shall be reset to the origin before other operations. Turn the three-position knob "M/A" knob to "M" state. Press "RESET" button. The blades I, II and III return to the origin in turn. If the three blades are all at the origin, the blades will not act after pressing "RESET" button. If one or two blades are not at the origin, the blades return to the origin in a certain sequence. Press "STOP" button during operation, blade stops reset operation. When the three blades return to the origin, the indicator lights of the three blade sensors will be on, KA14 relay will act for 1s, the absolute value encoder of blade shall be reset, the signals I2. 0-I2. 7 of blade acquisition signals shall be cleared, and the corresponding indicator light on PLC shall be off. Blade reset is completed.

(2) Blade manual forward and reverse control

Turn the three-position knob "M/A" knob to "M" state. Turn the three-position knob "P: FWD/REV" knob to "FWD" and blade rotates clockwise. Turn the three-position knob "P: FWD/REV" knob to "REV" and blade rotates anticlockwise.

(3) Automatic forward and reverse control of blade

Turn the three-position knob "M/A" knob to "A" state. When the input angle is 0-180°, press the "RESET" button first, and the three blades will reset to the origin. Press the "confirm" button and then press the "Start" button after completing the angle input in the main control computer or PLC programming software. The blades rotate forward to the set angle. When the next input value is less than the current angle, press the "Start" button of the upper computer, and the blade rotate reverse to the input angle.

When the input angle is 180°-360°, the operation is the same as that described above, but special attention shall be paid to: After completing the first pitch, if you want the next input angle value to be greater than the current value (e. g. 270° for the current value and 300° for the input value), press the "reset" button first, and then input the set angle (300° as mentioned above) after the blade returns to the origin, which is different from 0-180° pitch.

3. Write training report

Project Work

Describe the realization of pitch function with words or flow chart (divided into two situations: the pitch of the training equipment and the pitch of actual engineering unit).

Project Ⅵ Working Principle Operational Training of Grid-connected Inverter

Project Description

Consult the working principle and structure composition of grid-connected inverter of wind power generating unit, and complete the working principle operational training of 1.5MW wind power generating unit simulation grid-connected inverter with knowledge learned and relevant installation manuals. Observe safe utilization of electricity specifications and precautions.

Competency Objectives:

① Understand the working principle of grid-connected inverter.

② Master the use and function of grid-connected inverter.

Project Environment

This training task mainly involves grid-connected inverter of this training platform. To complete this task, refer to the equipment instruction manual of THWPWG-3B large-scale wind power generation system operational training platform to know and understand the principle and structure of grid-connected inverter, as well as the usage specifications and precautions of relevant tools.

Project Principle and Basic Knowledge

In general, the process of converting AC energy into DC energy is called rectification,

the circuit which completes the rectification function is called rectifier circuit, and the device realizing the rectification process is called rectifier equipment or rectifier. Correspondingly, the process of converting DC energy into AC energy is called inverter, the circuit which completes the inverter function is called inverter circuit, and the device realizing the inversion process is called inverter equipment or inverter.

Ⅰ. Control Mode of Inverter

The grid-connected inverters can be divided into voltage source voltage control, voltage source current control, current source voltage control and current source current control according to the control mode. However, because of the large inductance in the inverter loop, the dynamic response of the system is poor, so most of the inverters in the world are mainly voltage source input.

The output control of parallel operation of inverter and electric supply can be divided into voltage control and current control. The electric supply system can be regarded as a constant AC voltage source with infinite capacity. If the output of grid connected inverter adopts voltage control, it is actually a system in which voltage sources operate in parallel, which makes it difficult to ensure stable operation of the system; if the output of grid-connected inverter adopts current control, only the output current of grid-connected inverter needs to be controlled to track the electric supply voltage to achieve the purpose of parallel operation.

The grid-connected inverter generally adopts the control mode of voltage source input and current source output.

Ⅱ. Basic Structure of Inverter

The functional block diagram of grid-connected inverter is shown in Figure 3.6.1. The main circuit topology consists of DC/DC (Boost circuit) + DC/AC (single-phase grid-connected inverter) + filter, and the control loop consists of bus bar voltage detection + output current detection + grid voltage detection + isolation drive + DSP control circuit + keyboard + LCD screen.

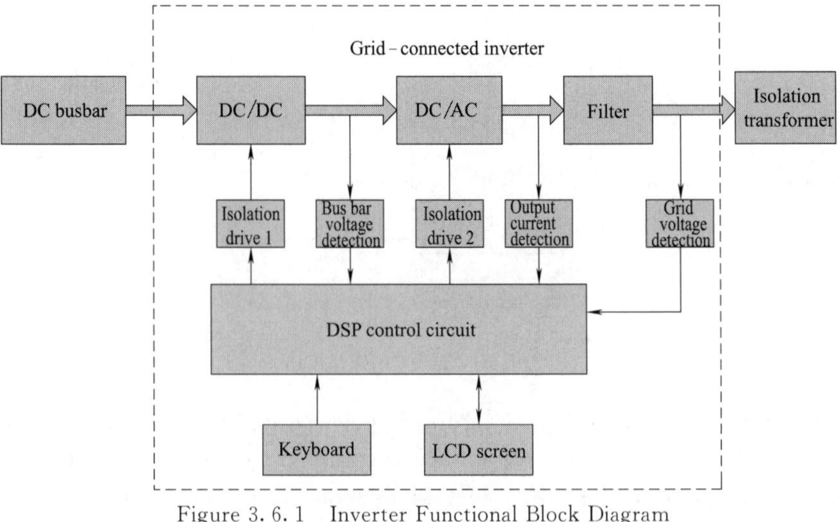

Figure 3.6.1 Inverter Functional Block Diagram

Description of grid-connected inverter module:

① DC/DC module (Boost circuit): Boost circuit mainly converts DC busbar output DC

voltage U_{bus} into DC bus bar voltage U that can meet grid connection requirements.

② DC/AC module: DC/AC inverts the DC bus bar voltage to sinusoidal AC of same frequency, same phase and same amplitude with grid voltage through DC/AC to realize grid connection with power grid.

③ Filter: The high frequency PWM harmonic current output by the inverter can be filtered, the high frequency circulating current in the grid current can be reduced, and the energy can be transferred between the inverter and the power grid, so that the grid-connected inverter can obtain certain damping characteristics and reduce the impulse current, which is beneficial to the stable operation of the system.

④ Bus bar voltage detection: Complete voltage closed loop and protection.

⑤ Output current detection: Complete current closed loop and protection.

⑥ Grid voltage detection: Complete grid voltage phase lock, voltage feedforward and protection.

⑦ Isolation drive 1, 2: Complete the isolation drive function on the switching tube.

⑧ DSP control circuit: Execute software algorithm function of grid-connected inverter.

⑨ Keyboard, LCD screen: Display grid connection parameters and set the parameters affecting grid connection current quality.

Ⅲ. Working Principle of Inverter

The working principle of single-phase voltage full-controlled PWM inverter is shown in Figure 3. 6. 2, which is a single-phase output full bridge inverter main circuit commonly used; wherein, the AC components adopt IGBT tubes Q11, Q12, Q13 and Q14. PWM pulse width modulation controls the connection or disconnection of IGBT tube.

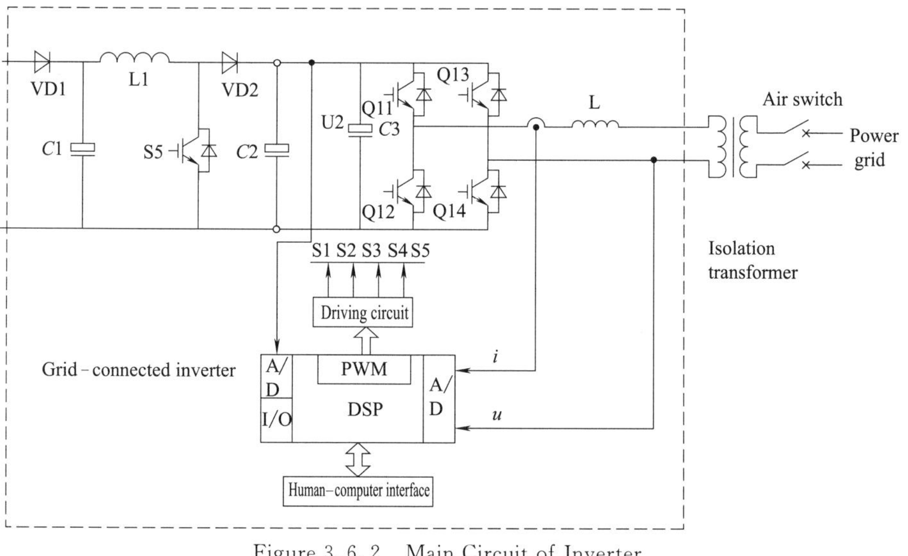

Figure 3. 6. 2 Main Circuit of Inverter

When the inverter circuit is connected to DC power supply, Q11 and Q14 are connected first, and Q12 and Q13 are disconnected. Then the current is output from the positive electrode of DC power supply and returns to Q14 negative electrode of power supply through Q11, inductor L and transformer primary coil. When Q11 and Q14 are disconnected, Q12 and Q13 are connected, and the current returns to Q12 negative electrode from the positive electrode of the power supply through Q13, transformer primary coil and inductor. At this

time, positive and negative alternating square waves have been formed on the primary coil of the transformer. With high frequency PWM control, two pairs of IGBT tubes are alternately repeated to generate AC voltage on the transformer. Due to function of the LC AC filter, a sine wave AC voltage is formed at the output terminal.

Ⅳ. Operation Instruction for Parameter Setting of Grid-connected Inverter

The controller is powered on, the system is initialized, the working indicator light and fault indicator light of the interface module are all on, and the LCD displays "initialization..." . After 3-5s, the indicator light goes out, and the LCD displays the interface as shown in Figure 3.6.3.

1. Parameter setting steps

① Click the "OK" key on the keyboard to display the parameter setting interface, as shown in Figure 3.6.4.

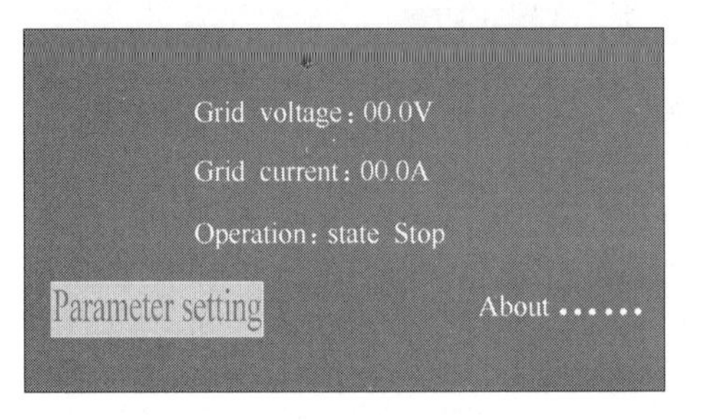

Figure 3.6.3 Default System Interface

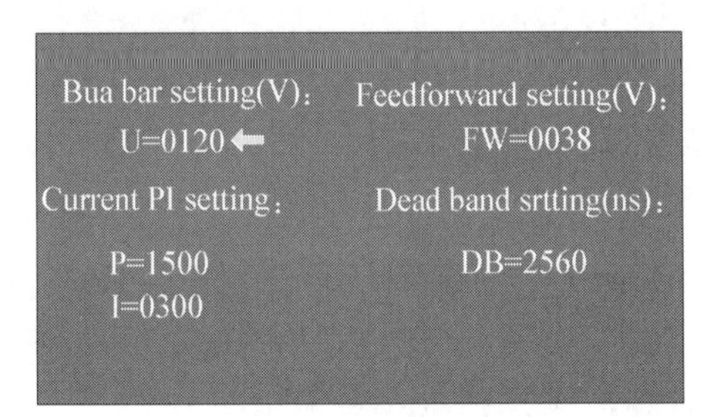

Figure 3.6.4 Parameter Setting Interface (1)

② Move the cursor on LCD screen by pressing "▼" or "▲" on the keyboard, select parameters to be set, select bus bar setting by default, click "set" key, and the interface as shown in Figure 3.6.5 will appear.

③ Set the required value through the numeric keys on the keyboard, for example, input "1", "0", "0", display U=100, click "ok" to complete the parameter setting, as shown in Figure 3.6.6.

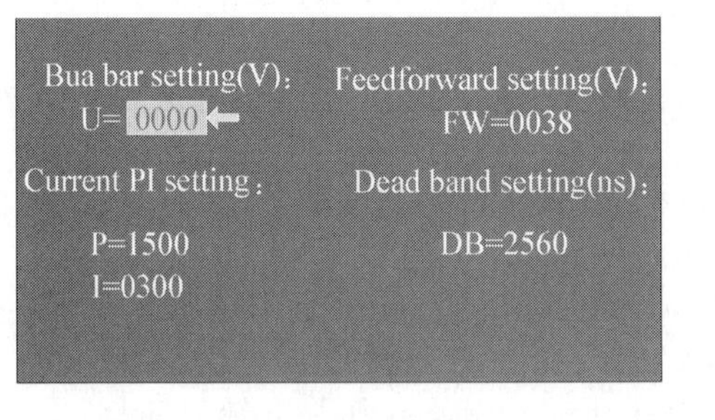

Figure 3.6.5 Parameter Setting Interface (2)

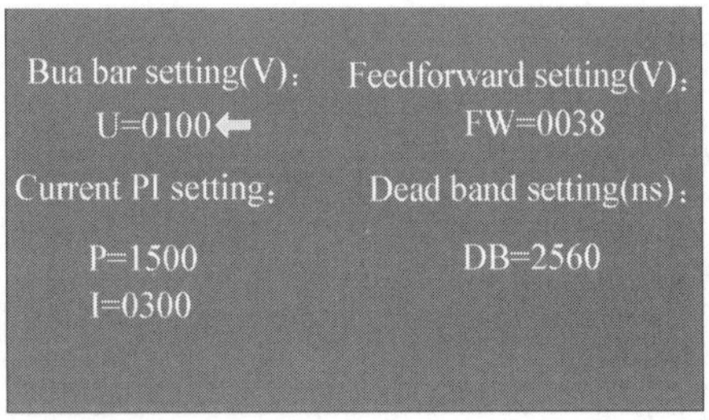

Figure 3.6.6 Parameter Setting Interface (3)

④ The setting of other parameters is similar to that of bus bar, and the schematic diagram of operation function is as shown in Figure 3.6.7.

2. Parameter setting instruction

in order to ensure safe operation of the system, the effective range of parameter setting is limited as follows:

① Effective setting range of bus bar voltage: 60-120V;

② Effective setting range of current loop scale factor P: 150-1500;

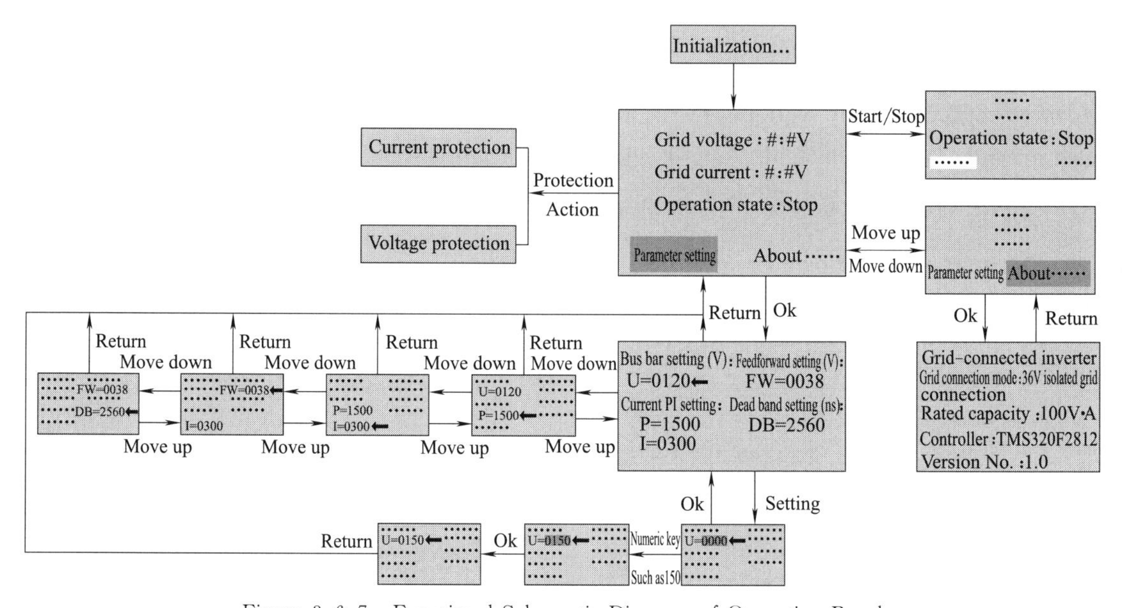

Figure 3. 6. 7 Functional Schematic Diagram of Operation Panel

③ Effective setting range of current loop integral coefficient I: 30-300;

④ Effective setting range of feedforward voltage: 0-50V;

⑤ Effective value of dead band setting: 2560ns, 2780ns, 2990ns, 3200ns.

Project Implementation

Ⅰ. List of Instruments, Equipment and Tools

① THWPWG-3B large-scale wind power generation system operational training platform.

② Multimeter.

Ⅱ. Safe Operation Specification

① Check whether each power supply is normal before use.

② There is AC 220V access point in the control cabinet. Pay attention to safety during training.

③ Always keep the training platform clean and tidy, and don't place sundries at will to avoid faults such as short circuit.

④ After the training is completed, the power switch shall be turned off in time and the training platform shall be cleaned in time.

⑤ Power on and off the system in strict accordance with correct operation steps to avoid damage to the system caused by misoperation.

⑥ In the process of operating the system, after the battery switch of the energy conversion storage control system is turned on, there is a process of waiting for the self-test initialization of the intelligent charge and discharge controller; the next operation can be carried out only after the "red light" of the intelligent charge and discharge controller goes off.

⑦ In the process of training, pay attention to the safety of strong current where there is "danger" sign.

Ⅲ. Operational Training Steps

① Turn on the main power switch of each control system; the power indicator light dis-

plays.

② Set the brake of charge and discharge controller to "RELEASE" state, close the "battery" air switch of the "energy conversion storage control system", connect the battery and supply power to charge and discharge controller at the same time. At this time, charge and discharge controller is initialized and the red indicator light is on (working in braking state) . Do not proceed until the red indicator goes out (exit the braking state).

③ Close the "fan output" and "charge and discharge controller" circuit breaker on the "energy storage control unit", and the "handle" switch is centered.

④ Start the fan and keep the rotating speed constant.

⑤ Turn on the "controller" switch of "grid-connected inverter control system", power on the grid-connected inverter controller and initialize the LCD screen.

⑥ Turn on the switches of "battery" and "grid-connected power generation" in turn, and partial input and output voltage of inverter have values.

⑦ Operate human-computer interface, move cursor through keyboard to select "parameter setting", bus bar voltage $U=120V$, current loop scale factor $P=1500$, current loop integral coefficient $I=300$, feedforward voltage FW=38V, dead band time DB=2560ns; click "return" key to return to initial interface, then click "start/stop" key to start inverter; record the value of each electricity meter before and after grid connection of inverter:

Serial Number	Project	Inverter input		Inverter Output	
		U/V	I/A	U/V	I/A
1	Before Grid Connection				
2	After Grid Connection				

⑧ After the inverter works normally, record the parameters of the inverter output coulombmeter under no-load, DC load and AC load respectively:

Serial Number	Project	Inverter Output Coulombmeter				
		U/V	I/A	P/kW	Q/kV·A	PF
1	After Grid Connection(no-load)					
2	After Grid Connection(motor)					
3	After Grid Connection(LED light)					

⑨ After training, click "start/stop" key to stop inverter and fan.

⑩ Turn off "grid-connected power generation" "DC load" "AC load" "battery" and "controller" air switch of "grid-connected inverter control system" and "fan output", "charge and discharge controller" and "battery" air switch of "energy conversion storage control system" in turn, and finally turn off "main power" switch of each control system.

Project Work

After the inverter works normally, change the value of bus bar voltage U to observe the value of each electricity meter.

Project Ⅶ Parameter Setting and Power Quality Analysis of Grid-connected Inverter

Project Description

Complete this training task and learn parameter setting and power quality analysis of

grid-connected inverter based on THWPWG-3B large-scale wind power generation system operational training platform.

Competency Objectives:

① Understand the working principle of the grid-connected inverter.

② Master the usage and function of grid-connected inverter.

③ Understand power quality and harmonic control.

Project Environment

With the development and application of new energy, the research of grid-connected inverter is paid more and more attention. It is a device to convert the DC electric energy generated by the power generation system to the AC electric energy suitable for the power grid. With the increasing number of grid-connected inverter devices put into use, the pollution to the grid by output grid current harmonic cannot be neglected. The total harmonic distortion (THD) of grid current is usually used to describe the power quality. In addition to causing additional loss to the grid, the harmonic of grid current may also damage the electric equipment in the power grid.

The grid-connected inverter control of THWPWG-3B large-scale wind power generation system operational training platform adopts voltage PWM technology, which completes phase lock of power grid, DC bus bar regulation and current regulation to realize electric energy grid-connected and on-load operation, featuring sine wave current output and unit power factor, effectively solving the problems of harmonic pollution and low power factor of grid-connected devices. The grid-connected inverter block diagram of THWPWG-3B large-scale wind power generation system operational training platform is shown in Figure 3. 7. 1, mainly composed of grid-connected inverter, inverter output coulombmeter, isolation transformer, DC load, AC load, etc.

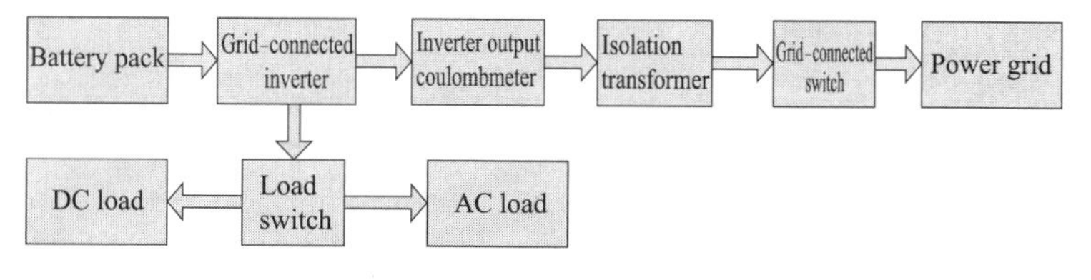

Figure 3. 7. 1 Block Diagram of Grid-connected Inverter

In grid-connected inverter control unit, the function of grid-connected inverter is to convert DC power from battery pack into AC power output. Inverter output coulombmeter has the function of monitoring electric quantity parameters, such as voltage, current, active power, reactive power, power factor, voltage/current harmonic distortion rate, etc. The isolation transformer can boost the inverted AC power to 220V and connect to the power grid to supply power to the grid and play a protective role.

Inverter output coulombmeter is the main component related to power quality analysis in THWPWG-3B large-scale wind power generation system operational training platform.

Ⅰ. Introduction of Inverter Output Coulombmeter

ACR series harmonic meter has full-scale functions of AC electric quantity measurement, multi-rate electric energy calculation, harmonic analysis, remote signaling input, remote control output, network communication, etc; meanwhile, it also has the function of

Figure 3.7.2 Inverter Output Coulombmeter

real-time tracking display of grid waveform and SOE event recording, which is mainly used for comprehensive monitoring and diagnosis of power supply quality of power grid and electric energy management.

Inverter output coulombmeter (Figure 3.7.2) is mainly used for monitoring grid-connected current, voltage, frequency, power, power factor, THD value of voltage and current, harmonic waveform of voltage and current, etc.

Ⅱ. Model and Performance Index of Inverter Output Coulombmeter

In this platform, ACR230ELH inverter output coulombmeter is selected. The inverted output electric quantity includes single-phase current, voltage input, LCD dot matrix display and a RS485 communication port. The performance indexes are as follows:

① Input network: Single-phase 2-wire.

② Input frequency: 45-65Hz.

③ Input voltage: AC 100V; overload 1.2 rated value (continuous), 2 times rated value for 1s; power consumption less than 0.2V • A.

④ Input current: AC 5A; overload 1.2 rated value (continuous), 10 times rated value for 1s; power consumption less than 0.2V • A.

⑤ Output electric energy: Optocoupler pulse of open collector.

⑥ Communication: RS485 interface, Modbus-RTU protocol.

⑦ Digital input: Contact input, built-in power supply.

⑧ Digital output: Relay normally open contact output; contact capacity: AC: 250V 3A, DC: 30V 3A.

⑨ Measurement accuracy: Frequency: 0.05Hz; reactive electricity: Class 1; others: Class 0.5.

⑩ Power supply: AC/DC 85-270V, power consumption less than 6V • A.

⑪ Safety: Power frequency withstand voltage: AC 2kV/1min between power supply, voltage and current input loop, between each input terminal (parallel connection) and enclosure; AC 1.5kV/1min between power supply and digital input loop, communication loop, transmitting output loop, and between each input terminal and each output terminal. Insulation: The input and output terminals to the enclosure shall be greater than 100MΩ.

It is divided into DC load and AC load according to different application of functions.

Ⅲ. Wiring and Operation of Inverter Output Coulombmeter

The wiring diagram of inverter output coulombmeter is shown in Figure 3.7.3.

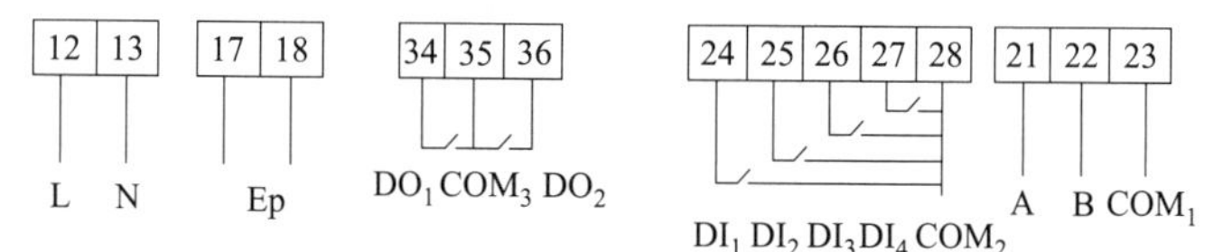

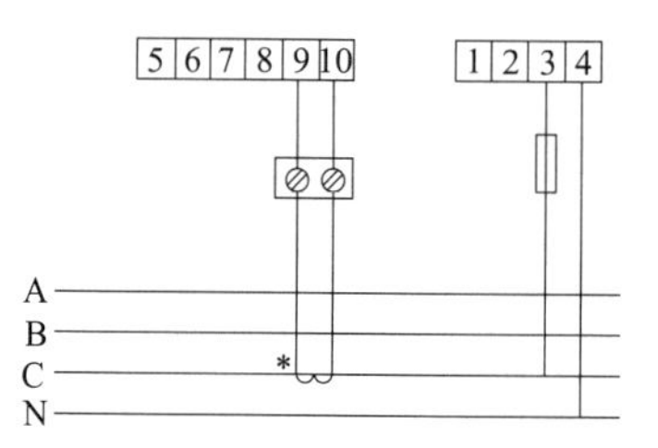

Figure 3. 7. 3 Wiring Diagram of Inverter Output Coulombmeter

Press SET key: Return to the previous menu. Press the left key: Page up. Press the right key: Page down. Press enter key: Confirm the selection of items and modification of parameters. The port definition is shown in Table 3. 7. 1.

Table 3. 7. 1 Port Description of Inverter Output Coulombmeter

Number	Definition	Description
1	U+	Measured voltage input
2	U−	
3	I+	Measured current input
4	I−	
8	A	RS485
9	B	
17	L	AC 220V power input
18	N	

Project Principle and Basic Knowledge

Ⅰ. Current Control Strategy

Using voltage outer loop and current inner loop control, first establish the mathematical model of the inverter in two-phase synchronous rotating coordinate system. Based on this, the current closed-loop control strategy based on space vector modulation is proposed, and the independent control of active component and reactive component of grid-connected current is realized. The current inner loop controls the phase and amplitude of grid-connected current, i. e. tracking grid-connected current command i_{ref}. The current inner loop generally adopts PI controller that joins the grid voltage feedforward. The current feedback sampling value is compared with the current command value. The error is output by PI control, and the output command type is voltage. After adding with the grid feedforward voltage, the required grid-connected voltage command is obtained. Adding the grid voltage feedforward actually offsets the grid voltage and make out the output value of current PI loop, i. e. inductance voltage, and then fine tuning control the grid-connected current, which is a kind of leading control. In voltage feedforward PI control, the actual output current cannot be the same as the output current command because the PI controller inevitably has steady state error when tracking sinusoidal signal.

The current inner loop control generally includes current hysteresis tracking control and

constant switching frequency current control. The schematic block diagram of constant switching frequency current control is shown in Figure 3. 7. 4. The error amount obtained by comparing the sinusoidal current reference value i_{ref} and the output instantaneous current i_o enters into the controller for regulation and add the feedforward voltage to send to the comparator. SPWM signal is obtained by comparison with triangular wave to control the connection and disconnection of the main circuit power tube. Unipolar frequency doubling sinusoidal pulse width modulation (SPWM) mode is selected for the platform. The harmonic frequency of SPWM waveform is increased without increasing the switching frequency, so that the harmonic component of output voltage can be effectively controlled.

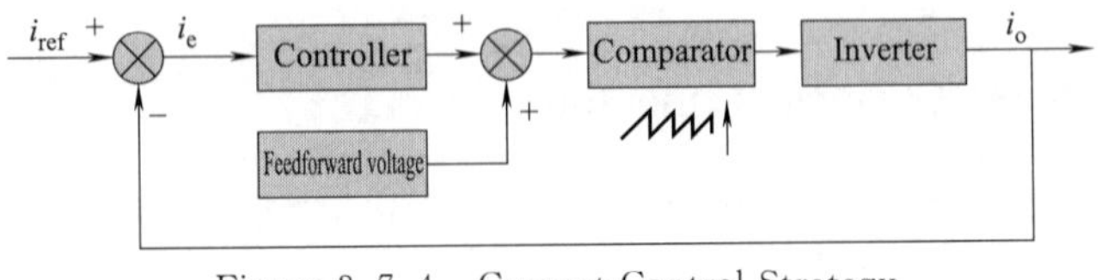

Figure 3. 7. 4 Current Control Strategy

The function of the voltage outer loop is to track the DC voltage command U_{dc} through the PI controller and provide the current inner loop command i_{ref} at the same time. The DC voltage command U_{dc} is a fixed constant, and the current given command i_{ref} is generated by comparing with the bus bar voltage U provided by the front-end Boost circuit.

Ⅱ. Other Parameters Affecting the Quality of Current Grid Connection

The power quality of grid-connected inverter is not only related to the control strategy, but also affected by the grid current circulation generated when the grid-connected switch is closed, the dead band time of the switch, DC voltage, the power system disturbance and the saturation nonlinearity characteristics of the isolation transformer iron core, etc.

Before the grid-connected switch is closed, the output voltage of the inverter is inconsistent with the grid voltage, including amplitude, phase, frequency and DC component, etc. Even if the given value of grid current is 0 at this time, there will be current flowing between the inverter and the power grid, which is called current circulation here. Current circulation includes DC circulation, fundamental wave circulation and harmonic circulation. Circulation accounts for a large proportion under light load and decreases under heavy load. Because the equivalent output impedance and line impedance of the inverter are very small, the tiny voltage difference between the inverter output voltage and the grid voltage will bring a large circulation, and there is inductance between the inverter and the grid, which makes the circulation lags voltage by 90°. Therefore, the existence of the circulation seriously affects the quality of the grid-side current. PI regulation is adopted for inverter control, but PI regulation cannot realize tracking without static error, so there is amplitude difference and phase difference between inverter output voltage and grid voltage, so the problem of reducing circulation is transformed into the problem of reducing steady state error of PI regulation. The influence of dead band time on output current waveform cannot be ignored when frequency of inverter control switch is high.

Power grid voltage disturbance is a transient process, its influence is not continuous, but the transformer is affected by manufacturing and operating conditions and other factors. In actual operation, the nonlinear saturation of iron core often produces strong harmonic current disturbance, and the disturbance is continuous, which has great influence on the

quality of output current waveform.

The phase lock accuracy adopts digital phase lock method, the sine table is made up of 4000 points, so the phase lock accuracy is 360/4000=0.09, cos0.09=0.99999, so this phase lock method can obtain very good phase lock result, thus has higher PF value. Phase lock is realized by tracking the voltage synthesized vector of power grid in rotating coordinate, that is, the space position angle of voltage synthesized vector of power grid is obtained, as the given angle of coordinate transformation used in control strategy, so that the tracking of phase and phase sequence of power grid can be realized.

Project Implementation

Ⅰ. List of Instruments, Equipment and Tools

① THWPWG-3B large-scale wind power generation system operational training platform.

② Multimeter.

Ⅱ. Safe Operation Specification

① Check whether each power supply is normal before use.

② There is AC 220V access point in the control cabinet. Pay attention to safety during training.

③ Always keep the training platform clean and tidy, and don't place sundries at will to avoid faults such as short circuit.

④ After the training is completed, the power switch shall be turned off in time and the training platform shall be cleaned in time.

⑤ Power on and off the system in strict accordance with correct operation steps to avoid damage to the system caused by misoperation.

⑥ In the process of operating the system, after the battery switch of the energy conversion storage control system is turned on, there is a process of waiting for the self-test initialization of the intelligent charge and discharge controller; the next operation can be carried out only after the "red light" of the intelligent charge and discharge controller goes off.

⑦ In the process of training, pay attention to the safety of strong current where there is "danger" sign.

Ⅲ. Operational Training Steps

① This training mainly observes energy conversion storage control system and grid-connected inverter control system.

② Turn on the main power supply and the power indicator light is on.

③ Insert one end of connecting wire into terminal A of connection module of energy conversion storage control system cabinet.

④ The other end is inserted into terminal A of the connection module of control cabinet of grid-connected inverter control system.

⑤ Set the brake of charge and discharge controller to "Release" state.

⑥ Close the "battery" air switch.

⑦ At this time, the charge and discharge controller initializes.

⑧ The red indicator light is on.

⑨ The next operation can be carried out only after the red indicator light goes off.

⑩ Close the "fan output" and "charge and discharge controller" circuit breaker on "energy storage control unit" .

⑪ The "handle" switch is centered.

⑫ Input the starting signal Q0. 0=1 of the spindle frequency converter in the program, and the given analog quantity is 10000.

⑬ Start the fan and keep the rotating speed constant.

⑭ Turn on the "controller" switch of "grid-connected inverter control system" .

⑮ Power on grid-connected inverter controller and initialize LCD screen.

⑯ Turn on "battery" and "grid-connected power generation" switch in turn.

⑰ Click "ok" key on the keyboard.

⑱ Enter the human-computer interface for parameter setting.

⑲ Click "return" key to return to the initial interface.

⑳ Click "start/stop" key to start inverter.

㉑ Change PID parameter setting of the current loop.

㉒ Record the value of inverter output coulombmeter in the table below.

Number	Project	Inverter Output Coulombmeter				
		U/V	I/A	PF	Voltage THD	Current THD
1	P=150; I=30					
2	P=500; I=100					
3	P=800; I=180					
4	P=1200; I=250					
5	P=1500; I=300					

㉓ Draw corresponding harmonic waveform.

㉔ Click "start/stop" key to stop the inverter.

㉕ Click "ok" key on the keyboard.

㉖ Enter the human-computer interface for parameter setting.

㉗ Move cursor through keyboard to select "parameter setting" .

㉘ Set dead band time DB=2780.

㉙ Click "return" key to return to the initial interface.

㉚ Click "start/stop" key to start inverter.

㉛ Change DB parameter setting of the dead band time.

㉜ Record the value of inverter output coulombmeter in the table below.

Number	Project	Inverter Output Coulombmeter				
		U/V	I/A	PF	Voltage THD	Current THD
1	DB=2560ns					
2	DB=2780ns					
3	DB=2990ns					
4	DB=3200ns					

㉝ After training, click "start/stop" key to stop the inverter.

㉞ Input Q0. 0=0 in the program, and the given analog quantity is 0, stop the fan.

㉟ Close each circuit breaker of "grid-connected inverter control system" in turn.

㊱ Close each circuit breaker of "energy conversion storage control system" in turn and finally turn off "main power" of each control system.

Project Work

To complete the task of this project, it is necessary to complete the preparation before training, learn the safety precautions, complete the project task of "After the inverter works normally, observe the value of total harmonic distortion (THD) by changing the bus bar voltage value", then write the project report. The project report is a summary and feelings and experiences of operational training based on the observed problems found during the project implementation through self-analysis and research or analysis and discussion among the team members, which shall be concise, clear in writing and conclusion.

Project Ⅷ Installation and Commissioning of Grid-connected Inverter Control System

Project Description

Consult the installation manual of grid-connected inverter control system of wind power generating unit to understand the circuit principle, structure composition and relevant wiring process of grid-connected inverter control, and complete the installation and commissioning of simulation grid-connected inverter control system of wind power generating unit with knowledge learned and relevant installation manuals.

Competency Objectives:

① Grasp the grid-connected inverter control principle.

② Be able to read the circuit diagram of grid-connected inverter control unit.

③ Be able to connect the main control circuit according to the given control wiring diagram and port distribution table.

Project Environment

This training task mainly involves grid-connected inverter control part of this training platform. To complete this task, it is necessary to refer to the equipment instruction manual of THWPWG-3B large-scale wind power generation system operational training platform to know and understand the principle, components and structure composition of grid-connected inverter control system, as well as the usage specifications and precautions of relevant tools.

Ⅰ. Composition and Function of Grid-connected Inverter Control System

The block diagram of grid-connected inverter control system is shown in the Figure 3.8.1, mainly composed of grid-connected inverter, inverter output coulombmeter, isolation transformer, DC load, AC load, etc.

Description of grid-connected inverter control unit module:

① Grid-connected inverter: Convert the DC power from the battery pack into AC output.

② Inverter output coulombmeter: Monitor electric quantity parameters such as voltage, current, active power, reactive power, power factor, voltage/current harmonic distortion rate, etc.

③ Isolation transformer: Boost the inverted AC to 220V and combine into the power grid to supply power to the grid and play a protective role at the same time.

④ DC load: 24V DC motor for local load.

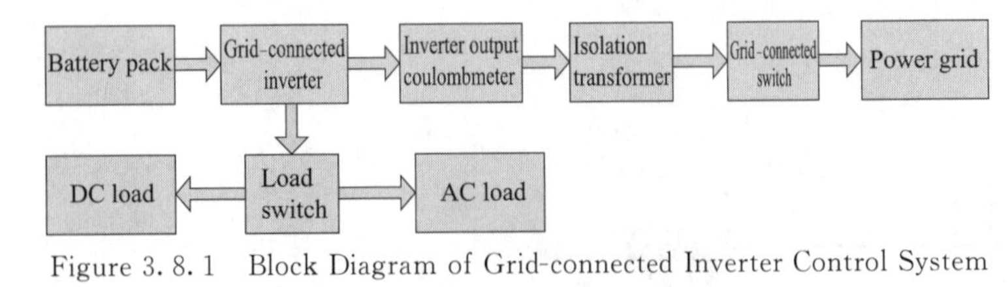

Figure 3. 8. 1 Block Diagram of Grid-connected Inverter Control System

⑤ AC load: 36V AC indicator light for local load.

Function: The DC voltage from the battery is inverted by grid-connected inverter into sinusoidal AC of the same frequency, same phase and same amplitude with the grid voltage to realize grid connection with the power grid; the electricity monitor is used to monitor the electric energy parameters at the output terminal of the inverter; part of the AC voltage output from the inverter is boosted to 220V through the isolation transformer and connected to the power grid, and part is supplied to the local load.

Ⅱ. Structural Composition of Grid-connected Inverter Control System

Grid-connected inverter control system mainly consists of core module, interface module, LCD display module, keyboard interface module, driving module, DC voltage booster module, DC voltage sampling module, AC voltage sampling module, AC current sampling module, temperature alarm module, communication module, switching power supply, DC motor, square indicator light, DC voltmeter, DC ammeter, multi-functional digital display meter, transformer, etc. See Table 3. 8. 1. Main technical parameters are as follows:

① Rated input voltage: DC24V.

② Rated output voltage: 220V±10%, 50Hz±1Hz.

③ Rated power: 100V • A.

④ Output power factor: ≥0. 80 (inductive load, capacitive load).

⑤ Inverter efficiency: ≥80%.

Table 3. 8. 1 Equipment of Grid-connected Inverter Control System

Number	Name	Main Component, Device and Specification	Quantity	Remark
1	Grid-connected inverter	Power: 100W Rated input voltage: DC 24V Rated output voltage: AC 36V	1 set	
2	Switching power supply	Rated input voltage: AC 220V Rated output voltage: DC 24V Rated power: 35W	1 set	
3	DC motor	Rated voltage: DC 24V	1 set	
4	Square indicator light	Rated voltage: AC 36V	1 set	
5	DC ammeter	Input current range: 0-5A Accuracy: 0. 5%±5 bits Communication: RS485 communication interface	1 piece	
6	Multi-functional digital display meter	Measured voltage range: AC0-250V Measured current range: AC0-5A Communication: RS485 communication interface	1 piece	
7	Transformer	Transformation ratio: 36/220	1 piece	

Ⅲ. Schematic Block Diagram of Grid-connected Inverter Control System

The schematic block diagram of grid-connected inverter control system is shown in Figure 3. 8. 2.

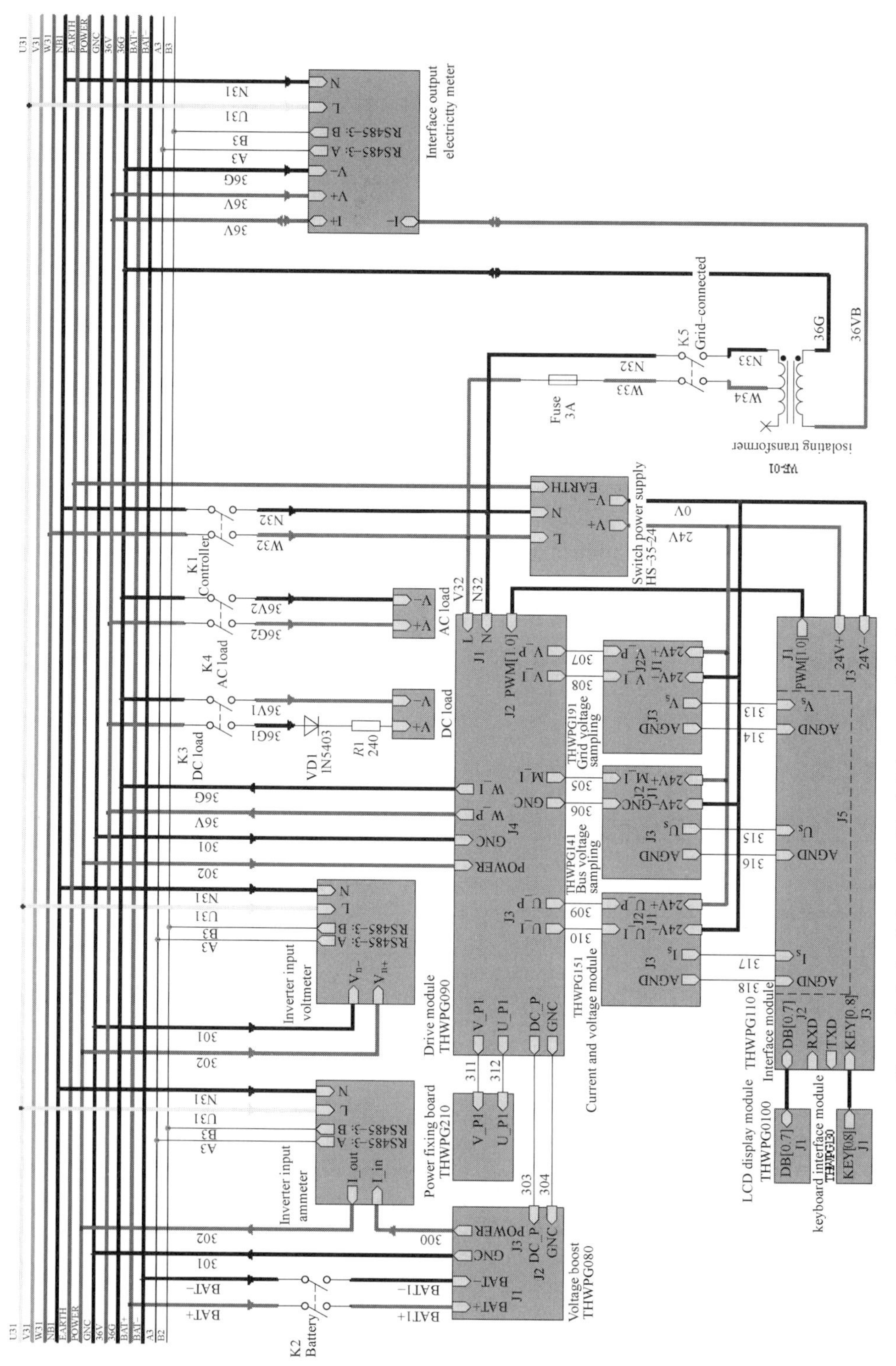

Figure 3. 8. 2 Schematic Block Diagram of Grid-connected Inverter Control System

Project Principle and Basic Knowledge

Ⅰ. Description of Grid-connected Inverter

The functional block diagram of grid-connected inverter is shown in the Figure 3. 8. 3. The main circuit topology consists of DC/DC (Boost circuit module) + DC/AC (driving circuit module) + filter (filter board). The control circuit consists of bus bar voltage sampling module + current sampling module + grid voltage sampling module + temperature alarm module + isolation drive + DSP control circuit + keyboard interface module + LCD display module.

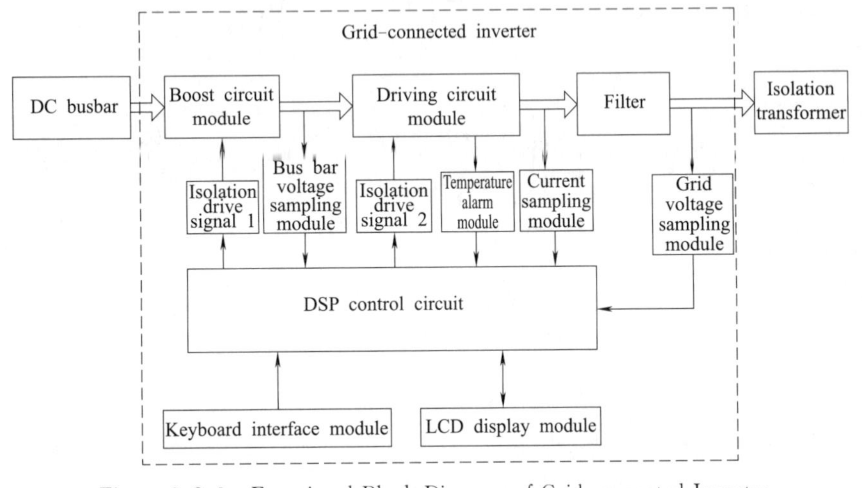

Figure 3. 8. 3 Functional Block Diagram of Grid-connected Inverter

① Boost circuit module: Boost circuit mainly converts the output DC voltage of DC busbar into bus bar voltage which can meet the requirements of grid connection.

② Driving circuit module: The driving circuit inverts the DC bus bar voltage into sinusoidal AC of the same frequency, same phase and same amplitude with the grid voltage through DC/AC to realize grid connection with the power grid.

③ Filter (filter board): The high frequency PWM harmonic current output by the inverter can be filtered, the high frequency circulating current in the grid current can be reduced, and the energy can be transferred between the inverter and the power grid, so that the grid-connected inverter can obtain certain damping characteristics and reduce the impulse current, which is beneficial to the stable operation of the system.

④ Bus bar voltage sampling module: Bus bar voltage detection, complete voltage closed loop and protection.

⑤ Current sampling module: Output current detection, complete current closed loop and protection.

⑥ Grid voltage sampling module: Grid voltage detection, complete grid voltage phase lock, voltage feedforward and protection.

⑦ Isolation drive signal 1: Complete the isolation drive function on switching tube of Boost circuit module.

⑧ Isolation drive signal 2: Complete the isolation drive function on IPM intelligent module of driving circuit.

⑨ DSP control circuit: Execute software algorithm function of grid-connected inverter.

⑩ Keyboard interface module: Set the parameters affecting the quality of grid-connected current.

⑪ LCD display module: Display grid connection parameters.

Ⅱ. Boost Circuit Module

Boost circuit module mainly converts 24V DC voltage output by battery pack into DC bus bar voltage that can meet grid connection requirements as shown in Figure 3. 8. 4.

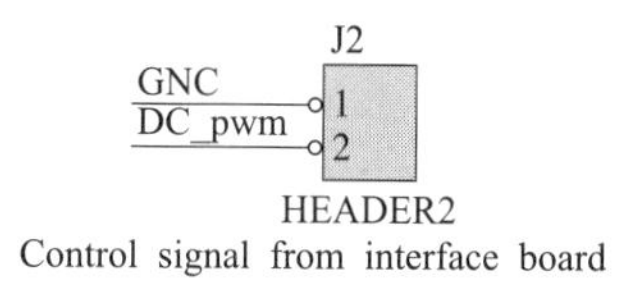

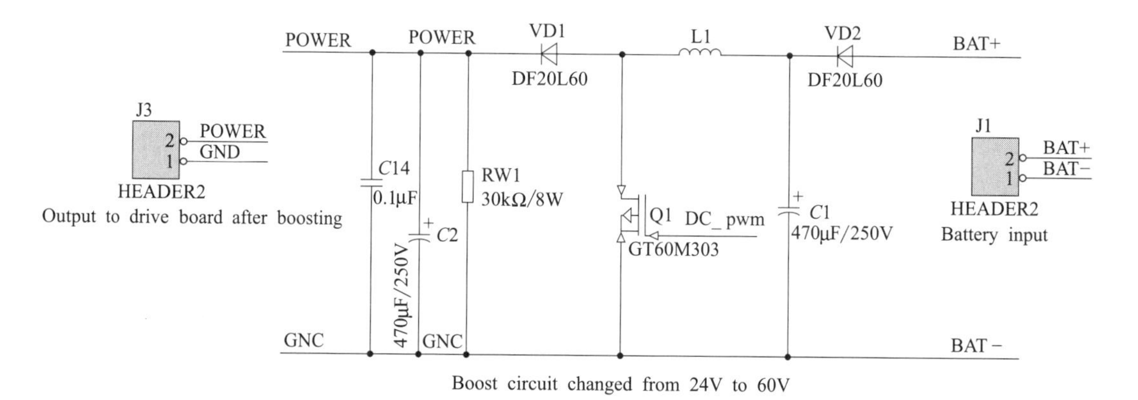

Figure 3. 8. 4 Schematic Diagram of Boost Circuit Module

The physical picture of Boost circuit module is shown in Figure 3. 8. 5.

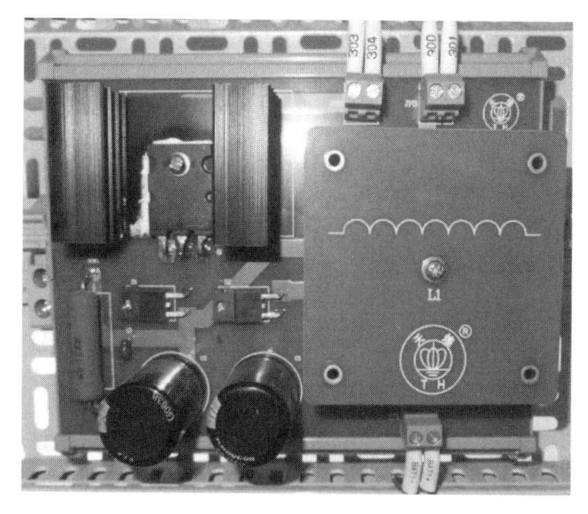

Figure 3. 8. 5 Boost Circuit Module

Ⅲ. Driving Circuit Module

The driving circuit module boosts the battery voltage through Boost, then inverts it to sinusoidal AC of the same frequency, same phase and same amplitude with the grid voltage

to realize grid connection with the power grid; it is mainly the application of Mitsubishi IPM intelligent module. The principle of driving circuit module is shown in Figure 3. 8. 6.

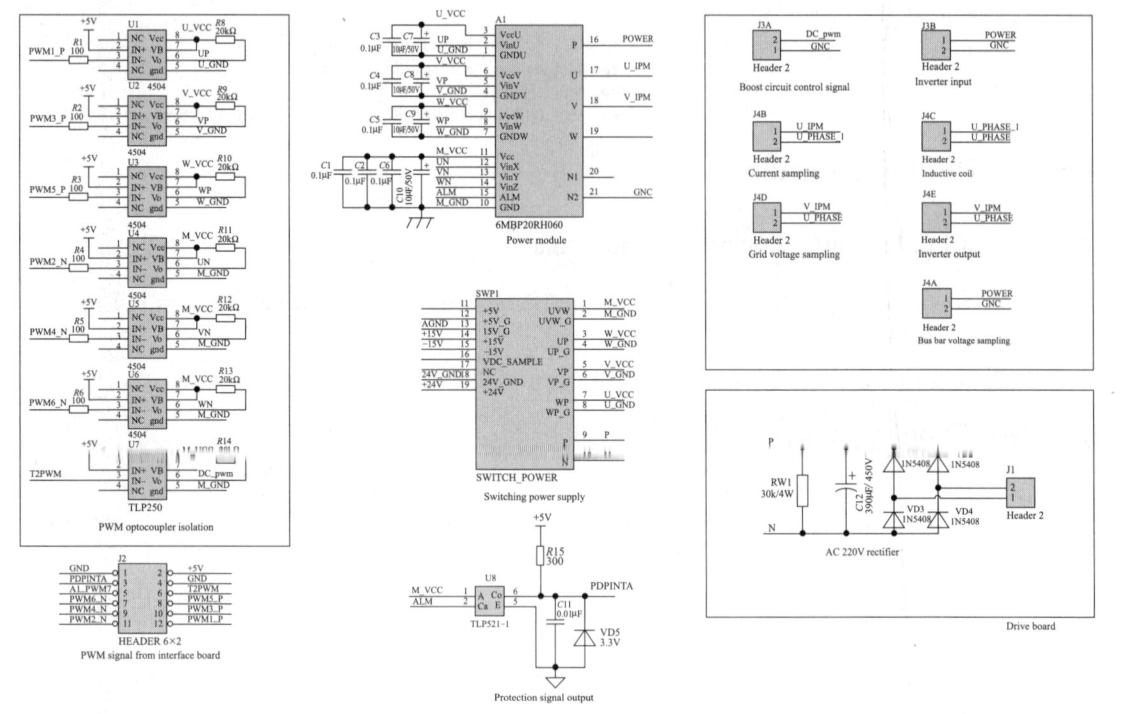

Figure 3. 8. 6 Schematic Diagram of Driving Circuit Module

The physical Picture is shown in Figure 3. 8. 7.

Figure 3. 8. 7 Driving Circuit Module

Terminal Line Port Definition is shown in Figure 3. 8. 8.

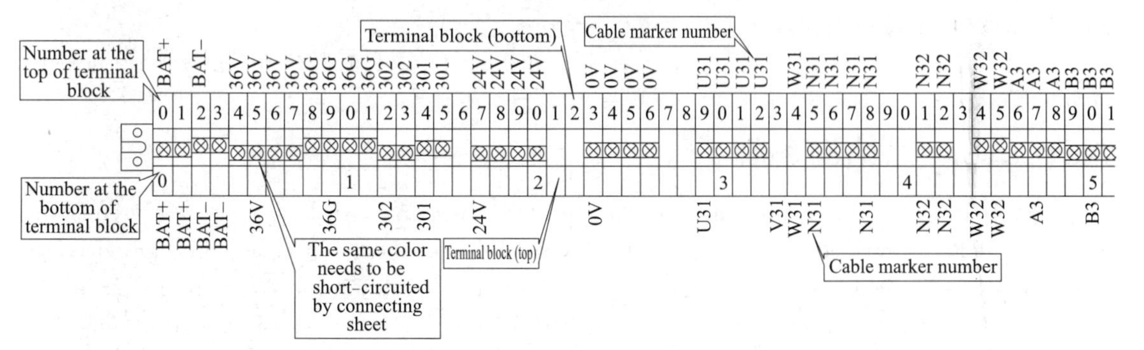

Figure 3. 8. 8 Terminal block Port Definition

Terminal line Number Definition

Name	Number

Terminal line(X) XT X : X

X (last) — Represent the number at the top of terminal line
X (first) — Represent the number at the bottom of terminal line
XT — Represent terminal line
Terminal line(X) — Represent the connection position [terminal line (top/bottom)]

Project Implementation

Ⅰ. List of Instruments, Equipment and Tools

① Wind turbine simulator of THWPWG-3B large-scale wind power generation system operational training platform.

② Wire stripper, crimping pliers, flat Screwdriver, cross screwdriver.

Ⅱ. Safe Operation Specification

① Check whether each power supply is normal before use.

② Be sure to be familiar with the function and wiring position of each unit module of the device before wiring.

③ The main power supply must be disconnected before the training wiring. Live wiring is strictly prohibited.

④ After wiring is completed and checked without error, power on can be conducted.

⑤ Proficient in control principle and program control method of yaw pitch motor.

⑥ There is AC 220V access point in the control cabinet. Pay attention to safety during training.

⑦ Always keep the training platform clean and tidy, and don't place sundries at will to avoid faults such as short circuit.

⑧ After the training is completed, the power switch shall be turned off in time and the training platform shall be cleaned in time.

⑨ Power on and off the system in strict accordance with correct operation steps, so as to avoid damage to the system caused by misoperation.

⑩ In the process of operating the system, after the battery switch of the energy conversion storage control system is turned on, there is a process of waiting for the self-test initialization of the intelligent charge and discharge controller; the next operation can be carried out only after the "red light" of the intelligent charge and discharge controller goes off.

⑪ In the process of training, pay attention to prevent falling from high place and squeezing injury during equipment installation.

⑫ In the process of training, pay attention to the safety of strong current where there is "danger" sign.

Ⅲ. Operational Training Steps

① Get to know the equipment. Grid-connected inverter control system mainly consists of core module, interface module, LCD display module, keyboard interface module, driving module, DC voltage booster module, DC voltage sampling module, AC voltage sampling module, AC current sampling module, temperature alarm module, communication module, switching power supply, DC motor, square indicator light, DC voltmeter, DC ammeter,

multi-functional digital display meter, transformer, etc. The layout of the equipment is shown in Figure 3.8.9.

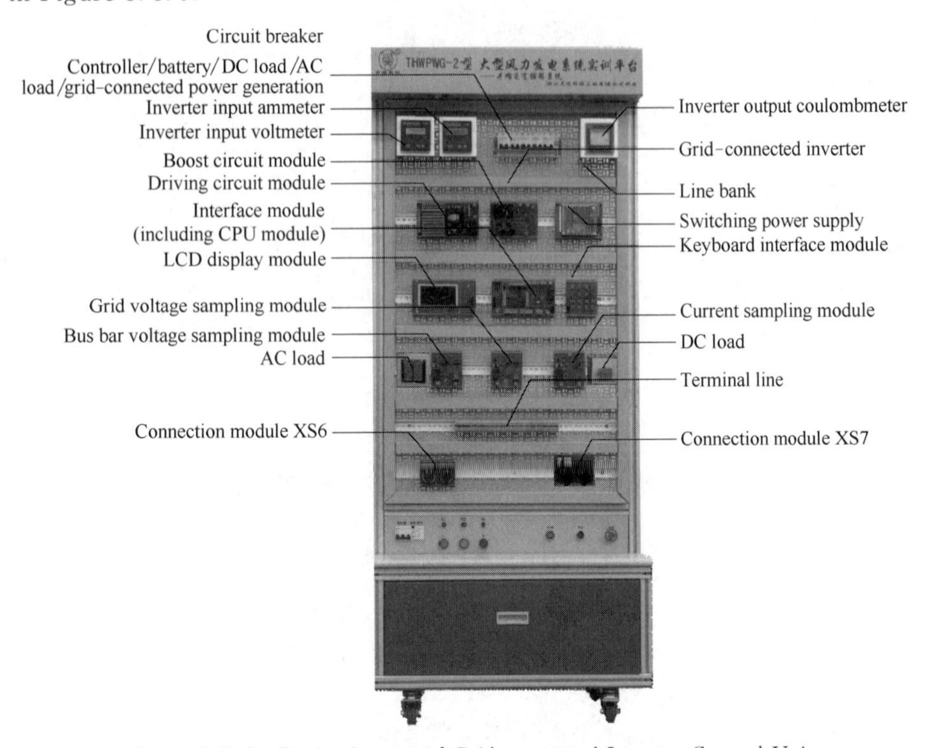

Figure 3.8.9 Device Layout of Grid-connected Inverter Control Unit

Please use the mobile phone and other devices to take pictures of the corresponding equipment, and put the photos in the appearance part of Table 3.8.2.

Table 3.8.2 Equipment of Grid-connected Control System

Number	Name	Main Component, Device and Specification	Quantity	Appearance
1	Grid-connected inverter	Power: 100W Rated input voltage: DC 24V Rated output voltage: AC 36V	1 set	
2	Switching power supply	Rated input voltage: AC 220V Rated output voltage: DC 24V Rated power: 35W	1 set	
3	DC motor	Rated voltage: DC 24V	1 set	
4	Square indicator light	Rated voltage: AC 36V	1 set	
5	DC ammeter	Input current range: 0-5A Accuracy: 0.5%±5 bits Communication: RS485 communication interface	1 piece	
6	Multi-functional digital display meter	Measured voltage range: AC 0-250V Measured current range: AC 0-5A Communication: RS485 communication interface	1 piece	
7	Transformer	Transformation ratio: 36/220	1 piece	

② Draw and simulate Boost circuit and describe the phenomenon.

③ Refer to Figure 3.8.10 circuit block diagram to describe the functions of each module in the circuit block diagram of grid-connected inverter control unit.

④ Install the driving circuit module. Attach aluminum guide rails, driving circuit module, etc. to the mesh plate to prepare for wiring according to Figure 3.8.11.

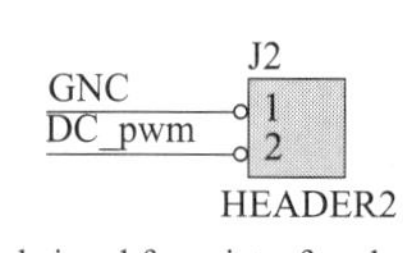

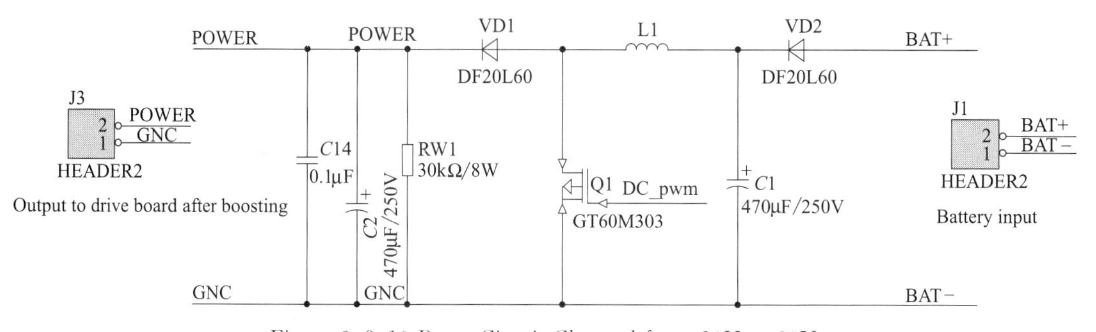

Figure 3. 8. 10 Boost Circuit Changed from 24V to 60V

Figure 3. 8. 11 Driving Circuit Module

Complete wiring according to driving circuit module wiring list, as shown in Table 3. 8. 3.

Table 3. 8. 3 Wiring List of Driving Circuit Module

Number	Start Port Position	End Port Position		Cable marker number	Line Type
	Driving circuit module	Name	Number		
1	J1:AC-220V	Terminal line (top)	XT4:4	W32	23 red
2	J1:AC-220V		XT4:1	N32	23 black
3	J2	Interface module	J1		14P cable
4	J3: DC_P	Boost circuit module	J2: DC_P	303	12 blue
5	J3: GNC		J2: GNC	304	12 blue
6	J3: GNC	Bus bar voltage Sampling module	J2:GNC	306	42 black
7	J3: M_I		J2:M_I	305	42 red
8	J3: W_I	Terminal line (bottom)	XT0:9	36G	42 black
9	J3: W_P		XT0:5	36V	42 red
10	J4: V_I	Grid voltage sampling module	J2:V_I	308	42 black
11	J4: V_P		J2:V_P	307	42 red
12	J4: V_P1	Filter board	J1:1	311	42 red
13	J4: U_P1		J1:2	312	42 red
14	J4: U_P	Current sampling module	J2:U_P	309	42 red
15	J4: U_I		J2:U_I	310	42 red
16	J4: GNC	Terminal line (top)	XT1:5	301	42 black
17	J4: POWER		XT1:3	302	42 red

⑤ Installation of Boost circuit module. Attach aluminum guide rails, Boost circuit mod-

ule (Figure 3. 8. 12), etc. to the mesh plate to prepare for wiring according to the device layout in Figure 3. 8. 9.

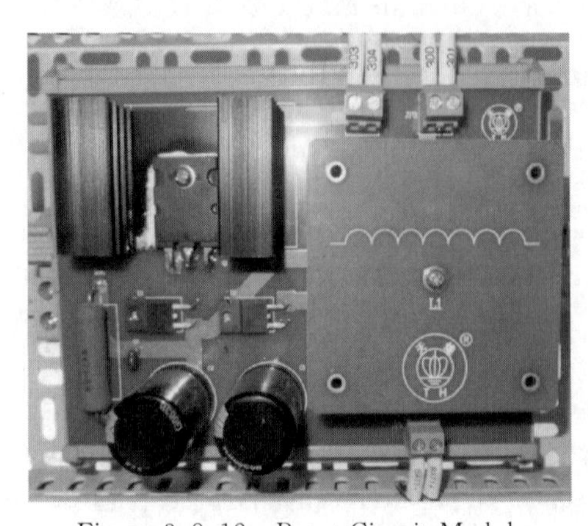

Figure 3. 8. 12 Boost Circuit Module

Complete wiring according to Boost circuit module wiring list, see Table 3. 8. 4.

Table 3. 8. 4 Wiring List of Boost Circuit Module

Number	Start Port Position	End Port Position		Cable marker number	Line Type
	Boost circuit module	Name	Number		
1	J2: DC_P	Driving circuit module	J3: DC_P	303	12 blue
2	J2: GNC		J3: GNC	304	12 blue
3	J1:BAT+	Battery circuit breaker	Bottom left	BAT1+	42 red
4	J1:BAT−		Bottom right	BAT1−	42 black
5	J3: POWER	Line bank of inverter input ammeter (bottom)	I+	300	42 red
6	J3: GNC	Terminal line (bottom)	XT1:4	301	42 black

Project Work

Ⅰ. Describe the problems encountered in wiring of driving module and the solutions

Ⅱ. Describe the main functions of driving module and booster module.

Project Ⅸ Overall Operation of Power Generation System

Project Description

Learn the key contents of wind power generation technology, understand the basic principles of wind power generation, and complete the operation and commissioning of wind power generation system through the operational training platform.

Competency Objectives:

① Grasp the principle of wind power generation system.

② Grasp the functions of various components of the wind power generation system.

Project Environment

The energy conversion storage control system is mainly composed of DC voltage and current sampling module, temperature alarm module, PWM driving module, CPU core module, human-computer interaction module, communication module, lightning protector,

intelligent charge and discharge controller, battery pack, switching power supply, DC voltmeter and DC ammeter, etc. as shown in Figure 3. 3. 1.

Project Principle and Basic Knowledge

The wind power generation system mainly consists of wind power generating unit, rectifier, controller, battery and inverter, DC load and AC load.

① Wind power generating unit. Wind power generating unit mainly converts wind energy into electric energy, and the output electric energy is sent to battery for storage after rectification, and can also be sent to power grid through grid-connected inverter.

② Battery. The battery is mainly used to store the electric energy generated by the wind power generating unit and supply power to the load at any time. The basic requirements for battery of wind power generation system are as follows: Low self-discharge rate, long service life, high charge efficiency, strong deep discharge capacity, wide operation temperature range, low maintenance or maintenance-free and low price. At present, maintenance-free lead-acid batteries are mainly used for wind power generation systems. In small and micro systems, Ni-MH batteries, Ni-Cd batteries, lithium batteries or super capacitors can also be used. If large amount of electric energy storage is required, a plurality of batteries needs to be connected in series or in parallel to form a battery pack.

③ Controller. The controller is used to control the working state of the whole system. Its functions mainly include: protection against over-charging and over-discharging of battery, short circuit protection of system, anti-reverse charge protection at night, etc.

④ Load. The load includes DC load and AC load. The DC load power supply of the wind power generation system directly comes from the controller, while the power supply of the AC load comes from the inverter. The AC inverter is the equipment used to convert the DC output from the battery into AC power and supply it to the power grid or AC load. Inverter can be divided into independent operation inverter and grid-connected inverter according to operation mode. Independent operation inverter is used for independent operation wind power generation system to supply independent loads. Grid-connected inverter is used for grid-connected wind power generation system.

Working Principle of Wind Power Generation System is as follows.

Wind power generation system can be divided into independent (off-grid) wind power generation system and grid-connected wind power generation system. Figure 3. 9. 1 is the schematic diagram of the working principle of independent operation wind power generation system. The core component of wind power generation is the wind power generating unit, which converts the wind energy directly into electric energy and stores it in the battery. When the load is powered by electricity, the electric energy in the battery is reasonably distributed to each load through the controller. The electric energy generated by the wind power generating unit is variable voltage variable frequency AC electric energy, which is converted into DC power by controller and charged to the battery. The electric energy is stored by the battery and used when necessary.

Figure 3. 9. 2 is the schematic diagram of the working principle of grid-connected wind power generation system. Grid-connected wind power generation system directly converts wind energy into electric energy and enters into grid-connected inverter through distribution box. Some grid-connected wind power generation systems shall be equipped with battery to

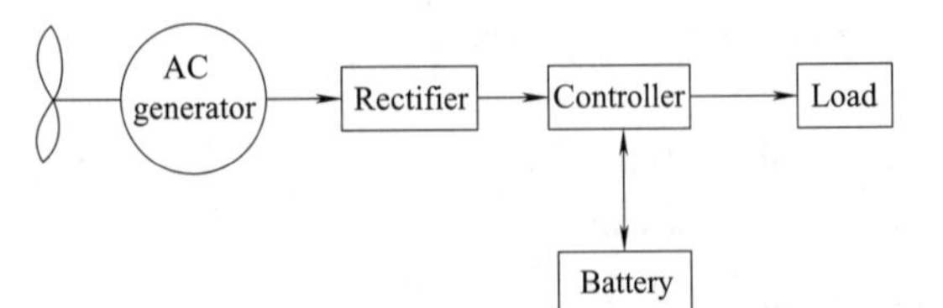

Figure 3. 9. 1 Working Principle of Independent Operation Wind Power Generation System

store DC electric energy. The grid-connected inverter consists of charge and discharge control, power regulation, AC inverter, grid-connected protection switching, etc. The AC power output by the inverter is used by the load, and the excess electric energy is fed into the public power grid (can be called selling electricity) through equipment such as power transformer. When the wind power generation system does not generate enough power due to weather reasons or its own large power consumption, the AC load can be supplied by the public power grid (called purchasing electricity).

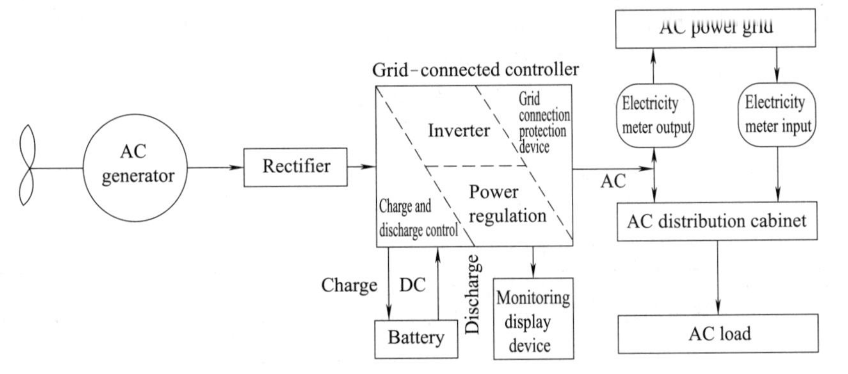

Figure 3. 9. 2 Working Principle of Grid-connected Wind Power Generation System

Project Implementation

Ⅰ. List of Instruments, Equipment and Tools

Wind turbline simulator of THWPWG-3B large-scale wind power generation system operational training platform fan object model and control cabinet of energy control. monitoring management. weather station.

Ⅱ. Safe Operation Specification

① Proficient in power generation principle of wind power generating unit and program control method;

② Before training, carefully read the instruction manual to get familiar with the parts related to power generation operation, carefully read the relevant operation instructions to ensure that the power supply of each system control cabinet is disconnected, and get familiar with the operation steps of this training according to relevant contents in the training instruction;

③ Power on and off the system in strict accordance with correct operation steps, so as to avoid damage to the system caused by misoperation;

④ Check whether each power supply and equipment are normal first, and then formally start the training task after ensuring they are normal;

⑤ In the process of training, always keep the training platform clean and tidy, do not

place sundries at will, so as to avoid short circuit and other faults, and pay attention to safety when there is " danger" sign;

⑥ After the training is completed, the power switch shall be turned off in time and the training platform shall be cleaned in time.

Ⅲ. Operational Training Steps

① Close the "main power" switch on the "energy storage control unit", the system is powered on and the three-phase power indicator light is on.

② Set the brake of charge and discharge controller to "RELEASE" state, close the "battery" circuit breaker of the "energy storage control unit", connect the battery and supply power to charge and discharge controller at the same time. At this time, charge and discharge controller is initialized and the red indicator light is on (working in braking state) . Do not proceed until the red indicator goes out (exit the braking state) .

③ Close the "fan output" and "MPPT" circuit breaker on the "energy storage control unit", and the "handle" switch is downward.

④ Press reset button K1 on "CPU core module" to reset the system.

⑤ Start the fan and keep the rotating speed constant.

⑥ Press "ENTER" key on "human-computer interaction module" to enter into manual adjustment interface, then press "UP" and "DOWN" keys to manually adjust duty ratio, measure output current and voltage of multiple groups of fans and input current and voltage of the battery. Record in the table below.

Number	Duty Ratio (number on LCD)	Fan Output			Battery		
		Voltage/V	Current/A	Power/W	Voltage/V	Current/A	Power/W
1							
2							
3							
4							
5							
6							
7							
8							
9							
10							

⑦ Based on the recorded voltage and current data, calculate the power corresponding to each voltage and current.

⑧ After completion, turn off the "fan output" "MPPT" "battery" circuit breaker and "main power" switch of "energy storage control unit" in turn.

Project Work

Draw the working principle block diagram of the wind power generation system of the training platform, and write out the energy transfer process.